Mechanical and Corrosion-Resistant Properties of Plastics and Elastomers

CORROSION TECHNOLOGY

Editor
Philip A. Schweitzer, P.E.
Consultant
Fallston, Maryland

ADDITIONAL VOLUMES IN PREPARATION

Environmental Degradation of Metals, U. K. Chatterjee, S. K. Bose, and S. K. Roy
Corrosion-Resistant Linings and Coatings, Philip A. Schweitzer

Mechanical and Corrosion-Resistant Properties of Plastics and Elastomers

Philip A. Schweitzer, P.E.
Consultant
Fallston, Maryland

CRC Press
Taylor & Francis Group
Boca Raton London New York

CRC Press is an imprint of the
Taylor & Francis Group, an **informa** business

CRC Press
Taylor & Francis Group
6000 Broken Sound Parkway NW, Suite 300
Boca Raton, FL 33487-2742

First issued in paperback 2019

© 2000 by Taylor & Francis Group, LLC
CRC Press is an imprint of Taylor & Francis Group, an Informa business

No claim to original U.S. Government works

ISBN-13: 978-0-8247-0348-6 (hbk)
ISBN-13: 978-0-367-39875-0 (pbk)

Visit the Taylor & Francis Web site at
http://www.taylorandfrancis.com

and the CRC Press Web site at
http://www.crcpress.com

Preface

Plastic and elastomeric materials have become an integral part of everyone's lives. There isn't a home, an industry, or a commercial establishment that does not make use of these materials in some form.

There have been instances where plastic materials have failed to perform satisfactorily. This is not the fault of the material, but rather of a poor choice on the part of the designer in the selection of the specific polymer or composite—in other words, a misapplication.

Many basic polymeric materials are available, all of which can be modified through formulation to produce a product with the desired properties. When the proper choice is made, the resulting product will provide satisfactory service.

Each year new plastic and composite materials are being developed. With these new materials the opportunities for additional applications have increased. Included in this book are descriptions of all the most common plastic and elastomeric materials readily available.

It is the purpose of this book to provide a single source where the engineer and/or designer can find the basic information relating to the physical, mechanical, and corrosion-resistant properties of each plastic and elastomeric material. With this information, an intelligent selection may be made as to which polymer or composite is best suited for the proposed application.

Most other compilations of this nature ignore the corrosion-resistant properties of these materials. Failure to take these properties into account can have a detrimental effect on the performance of the final product. In addition, other compilations do not include all the materials that are available.

In order to properly make use of a polymeric material, a basic understanding of the nature of polymers is necessary. Chapter 1 covers the nature of plastics and provides a definition of the mechanical property terms as they relate to plastics. It is important that these terms be understood, since the same terms as used for metallic materials have a somewhat different meaning when applied to plastics.

Chapter 2 provides detailed information on the physical, mechanical, and corrosion-resistant properties of the 32 most common thermoplastic polymers, while Chapter 3 does likewise for the 20 most common thermosetting polymers. Information is also included on the reinforcing materials used, including their physical, mechanical, and corrosion-resistant properties.

Chapter 4 covers the 27 most common elastomeric materials and Chapter 5 covers thermoplastic piping systems, while Chapter 6 covers thermosetting piping systems. Chapter 7 provides details on several miscellaneous applications.

It is hoped that this information will be beneficial to the reader.

Philip A. Schweitzer, P.E.

Contents

v

Mechanical and Corrosion-Resistant Properties of Plastics and Elastomers

1

Polymers

1.1 INTRODUCTION TO POLYMERS

Plastics are an important group of raw materials for a wide array of manufacturing operations. Applications range from small food containers to large chemical storage tanks, from domestic water piping systems to industrial piping systems handling highly corrosive chemicals, from toys to boat hulls, and from plastic wrap to incubators, and a multitude of other products. When properly designed and applied, plastic products possess the advantages of light weight, sturdiness, economy, and corrosion resistance.

Plastics are in reality polymers. The term plastic is defined in the dictionary as "capable of being easily molded," such as putty or wet clay. The term plastics was originally adopted to describe the early polymeric materials since they could be easily molded. Unfortunately, many of today's polymers are quite brittle and once formed cannot be molded. In view of this, the term polymer will be used throughout the book.

There are three general categories of polymers: thermoplastic polymers commonly called thermoplasts, thermosetting polymers called thermosets, and elastomers, more commonly called rubbers. Thermoplasts are long-chain linear molecules that can be easily formed by heat and pressure at temperatures above a critical temperature referred to as the "glass temperature." This term was originally applied to glass and was the temperature at which glass became plastic and easily formed. The glass temperatures for many polymers are above room temperature, and therefore these polymers are brittle at room temperature. However, they can be reheated and re-formed into new shapes and therefore can be recycled.

Thermosets are polymers that assume a permanent shape or set when heated, although some will set at room temperature. The thermosets begin as powders or liquids which are reacted with a second material, or which through catalyzed polymerization result in a new product whose properties differ from those of either starting material. Examples of a thermoset that will set at room temperature are epoxies that result from combining an epoxy polymer with a curing agent or catalyst at room temperature. Rather than a long-chain molecule, thermosets consist of a three-dimensional network of atoms. Since they decompose on heating they cannot be re-formed or recycled. Thermosets are amorphous polymers.

Elastomers are polymeric materials whose dimensions can be changed drastically by applying a relatively modest force, but which return to their original values when the force is released. The molecules are extensively kinked so that when a force is applied they unkink or uncoil and can be extended in length by approximately 100% with a minimum force and return to their original shape when the force is released. Since their glass temperature is below room temperature, they must be cooled below room temperature to become brittle.

Polymers are the building blocks of "plastics." The term is derived from the Greek meaning "many parts." They are large molecules composed of many repeat units that have been chemically bonded into long chains. Wool, silk, and cotton are examples of natural polymers.

The monomeric building blocks are chemically bonded by a process known as polymerization, which can take place by one of several methods. In condensation polymerization the reaction between monomer units or chain-end groups releases a small molecule, usually water. This is an equilibrium reaction that will halt unless the by-product is removed. Polymers produced by this process will degrade when exposed to water and high temperatures.

In addition polymerization, a chain reaction appends new monomer units to the growing molecule one at a time. Each new unit creates an active site for the next attachment. The polymerization of ethylene gas (C_2H_4) is a typical example. The process begins with a monomer of ethylene gas in which the carbon atoms are joined by covalent bonds as shown below:

$$
\begin{array}{cc}
H & H \\
| & | \\
C & = & C \\
| & | \\
H & H
\end{array}
$$

Each bond has two electrons, which satisfies the need for the s and p levels to be filled. Through the use of heat, pressure, and a catalyst, the double bonds, which are said to be unsaturated, are broken to form single bonds as below:

$$
\begin{array}{cc}
\text{H} & \text{H} \\
| & | \\
-\text{C}-\text{C}- \\
| & | \\
\text{H} & \text{H}
\end{array}
$$

This resultant structure, called a mer, is now free to react with other mers forming the long chain molecule shown below:

$$
\begin{array}{cccccccc}
\text{H} & \text{H} & \text{H} & \text{H} & \text{H} & \text{H} & \text{H} & \text{H} \\
| & | & | & | & | & | & | & | \\
-\text{C}-\text{C}-\text{C}-\text{C}-\text{C}-\text{C}-\text{C}-\text{C}- \\
| & | & | & | & | & | & | & | \\
\text{H} & \text{H} & \text{H} & \text{H} & \text{H} & \text{H} & \text{H} & \text{H}
\end{array}
$$

Most addition polymerization reactions follow a method of chain growth, where each chain, once initiated, grows at an extremely rapid rate until terminated, and then once terminated cannot grow any more except by side reactions.

The year 1868 marks the beginning of the polymer industry with the production of celluloid which was produced by mixing cellulose nitrate with camphor. This produced a molded plastic material that became very hard when dried. Synthetic polymers appeared in the early twentieth century when Leo Bakeland invented Bakelite by combining the two monomers, phenol and formaldehyde. An important paper published by Staudinger in 1920 proposed chain formulas for polystyrene and polox-methylene. In 1953 he was awarded the Nobel prize for his work in establishing polymer science. It was demonstrated in 1934 by W. H. Carothers that chain polymers could be formed by condensation reactions which resulted in the invention of nylon through polymerization of hexamethylenediamine and adipic acid. Commercial nylon was placed on the market in 1938 by the DuPont Company. By the late 1930s polystyrene, polyvinyl chlorides (PVC), and polymethyl methacrylate (Plexiglas) were in commercial production.

Further development of linear condensation polymers resulted from the recognition that natural fibers such as rubber, sugars, and cellulose were giant molecules of high molecular weight. These are natural condensation polymers, and understanding their structure paved the way for the development of the synthetic condensation polymers, such as polyesters, polyamides, polyimides, and polycarbonates. The chronological order of the development of polymers is shown in Table 1.1

A relatively recent term, engineering polymers, has come into play. It has been used interchangeably with the terms "high-performance polymers" and "engineering plastics." According to the ASM Handbook, engineering plastics are defined as "synthetic polymers of a resin-based material that have load-bearing characteristics and high-performance properties, which permit them to be

TABLE 1.1 Chronological Development of Polymers

Year	Material
1868	Celluloid
1869	Cellulose nitrate, cellulose propionate, ethyl cellulose
1907	PF resin (Bakelite)
1912	Cellulose acetate, vinyl plastics
1919	Glass-bonded mica
1926	Alkyl polyester
1928	Polyvinyl acetate
1931	Acrylic plastics
1933	Polystyrene plastics, ABS plastics
1937	Polyester-reinforced urethane
1938	Polyamide plastics (nylon)
1940	Polyolefin plastics, polyvinyl aldehyde, PVC, PVC plastisols
1942	Unsaturated polyester
1943	Fluorocarbon resins (Teflon), silicones, polyurethanes
1947	Epoxy resins
1948	Copolymers of butadiene and styrene (ABS)
1950	Polyester fibers, polyvinylidene chloride
1954	Polypropylene plastic
1955	Urethane
1956	POM (Acetals)
1957	PC (polycarbonate)
1961	Polyvinylidene fluoride
1962	Phenoxy plastics, polyallomers
1964	Polyimides, (Polyphenylene oxide PPO)
1965	Polysulfones, methyl pentene polymers
1970	Polybutylene terephthalate (PBT)
1971	Polyphenylene sulfide
1978	Polyarylate (Ardel)
1979	PET–PC blends (Xenoy)
1981	Polyether block amides (Pebax)
1982	Polyetherether ketone (PEEK)
1983	Polyetheramide (Ultem)
1984	Liquid crystal polymers (Xydar)
1985	Liquid crystal polymers (Vectra)
1988	PVC–SMA blend

used in the same manner as metals and ceramics." Others have limited the term to thermoplasts only. Many engineering polymers are reinforced and/or alloy polymers (a blend of polymers). Polyethylene, polypropylene, polyvinyl chloride, and polystyrene, the major products of the polymer industry are not considered engineering polymers.

Reinforced polymers are those to which fibers have been added that increase the physical properties—especially impact resistance and heat deflection temperatures. Glass fibers are the most common additives, but carbon, graphite, aramid, and boron fibers are also used. In a reinforced polymer, the resin matrix is the continuous phase and the fiber reinforcement is the discontinuous phase. The function of the resin is to bond the fibers together to provide shape and form and to transfer stresses in the structure from the resin to the fiber. Only high-strength fibers with high modulus are used. Because of the increased stiffness resulting from the fiber reinforcement, those polymers that are noted for their flexibility are not normally reinforced.

Virtually all thermosetting polymers can be reinforced with fibers. Polyester resins are particularly useful in reinforced polymers. They are used extensively in manufacturing very large components such as swimming pools, large tankage, boat hulls, shower enclosures, and building components. Reinforced molding materials such as phenolic, alkyd, or epoxy are used extensively in the electronics industry.

Many thermoplastic polymers are reinforced with fibers. Reinforcement is used to improve physical properties—specifically heat deflection temperature. Glass fibers are the most commonly used reinforcing material. The wear resistance and abrasion resistance of thermoplastic polymers are improved by the use of aramid reinforcing. Although fibers can be used with any thermoplastic polymer, the following are the most important:

1. Polyamide polymers use glass fibers to control brittleness. Tensile strengths are increased by a factor of three, and heat deflection temperature increases from 150 to 500°F (66 to 260°C).
2. Polycarbonate compounds using 10, 20, 30, and 40% glass fiber loading have their physical properties greatly improved.
3. Other polymers benefiting from the addition of glass fibers include polyphenylene sulfide, polypropylene, and polyether sulfone.

Polymers chosen for structural application are usually selected as a replacement for metal. Usually a like replacement of a polymer section for a metallic section will result in a weight saving. In addition, polymers can be easily formed into shapes that are difficult to achieve with metals. By using a polymer, the engineer can design an attractive shape that favors plastic forming and achieve a saving in cost and weight and a cosmetic improvement. An additional cost saving is realized since the polymer part does not require painting for corrosion protection as would the comparable metal part. Selection of the specific polymer will be based on mechanical requirements and the temperature and chemical end-use environment.

The following chapters will provide the necessary data to permit making an informed selection.

A. Physical and Mechanical Properties of Polymers

In order to properly apply any material, including polymers, it is necessary to know certain physical and mechanical properties of the material. The following chapters will include detailed discussions of each specific polymer, including a listing of their physical and mechanical properties. Presented here are the definitions of the various terms used in the tables.

Polymers have the characteristics of both a viscous liquid and a springlike elastomer, traits known as viscoelasticity. These characteristics are responsible for many of the peculiar mechanical properties displayed by polymers. Under mild loading conditions, such as short-term loading with low deflection and small loads at room temperature, polymers usually react like springs, returning to their original shape after the load is removed. During this purely elastic behaviour, no energy is lost or dissipated. Stress versus strain remains a linear function. Refer to Figure 1.1. As the applied load increases, there is a proportional increase in the parts deflection.

Under long-term heavy loads or elevated temperatures many polymers exhibit viscous behaviour. Polymers will deform and flow similarly to highly viscous liquids, although still solid. This viscous behavior is explained by two terms, strain (ε) and stress (σ). Strain is measured as percent elongation while stress is measured in load per area. Under tensile loading, viscous behavior shows that the strain resulting from a constant applied stress increases with time as a nonlinear function. Refer to Figure 1.2. This time- and temperature-dependent

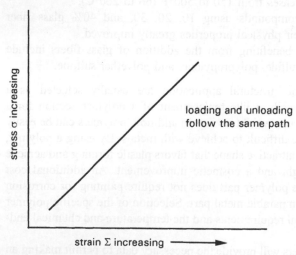

FIGURE 1.1   Strain–stress behavior linear relationship.

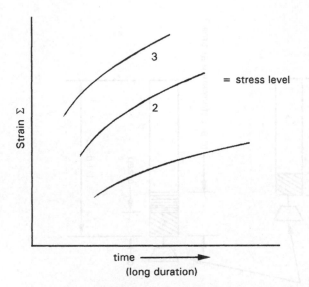

FIGURE 1.2 Viscous behavior of polymers with varying stress levels over time.

behavior takes place since the polymer chains in the part slip and do not return to their original position when the load is removed.

1. Creep

Creep is the deformation that occurs over time when a material is subjected to constant stress at constant temperature. This is the result of the viscoelastic behavior of polymers. Under these conditions, the polymer chains slowly slip past each other. Since some of this slippage is permanent, only a portion of the creep deformation can be recovered when the load is removed. Figure 1.3 demonstrates the phenomenon of creep. A weight hung from a polymer tensile bar will cause the initial deformation "*d*." Over an extended period of time the weight increases the bars elongation or creep "*C*."

2. Stress Relaxation

Stress relaxation is another viscoelastic phenomenon. It is defined as a gradual decrease in stress at constant temperature. Stress relaxation occurs as a result of the same polymer chain slippage found in creep. It takes place in simple tension as well as in parts subjected to multiaxial tension, bending, shear, and compression. Load, load duration, temperature, and other factors determine the degree of stress relaxation. Figure 1.4 demonstrates that a large weight initially produces an elongation "*d*" and a strain d/L (L = original length). Over time, less weight is

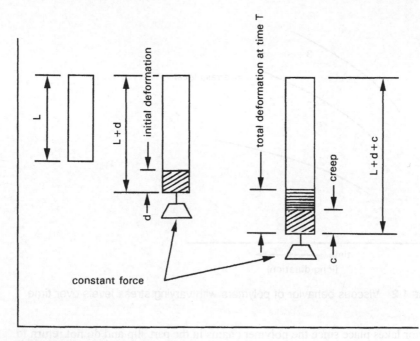

FIGURE 1.3   Creep phenomenon.

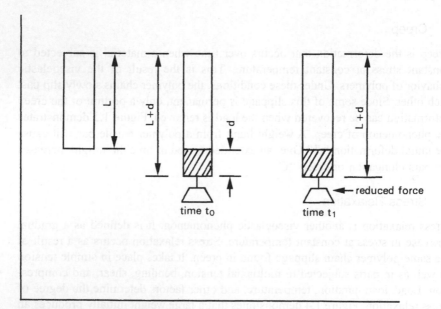

FIGURE 1.4   Stress relaxation.

needed to maintain the same elongation and strain in the test bar because of stress relaxation.

When designing parts that will be subject to constant strain, stress relaxation must be taken into account. A typical press fit, such as a metal insert in a polymer boss, relies upon stresses from the imposed strain of an interference fit to hold the insert in place. Polymer-chain slippage can relax these stresses and reduce the insert retention strength over time.

3. Recovery

Recovery is the degree to which a polymer returns to its original shape after a load is removed. Refer to Figure 1.5 where strain is plotted against time. The initial strain from an applied load is shown at point A. Over time, creep causes the strain to increase to point B. The strain immediately drops to point C when the load is removed. If 100% recovery were possible, the polymer part would return to its original size at point E. More commonly, the polymer retains some permanent deformation, as shown by point D.

4. Specific Gravity

The ratio of the weight of any volume to the weight of an equal volume of some other substance taken as the standard at a stated temperature. For polymers, the standard is water.

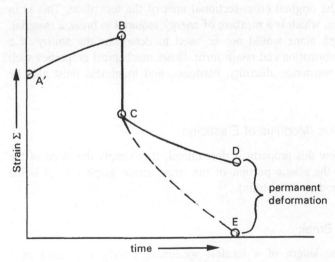

FIGURE 1.5 Load and recovery behavior.

5. Water Absorption

The ratio of the weight of water absorbed by a material to the weight of the dry material, expressed as a percentage. Many polymers are hygroscopic, meaning that over time they absorb water. Through a process called hydrolysis, water in a polymer severs the polymer chains, reduces molecular weight, and decreases mechanical properties. Longer exposure at elevated temperature and/or loads increases hydrolytic attack.

Water absorption can also change the physical properties of polyamides without degrading them. Some polyamides absorb large amounts of water causing them to swell. As the moisture contact increases, other mechanical and electrical properties may also change. These changes are reversible. When the polymer is dried, the mechanical properties return to their original values.

6. Dielectric Strength

The voltage that an insulating material can withstand before breakdown occurs, usually expressed as a voltage gradient (such as volts per mil). The value obtained will depend on the thickness of the material and on the method and conditions of test.

7. Tensile Strength at Break

This is a measure of the stress required to deform a material prior to breakage. It is calculated by dividing the maximum load applied to the material before its breaking point by the original cross-sectional area of the test piece. This is in contrast to toughness, which is a measure of energy required to break a material.

Tensile strength alone would not be used to determine the ability of a polymer to resist deformation and retain form. Other mechanical properties such as elasticity, creep resistance, ductility, hardness, and toughness must also be taken into account.

8. Tensile Modulus (Modulus of Elasticity)

Figure 1.6 shows how this property is determined. It is simply the slope of the line that represents the elastic portion of the stress–strain graph (i.e., it is the stress required to produce unit strain).

9. Elongation at Break

The increase in the length of a tension specimen, usually expressed as a percentage of the original length of the specimen.

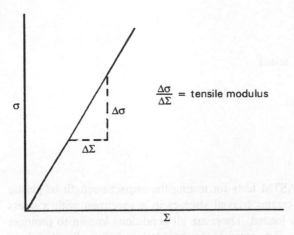

$$\frac{\Delta\sigma}{\Delta\Sigma} = \text{tensile modulus}$$

Elastic stress-strain relationship showing
how the tensile modulus is determined

FIGURE 1.6   Elastic stress–strain relationship showing how the tensile modulus is determined.

10. Compressive Strength

The maximum compressive stress a material is capable of sustaining. For materials that do not fail by a shattering fracture, the value is arbitrary, depending on the distortion allowed.

11. Flexural Strength

The strength of a material in bending expressed as the tensile stress of the outermost fibers of a bent test sample at the instant of failure.

12. Flexural Modulus

The ratio, within the elastic limit, of stress to the corresponding strain. It is calculated by drawing a tangent to the steepest initial straight-line portion of the load-deformation curve and calculating by the following equation:

$$E_{\mathrm{B}} = \frac{L^3 m}{4bd^3}$$

where

E_B = modulus
  b = width of beam tested
  d = depth of beam
  m = slope of tangent
  L = span, inches

13. Izod Impact

One of the most common ASTM tests for testing the impact strength of plastic materials. The impact test creates triaxial stresses in a specimen with a stress concentration that is rapidly loaded. These are all conditions known to promote brittle failure. Izod tests give data to compare the relative ability of materials to resist brittle failure as the service temperature decreases. The procedure is described in ASTM D256 for plastics.

14. Hardness, Rockwell Number

The number derived from the net increase in depth of impression as the load on a penetrator is increased from a fixed minimum load to a high load and then returned to a minimum load. Penetrators include steel balls of several specified diameters and a diamond-cone penetrator.

15. Coefficient of Thermal Expansion

The change in unit length or volume resulting from a unit change in temperature. Commonly used units are 10^{-6} in./in./°F or 10^{-6} cm/cm/°C.

16. Thermal Conductivity

The ability of a material to conduct heat; a physical constant for the quantity of heat that passes through a unit cube of a material in a unit of time when the difference in temperature of two faces is 1°C.

17. Deflection Temperature

The heat deflection (distortion) temperature (HDT) test is one in which a bar of the polymer in question is heated uniformly in a closed chamber while a load of 66 psi or 264 psi is placed at the center of the horizontal bar. The HDT is the temperature at which a slight deflection of 0.25 mm is noted at the center. The HDT indicates how much mass (weight) the object must be constructed of to maintain the desired form stability and strength rating and provides a measure of the rigidity of the polymer under a load as well as temperature.

18. Limiting Oxygen Index

This is a measure of the minimum oxygen level required to support combustion of the polymer.

19. Flame Spread Classification

These ratings are based on common tests outlined by the Underwriters' Laboratories and are defined as follows:

Flame spread rating	Classification
0-26	Noncombustible
25-50	Fire retardant
50-75	Slow-burning
75-200	Combustible
over 200	Highly combustible

Various properties of polymers may be improved by the use of additives. In some instances the use of additives to improve a specific property may have an adverse effect on certain other properties. Corrosion resistance is a property most often affected by the use of additives. In many cases the corrosion resistance of a polymer is reduced as a result of additives being used. Common additives used to improve the performance of thermoplastic polymers are listed here:

Antioxidants	Protect against atmospheric oxidation
Colorants	Dyes and pigments
Coupling agents	Used to improve adhesive bonds
Filler or extenders	Minerals, metallic powders, and organic compounds used to improve specific properties or to reduce costs
Flame retardants	Change the chemistry/physics of combustion
Foaming agents	Generate cells or gas pockets
Impact modifiers	Materials usually containing an elastomeric component to reduce brittleness
Lubricants	Substances that reduce friction, heat, and wear between surfaces
Optical brighteners	Organic substances that absorb UV radiation below 3000 A° and emit radiation below 5500 A°
Plasticizers	Increase workability
Reinforcing fibers	Increase strength, modulus, and impact strength
Processing aids	Improve hot processing characteristics
Stabilizers	Control for adjustment of deteriorative and physicochemical reactions during processing and subsequent life

TABLE 1.2 Fillers and Their Property Contribution to Polymers

Filler	Chemical resistance	Heat resistance	Electrical insulation	Impact strength	Tensile strength	Dimensional stability	Stiffness	Hardness	Electrical conductivity	Thermal conductivity	Moisture resistance	Hardenability
Alumina powder											x	x
Alumina tetrahydrate			x								x	x
Bronze									x	x		
Calcium carbonate		x					x	x		x		
Calcium silicate		x					x	x		x		
Carbon black		x					x	x	x	x		x
Carbon fiber							x		x	x		
Cellulose				x								
Alpha cellulose				x	x							
Coal, powdered	x										x	
Cotton, chopped fibers				x			x	x				
Fibrous glass		x			x	x	x	x				
Graphite	x	x					x	x	x	x	x	
Jute							x	x				
Kaolin	x						x	x				
Mica	x	x	x			x	x	x				
Molybdenium disulfide							x					
Nylon, chopped fibers	x	x		x	x		x				x	
Orlon	x	x		x	x		x				x	
Rayon				x	x		x			x		
Silica, amorphous					x	x	x	x			x	x
TFE	x						x				x	x
Talc	x						x	x			x	x
Wood flour		x					x	x				x

A list of specific fillers and the properties they improve are given in Table 1.2. Many thermoplastic polymers have useful properties without the need for additives. However, other thermoplasts require additives to be useful: For example PVC benefits from all additives and is practically useless in the pure form. Examples of the effects of additives on specific polymers will be illustrated.

Impact resistance is improved in polybutylene terephthalates, polypropylene, polycarbonate, PVC, acetals (POM) and certain polymer blends by the use of additives. Figures 1.7 shows the increase in impact strength of nylon, polycarbonate, polypropylene, and polystyrene by the addition of 30 wt% of glass fibers.

Glass fibers also increase the strength and moduli of thermoplastic polymers. Figures 1.8 and 1.9 illustrate the effect on the tensile stress and flexural moduli of nylon, polycarbonate, polypropylene, and polystyrene when 30 wt% glass fiber additions have been made.

The addition of 30 wt% of glass fibers also increases the heat distortion temperature. Table 1.3 shows the increase in the HDT when glass fibers have been added to polymers.

20. Permeation

All materials are somewhat permeable to chemical molecules, but plastic materials tend to be an order of magnitude greater in their permeability than metals. Gases, vapors, or liquids will permeate polymers.

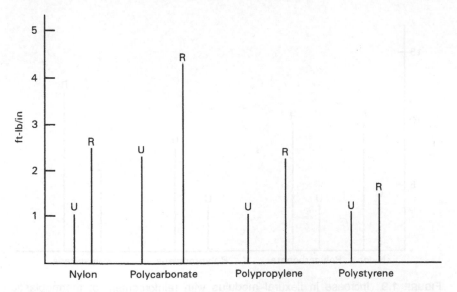

FIGURE 1.7 Izod impact change with glass reinforcement of thermoplastic polymers. U = unreinforced; R = reinforced.

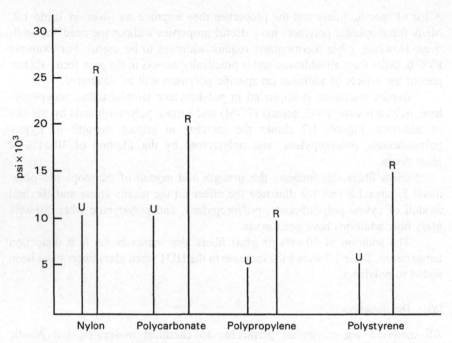

FIGURE 1.8 Increase in tensile strength with glass reinforcement of thermoplastic polymers. U = unreinforced; R = reinforced.

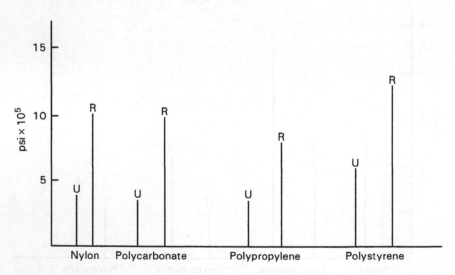

FIGURE 1.9 Increase in flexural modulus with reinforcement of thermoplastic polymers. U = unreinforced; R = reinforced.

TABLE 1.3 Increase of HDT with 20 wt% Glass Fiber Addition to the Polymer

Polymer	HDT at 264 psi 20% glass (°F/°C)	Increase over base polymer (°F/°C)
Acetal copolymer	325/163	95/52
Polypropylene	250/121	110/61
Linear polyethylene	260/127	140/77
Thermoplastic polyester	400/207	230/139
Nylon 6[a]	425/219	305/168
Nylon 6/6[a]	490/254	330/183
ABS	215/102	25/14
Styrene–acrylonitrile	215/102	20/12
Polystyrene	220/104	20/12
Polycarbonate	290/143	20/12
Polysulfone	365/185	20/12

[a] 30 wt% glass fibers.

Permeation is molecular migration through microvoids either in the polymer (if the polymer is more or less porous) or between polymer molecules. In neither case is there any attack on the polymer. This action is strictly a physical phenomenon. However, permeation can be detrimental when a polymer is used to line piping or equipment. In lined equipment, permeation can result in

1. Failure of the substrate from corrosive attack.
2. Bond failure and blistering, resulting from the accumulation of fluids at the bond, when the substrate is less permeable than the liner, or from corrosion/reaction products if the substrate is attacked by the permeant.
3. Loss of contents through substrate and liner as a result of the eventual failure of the substrate. In unbonded linings it is important that the space between the liner and support member be vented to the atmosphere, not only to allow minute quantities of permeant vapors to escape, but also to prevent expansion of entrapped air from collapsing the liner.

Permeation is a function of two variables, one relating to diffusion between molecular chains and the other to the solubility of the permeant in the polymer. The driving force of diffusion is the partial pressure of gases and the concentration gradient of liquids. Solubility is a function of the affinity of the permeant for the polymer.

All polymers do not have the same rate of permeation. In fact, some polymers are not affected by permeation. The fluoropolymers are particularly affected. Vapor permeation of PTFE is shown in Table 1.4, and Table 1.5 shows the vapor permeation of FEP. Table 1.6 provides permeation data of various gases in PFA and Table 1.7 gives the relative gas permeation into fluoropolymers.

There is no relationship between permeation and the passage of materials through cracks and voids, even though in both cases migrating chemicals travel through the polymer from one side to the other.

Some control can be exercised over permeation, which is affected by

1. Temperature and pressure
2. The permeant concentration
3. The thickness of the polymer

Increasing the temperature will increase the permeation rate since the solubility of the permeant in the polymer will increase, and as the temperature rises, polymer chain movement is stimulated, permitting more permeant to diffuse among the chains more easily. The permeation rates of many gases increase linearly with the partial pressure gradient, and the same effect is experienced with the concentration gradients of liquids. If the permeant is highly soluble in the polymer, the permeability increase may be nonlinear. The thickness will generally decrease permeation by the square of the thickness.

The density of a polymer, as well as the thickness, will have an effect on the permeation rate. The greater the density of the polymer, the fewer voids through which permeation can take place. A comparison of the density of sheets produced from different polymers does not provide any indication of the relative permea-

TABLE 1.4 Vapor Permeation into PTFE[a]

	Permeation g/100 in^2/24 h/mil	
Gases	73°F/23°C	86°F/30°C
Carbon dioxide		0.66
Helium		0.22
Hydrogen chloride, anh.		< 0.01
Nitrogen		0.11
Acetophenone	0.56	
Benzene	0.36	0.80
Carbon tetrachloride	0.06	
Ethyl alcohol	0.13	
Hydrochloric acid 20%	< 0.01	
Piperdine	0.07	
Sodium hydroxide 50%	5×10^{-5}	
Sulfuric acid 98%	1.8×10^{-5}	

[a] Based on PTFE having a specific gravity of >2.2.

TABLE 1.5 Vapor Permeation into FEP

	Permeation (g/100 in²/24 h/mil) at		
	73°F/23°C	93°F/35°C	122°F/50°C
Gases			
Nitrogen	0.18		
Oxygen	0.39		
Vapors			
Acetic acid		0.42	
Acetone	0.13	0.95	3.29
Acetophenone	0.47		
Benzene	0.15	0.64	
n-Butyl ether	0.08		
Carbon tetrachloride	0.11	0.31	
Decane	0.72		1.03
Ethyl acetate	0.06	0.77	2.9
Ethyl alcohol	0.11	0.69	
Hexane		0.57	
Hydrochloric acid 20%	< 0.01		
Methanol			5.61
Sodium hydroxide	4×10^{-5}		
Sulfuric acid 98%	8×10^{-6}		
Toluene	0.37		2.93

TABLE 1.6 Permeation of Gases into PFA

Gas	Permeation at 77°F/25°C (cc/mil thickness/100 in²/24 h/atm)
Carbon dioxide	2260
Nitrogen	291
Oxygen	881

TABLE 1.7 Relative Gas Permeation into Fluoropolymers[a]

Gas	PVDF	PTFE	FEP	PFA
Air	27	2,000	600	1150
Oxygen	20	1,500	2,900	—
Nitrogen	30	500	1,200	—
Helium	600	35,000	18,000	17,000
Carbon dioxide	100	15,000	4,700	7,000

[a] Permeation through a 100 μm film at 73°F/23°C. Units = cm³/m² deg bar.

tion rates. However, a comparison of the density of sheets produced from the same polymer will provide an indication of the relative permeation rates. The denser the sheet, the lower the permeation rate.

The thickness of a lining is a factor affecting permeation. For general corrosion resistance, thicknesses of 0.010–0.020 inch are usually satisfactory, depending on the combination of lining material and the specific corrodent. When mechanical factors such as thinning to cold flow, mechanical abuse, and permeation rates are a consideration, thicker linings may be required.

Increasing a lining thickness will normally decrease permeation by the square of the thickness. Although this would appear to be the approach to follow to control permeation, there are disadvantages. First, as thickness increases, the thermal stresses on the boundary increase, which can result in bond failure. Temperature changes and large differences in coefficients of thermal expansion are the most common causes of bond failure. Thickness and modulus of elasticity of the plastic are two of the factors that influence these stresses. Second, as the thickness of a lining increases, installation becomes difficult with a resulting increase in labor costs.

The rate of permeation is also affected by the temperature and temperature gradient in the lining. Lowering these will reduce the rate of permeation. Lined vessels, such as storage tanks, that are used under ambient conditions provide the best service.

Other factors affecting permeation consist of these chemical and physio-chemical properties:

1. Ease of condensation of the permeant. Chemicals that condense readily will permeate at higher rates.
2. The higher the intermolecular chain forces (e.g., Van der Waals hydrogen bonding) of the polymer, the lower the permeation rate.
3. The higher the level of crystallinity in the polymer, the lower the permeation rate.
4. The greater the degree of cross-linking within the polymer, the lower the permeation rate.
5. Chemical similarity between the polymer and permeant. When the polymer and permeant both have similar functional groups, the permeation rate will increase.
6. The smaller the molecule of the permanent, the greater the permeation rate.

21. Absorption

Polymers have the potential to absorb varying amounts of corrodents they come into contact with, particularly organic liquids. This can result in swelling, cracking, and penetration to the substrate of a lined component. Swelling can cause softening of the polymer, introduce high stresses, and cause failure of the bond on lined components. If the polymer has a high absorption rate, permeation will probably take place. An approximation of the expected permeation and/or absorption of a polymer can be based on the absorption of water. These data are usually available. Table 1.8 provides the water absorption rates for the more common polymers, Table 1.9 gives the absorption rates of various liquids by FEP, and Table 1.10 provides the absorption rates of representative liquids by PFA.

TABLE 1.8 Water Absorption Rates of Polymers

Polymer	Water absorption 24 h at 73°F/23°C (%)
PVC	0.05
CPVC	0.03
PP (Homo)	0.02
PP (Co)	0.05
EHMW PE	< 0.01
ECTFE	< 0.01
PVDF	< 0.04
PVCP (Saran)	Nil
PFA	< 0.03
ETFE	0.029
PTFE	< 0.01
FEP	< 0.01

TABLE 1.9 Absorption of Selected Liquids by FEP[a]

Chemical	Temp. (°F/°C)	Range of weight gains (%)
Aniline	365/185	0.3–0.4
Acetophenone	394/201	0.6–0.8
Benzaldehyde	354/179	0.4–0.5
Benzyl alcohol	400/204	0.3–0.4
n-Butylamine	172/78	0.3–0.4
Carbon tetrachloride	172/78	2.3–2.4
Dimethyl sulfide	372/190	0.1–0.2
Nitrobenzene	410/210	0.7–0.9
Perchlorethylene	250/121	2.0–2.3
Sulfuryl chloride	154/68	1.7–2.7
Toluene	230/110	0.7–0.8
Tributyl phosphate	392/200[b]	1.8–2.0

[a] Exposure for 168 hours at their boiling points.
[b] Not boiling.

TABLE 1.10 Absorption of Liquids by PFA

Liquid[a]	Temp. (°F/°C)	Range of weight gains (%)
Aniline	365/185	0.3−0.4
Acetophenone	394/201	0.6−0.8
Benzaldehyde	354/179	0.4−0.5
Benzyl alcohol	400/204	0.3−0.4
n-Butylamine	172/78	0.3−0.4
Carbon tetrachloride	172/78	2.3−2.4
Dimethyl sulfoxide	372/190	0.1−0.2
Freon 113	117/47	1.2
Isooctane	210/99	0.7−0.8
Nitrobenzene	410/210	0.7−0.9
Perchlorethylene	250/121	2.0−2.3
Sulfuryl chloride	154/68	1.7−2.7
Toluene	230/110	0.7−0.8
Tributyl phosphate[b]	392/200	1.8−2.0
Bromine, anhydrous	−5/−22	0.5
Chlorine, anhydrous	248/120	0.5−0.6
Chlorosulfonic acid	302/150	0.7−0.8
Chromic acid 50%	248/120	0.00−0.01
Ferric chloride	212/100	0.00−0.01
Hydrochloric acid 37%	248/120	0.00−0.03
Phosphoric acid, conc.	212/100	0.00−0.01
Zinc chloride	212/100	0.00−0.03

[a] Samples were exposed for 168 hours at the boiling point of the solvent. Exposure of the acidic reagents was for 168 hours.
[b] Not boiling.

The failure due to absorption can best be understood by considering the "steam cycle" test described in ASTM standards for lined pipe. A section of lined pipe is subjected to thermal and pressure fluctuations. This is repeated for 100 cycles. The steam creates a temperature and pressure gradient through the liner, causing absorption of a small quantity of steam, which condenses to water within the inner wall. Upon pressure release or on reintroduction of steam, the entrapped water can expand to vapor, causing an original micropore. The repeated pressure and thermal cycling enlarge the micropores, ultimately producing visible water-filled blisters within the liner.

In an actual process, the polymer may absorb process fluids, and repeated temperature or pressure cycling can cause blisters. Eventually, corrodent may find its way to the substrate.

Related effects can occur when process chemicals are absorbed, which may later react, decompose, or solidify within the structure of the polymer. Prolonged retention of the chemicals may lead to their decomposition within the polymer. Although it is unusual, it is possible for absorbed monomers to polymerize.

Several steps can be taken to reduce absorption. Thermal insulation of the substrate will reduce the temperature gradient across the vessel, thereby preventing condensation and subsequent expansion of the absorbed fluids. This also reduces the rate and magnitude of temperature changes, keeping blisters to a minimum. The use of operating procedures or devices that limit the ratio of process pressure reductions or temperature increases will provide added protection.

B. Painting of Polymers

Polymers are painted since this is frequently a less expensive process than precolored resins or molded-in coloring. They are also painted, when necessary, to provide UV protection. However, they are difficult to paint, and proper consideration must be given to the following:

1. Heat distortion point and heat resistance. This determines whether a bake-type paint can be used, and if so the maximum baking temperature the polymer can tolerate.

2. Solvent resistance. Since different polymers are subject to attack by different solvents, this will dictate the choice of paint system. Some softening of the surface is desirable to improve adhesion, but a solvent that attacks the surface aggressively and results in cracking or crazing must be avoided.

3. Residual stress. Molded parts may have localized areas of stress. A coating applied in these areas may swell the polymer and cause crazing. Annealing of the part prior to coating will minimize or eliminate the problem.

4. Mold-release residues. If excessive amounts of mold-release compounds remain on the part, adhesion problems are likely. To prevent such a problem, the polymer should be thoroughly rinsed or otherwise cleaned.

5. Plasticizers and other additives. Most polymers are formulated with plasticizers and chemical additives. These materials have a tendency to migrate to the surface and many even soften the coating and destroy the adhesion. The specific polymer formulation should be checked to determine whether the coating will cause short- or long-term softening or adhesion problems.

6. Other factors. The long-term adhesion of the coating is affected by properties of the polymer such as stiffness or rigidity, dimensional stability, and coefficient of expansion. The physical properties of the paint film must accommodate to those of the polymer.

C. Corrosion of Polymers

Corrosion of metallic materials takes place via an electrochemical reaction at a specific corrosion rate. Consequently, the life of a metallic material in a particular corrosive environment can be accurately predicted. This is not the case with polymeric materials.

Plastic materials do not experience specific corrosion rates. They are usually completely resistant to a specific corrodent or they deteriorate rapidly. Polymers are attacked either by chemical reaction or by solvation. Solvation is the penetration of the polymer by a corrodent which causes softening, swelling, and ultimate failure. Corrosion of plastics can be classified in the following ways as to attack mechanism:

1. Disintegration or degradation of a physical nature due to absorption, permeation, solvent action, or other factors
2. Oxidation, where chemical bonds are attacked
3. Hydrolysis, where ester linkages are attacked
4. Radiation
5. Thermal degradation involving depolymerization and possibly repolymerization
6. Dehydration (rather uncommon)
7. Any combination of the above

Results of such attacks will appear in the form of softening, charring, crazing, delamination, embrittlement, discoloration, dissolving, or swelling.

The common resistance of polymer matrix composites is also affected by two other factors, the nature of the laminate and in the case of thermoset resins, the cure. Improper or insufficient cure time will adversely affect the corrosion resistance, while proper cure time and procedures will generally improve corrosion resistance.

Polymeric materials in outdoor applications are exposed to weather extremes that can be extremely deleterious to the material. The most harmful weather component, exposure to ultraviolet (UV) radiation, can cause embrittlement, fading, surface cracking, and chalking. After exposure to direct sunlight for a period of years, most polymers exhibit reduced impact resistance, lower overall mechanical performance, and a change in appearance.

The electromagnetic energy from sunlight is normally divided into ultraviolet light, visible light, and infrared energy. Infrared energy consists of wavelengths longer than visible red wavelengths and starts above 760 manometers (nm). Visible light is defined as radiation between 400 and 760 nm. Ultraviolet light consists of radiation below 400 nm. The UV portion of the spectrum is further subdivided into UV-A, UV-B, and UV-C. The effects of the various wavelengths are shown in Table 1.11.

Since UV is easily filtered by air masses, cloud cover, pollution, and other factors, the amount and spectrum of natural UV exposure is extremely variable. Because the sun is lower in the sky during the winter months, it is filtered through a greater air mass. This creates two important differences between summer and

TABLE 1.11   Wavelength Regions of the UV

Region	Wavelength (nm)	Characteristics
UV-A	400–315	Causes polymer damage
UV-B	315–200	Includes the shortest wavelengths found at the earth's surface
		Causes severe polymer damage
		Absorbed by window glass
UV-C	280–100	Filtered out by the earth's atmosphere
		Found only in outer space

winter sunlight: changes in the intensity of the light and in the spectrum. During the winter months, much of the damaging short-wavelength UV light is filtered out. For example, the intensity of UV at 320 nm changes about 8 to 1 from summer to winter. In addition, the short-wavelength solar cutoff shifts from approximately 295 nm in summer to approximately 310 nm in winter. As a result, materials sensitive to UV below 320 nm would degrade only slightly, if at all, during the winter months.

Photochemical degradation is caused by photons or light breaking chemical bonds. For each type of chemical bond, there is a critical threshold wavelength of light with enough energy to cause a reaction. Light of any wavelength shorter than the threshold can break a bond, but longer wavelengths of light cannot break it. Therefore, the short wavelength cutoff of a light source is of critical importance. If a particular polymer is sensitive only to UV light below 295 nm (the solar cutoff point) it will never experience photochemical deterioration outdoors.

The ability to withstand weathering varies with the polymer type and within grades of a particular resin. Many resin grades are available with UV-absorbing additives to improve weatherability. However, the higher molecular weight grades of a resin generally exhibit better weatherability than the lower molecular weight grades with comparable additives. In addition, some colors tend to weather better than others.

Many of the physical property and chemical resistance differences of polymers stem directly from the type and arrangement of atoms in the polymer chains. The periodic table is shown in Figure 1.10.

In the periodic table, the basic elements of nature are placed into classes with similar properties, i.e., elements and compounds that exhibit similar behaviour. These classes are alkali metals, alkaline earth metals, transition metals, rare earth series, other metals, nonmetals, and noble (inert) gases.

IA	IIA	IIIB	IVB	VB	VIB	VIIB	VIII			IB	IIB	IIIA	IVA	VA	VIA	VIIA	0
1 H 1.00797																1 H 1.00797	2 He 4.0026
3 Li 6.939	4 Be 9.0122											5 B 10.811	6 C 12.01115	7 N 14.0067	8 O 15.9994	9 F 18.9984	10 Ne 20.183
11 Na 22.9898	12 Mg 24.312											13 Al 26.9815	14 Si 28.086	15 P 30.9738	16 S 32.064	17 Cl 35.453	18 Ar 39.948
19 K 39.102	20 Ca 40.08	21 Sc 44.956	22 Ti 47.90	23 V 50.942	24 Cr 51.996	25 Mn 54.938	26 Fe 55.847	27 Co 58.933	28 Ni 58.71	29 Cu 63.54	30 Zn 65.37	31 Ga 69.72	32 Ge 72.59	33 As 74.922	34 Se 78.96	35 Br 79.909	36 Kr 83.80
37 Rb 85.47	38 Sr 87.62	39 Y 88.905	40 Zr 91.22	41 Nb 92.906	42 Mo 95.94	43 Tc (98)	44 Ru 101.07	45 Rh 102.905	46 Pd 106.4	47 Ag 107.870	48 Cd 112.40	49 In 114.82	50 Sn 118.69	51 Sb 121.75	52 Te 127.60	53 I 126.904	54 Xe 131.30
55 Cs 132.905	56 Ba 137.34	57 La 138.91	72 Hf 178.49	73 Ta 180.948	74 W 183.85	75 Re 186.2	76 Os 190.2	77 Ir 192.2	78 Pt 195.09	79 Au 196.967	80 Hg 200.59	81 Tl 204.37	82 Pb 207.19	83 Bi 208.980	84 Po (210)	85 At (210)	86 Rn (222)
87 Fr (223)	88 Ra (226)	89 Ac (227)														71 Lu 174.97	103 Lw (257)

58 Ce 140.12	59 Pr 140.907	60 Nd 144.24	61 Pm (147)	62 Sm 150.35	63 Eu 151.96	64 Gd 157.25	65 Tb 158.924	66 Dy 162.50	67 Ho 164.930	68 Er 167.26	69 Tm 168.934	70 Yb 173.04
90 Th 232.038	91 Pa (231)	92 U 238.04	93 Np (237)	94 Pu (242)	95 Am (243)	96 Cm (247)	97 Bk (247)	98 Cf (249)	99 Es (254)	100 Fm (253)	101 Md (256)	102 No (254)

FIGURE 1.10   The periodic table.

Of particular importance and interest for thermoplasts is the category known as halogens, which are found within the nonmetal category. The elements included in this category are fluorine, chlorine, bromine, and iodine. Since these are the most electronegative elements in the periodic table, they are the most likely to attract an electron from another element and become part of a stable structure. Of all the halogens, fluorine is the most electronegative, permitting it to bond strongly with carbon and hydrogen atoms, but not well with itself. The carbon–fluorine bond, the predominant bond in PVDF and PTFE which gives it such important properties, is among the strongest known organic compounds. The fluorine acts as a protective shield for other bonds of lesser strength within the main chain of the polymer. The carbon–hydrogen bond of which such plastics as PPE and PP are composed is considerably weaker. This class of polymers are known as polyolefins.

The carbon–chlorine bond, a key bond of PVC is weaker yet.

The arrangement of the elements in the molecule, the symmetry of the structure, and the degree of branching of the polymer chains are as important as the specific elements contained in the molecule. Polymers containing the carbon–hydrogen bonds such as polypropylene and polyethylene, and the carbon–chlorine bonds such as polyvinyl chloride, ethylene chlorotrifluoroethylene and chlorotrifluorethylene, are different in the important property of chemical resistance from a fully fluorinated polymer such as polytetrafluorethylene. The latter has a much wider range of corrosion resistance.

The fluoroplastic materials are divided into two groups: fully fluorinated fluorocarbon polymers such as PTFE, FEP, and PFA called perfluoropolymers, and the partially fluorinated polymers such as ETFE, PVDF, and ECTFE which are called fluoropolymers. The polymeric characteristics within each group are similar, but there are important differences between the groups, as will be seen later.

2

Thermoplastic Polymers

2.1 INTRODUCTION TO THERMOPLASTIC POLYMERS

As discussed previously, thermoplastic polymers can be repeatedly re-formed by the application of heat, similar to metallic materials. They are long-chain linear molecules that are easily formed by the application of heat and pressure at temperatures above a critical temperature referred to as the "glass temperature." Because of this ability to be re-formed by heat, these materials can be recycled. However, thermal aging which results from repeated exposure to the high temperatures required for melting, causes eventual degradation of the polymer and limits the number of reheat cycles.

Polymers are formed as the result of a polymerization reaction of a monomer which is a single molecule or substance consisting of single molecules. Copolymers are long-chain molecules formed by the addition reaction of two or more monomers. In essence, they are chains in which one mer has been substituted with another mer. When the chain of a polymer is made up of a single repeating section, it is referred to as a homopolymer in contrast to a copolymer. Thermoplastic polymers can be either homopolymers or copolymers. Alloy polymers are blends of different polymers.

In general, thermoplastic materials tend to be tougher and less brittle than thermoset polymers so they can be used without the need for incorporating fillers. However, all thermoplasts do not fall into this category. Some tend to craze or crack easily, so each case must be considered on its individual merits. By virtue of their basic polymer structure, thermoplastics have been less dimensionally and thermally stable than thermosetting polymers. Therefore, thermosets have offered a performance advantage, although the lower processing costs for thermoplastics

29

have given the latter a cost advantage. Because of three major developments, thermosets and thermoplastics are now considered on the basis of performance. First, stability of thermoplastics has been greatly improved by the use of fiber reinforcement. Second has been the development of the so-called engineering, or high-stability, higher performance polymers, which can be reinforced with fiber filler to increase their stability even further. Third, to offset the gains in thermoplastics, has been the development of lower cost processing of thermo-setting polymers, specifically the screw-injection-molding technology.

The two most common materials used as reinforcement are glass fiber and carbon. When a reinforcing material is used in a thermoplast, the thermoplast is known as a composite. Compatibility of the reinforcing material with the corrodent to be encountered must be checked as well as the compatibility of the thermoplast. Table 2.1 provides the compatibilities of glass fibers with selected corrodents while Table 2.2 provides the compatibilities of carbon fibers with selected corrodents. Table 2.3 provides the compatibilities of imper-vious graphite with selected corrodents.

The engineering plastics are synthetic polymers of resin-based materials that have load-bearing characteristics and high-performance properties, which permit them to be used in the same manner as metals. The major products of the polymer industry which include polyethylene, polypropylene, polyvinyl chloride, and polystyrene are not considered engineering polymers because of their low strength. Many of the engineering plastics are copolymers or alloy polymers. Table 2.4 lists the abbreviations used for the more common thermoplasts.

One of the applications for plastic materials is to resist atmospheric corrosion. In addition to being able to resist attack by specific corrodents in plant operation, they are also able to resist corrosive fumes that may be present in the atmosphere. Table 2.5 provides the resistance of the more common thermo-plasts to various atmospheric pollutants, and Table 2.6 gives the allowable temperature range for thermoplastic polymers.

A. Joining of Thermoplastics

Thermoplastic materials are joined by either solvent cementing, thermal fusion, or by means of adhesives.

Solvent cementing is the easiest and most economical method for joining thermoplasts. Solvent-cemented joints are less sensitive to thermal cycling than joints bonded with adhesives, and they have the same corrosion resistance as the base polymer. It is possible to achieve a bond strength equivalent to 85 to 100% of the parent polymer. The major disadvantages of solvent cementing are the possibility of stress cracking of the part and the possible hazards of using low vapor point solvents. Adhesive bonding is generally recommended when two

TABLE 2.1 Compatibility of Borosilicate Glass Fiber with Selected Corrodents[a]

Chemical	Maximum temp. °F	Maximum temp. °C	Chemical	Maximum temp. °F	Maximum temp. °C
Acetaldehyde	450	232	Benzoic acid	200	93
Acetamide	270	132	Benzyl alcohol	200	93
Acetic acid 10%	400	204	Benzyl chloride	200	93
Acetic acid 50%	400	204	Borax	250	121
Acetic acid 80%	400	204	Boric acid	300	149
Acetic acid, glacial	400	204	Bromine gas, moist	250	121
Acetic anhydride	250	121	Bromine liquid	90	32
Acetone	250	121	Butadiene	90	32
Adipic acid	210	99	Butyl acetate	250	121
Allyl alcohol	120	49	Butyl alcohol	200	93
Allyl chloride	250	121	Butyric acid	200	93
Alum	250	121	Calcium bisulfite	250	121
Aluminum chloride, aqueous	250	121	Calcium carbonate	250	121
Aluminum chloride, dry	180	82	Calcium chlorate	200	93
Aluminum fluoride	x	x	Calcium chloride	200	93
Aluminum hydroxide	250	121	Calcium hydroxide 10%	250	121
Aluminum nitrate	100	38	Calcium hydroxide, sat.	x	x
Aluminum oxychloride	190	88	Calcium hypochlorite	200	93
Aluminum sulfate	250	121	Calcium nitrate	100	38
Ammonium bifluoride	x	x	Carbon bisulfide	250	121
Ammonium carbonate	250	121	Carbon dioxide, dry	160	71
Ammonium chloride 10%	250	121	Carbon dioxide, wet	160	71
Ammonium chloride 50%	250	121	Carbon disulfide	250	121
Ammonium chloride, sat.	250	121	Carbon monoxide	450	232
Ammonium fluoride 10%	x	x	Carbon tetrachloride	200	93
Ammonium fluoride 25%	x	x	Carbonic acid	200	93
Ammonium hydroxide 25%	250	121	Cellosolve	160	71
Ammonium hydroxide, sat.	250	121	Chloroacetic acid, 50% water	250	121
Ammonium nitrate	200	93	Chloroacetic acid	250	121
Ammonium persulfate	200	93	Chlorine gas, dry	450	232
Ammonium phosphate	90	32	Chlorine gas, wet	400	204
Ammonium sulfate 10–40%	200	93	Chlorine, liquid	140	60
Amyl acetate	200	93	Chlorobenzene	200	93
Amyl alcohol	250	121	Chloroform	200	93
Amyl chloride	250	121	Chlorosulfonic acid	200	93
Aniline	200	93	Chromic acid 10%	200	93
Antimony trichloride	250	121	Chromic acid 50%	200	93
Aqua regia 3:1	200	93	Citric acid 15%	200	93
Barium carbonate	250	121	Citric acid, concd	200	93
Barium chloride	250	121	Copper chloride	250	121
Barium hydroxide	250	121	Copper sulfate	200	93
Barium sulfate	250	121	Cresol	200	93
Barium sulfide	250	121	Cupric chloride 5%	160	71
Benzaldeyde	200	93	Cupric chloride 50%	160	71
Benzene	200	93	Cyclohexane	200	93
Benzenesulfonic acid 10%	200	93	Cyclohexanol		

(Continued)

TABLE 2.1 Continued

Chemical	Maximum temp. °F	°C	Chemical	Maximum temp. °F	°C
Dichloroacetic acid	310	154	Nitric acid, anhydrous	250	121
Dichloroethane (ethylene dichloride)	250	121	Oleum	400	204
Ethylene glycol	210	99	Perchloric acid 10%	200	93
Ferric chloride	290	143	Perchloric acid 70%	200	93
Ferric chloride 50% in water	280	138	Phenol	200	93
Ferric nitrate 10–50%	180	82	Phosphoric acid 50–80%	300	149
Ferrous chloride	200	93	Picric acid	200	93
Fluorine gas, dry	300	149	Potassium bromide 30%	250	121
Fluorine gas, moist	x	x	Silver bromide 10%		
Hydrobromic acid, dil.	200	93	Sodium carbonate	250	121
Hydrobromic acid 20%	200	93	Sodium chloride	250	121
Hydrobromic acid 50%	200	93	Sodium hydroxide 10%	x	x
Hydrochloric acid 20%	200	93	Sodium hydroxide 50%	x	x
Hydrochloric acid 38%	200	93	Sodium hydroxide, concd	x	x
Hydrocyanic acid 10%	200	93	Sodium hypochlorite 20%	150	66
Hydrofluoric acid 30%	x	x	Sodium hypochlorite, concd	150	66
Hydrofluoric acid 70%	x	x	Sodisum sulfide to 50%	x	x
Hydrofluoric acid 100%	x	x	Stannic chloride	210	99
Hypochlorous acid	190	88	Stannous chloride	210	99
Iodine solution 10%	200	93	Sulfuric acid 10%	400	204
Ketones, general	200	93	Sulfuric acid 50%	400	204
Lactic acid 25%	200	93	Sulfuric acid 70%	400	204
Lactic acid, concd	200	93	Sulfuric acid 90%	400	204
Magnesium chloride	250	121	Sulfuric acid 98%	400	204
Malic acid	160	72	Sulfuric acid 100%	400	204
Methyl chloride	200	93	Sulfurous acid	210	99
Methyl ethyl ketone	200	93	Thionyl chloride	210	99
Methyl isobutyl ketone	200	93	Toluene	250	121
Nitric acid 5%	400	204	Trichloroacetic acid	210	99
Nitric acid 20%	400	204	White liquor	210	99
Nitric acid 70%	400	204	Zinc chloride	210	99

[a] The chemicals listed are in the pure state or in a saturated solution unless otherwise indicated. Compatibility is shown to the maximum allowable temperature for which data are available. Incompatibility is shown by an x. A blank space indicates that data are unavailable.
Source: Schweitzer, PA. Corrosion Resistance Tables, 4th ed. Vols. 1–3. New York: Marcel Dekker, 1995.

TABLE 2.2   Compatibility of Carbon Fiber with Selected Corrodents[a]

Chemical	Maximum temp. °F	°C	Chemical	Maximum temp. °F	°C
Acetaldehyde	340	171	Benzene sulfonic acid 10%	340	171
Acetamide	340	171	Benzoic acid	350	177
Acetic acid 10%	340	171	Borax	250	121
Acetic acid 50%	340	171	Boric acid	210	99
Acetic acid 80%	340	171	Bromine gas, dry	x	x
Acetic acid, glacial	340	171	Bromine gas, moist	x	x
Acetic anhydride	340	171	Bromine, liquid	x	x
Acetone	340	171	Butadiene	340	171
Acetyl chloride	340	171	Butyl acetate	340	171
Acrylonitrile	340	171	Butyl alcohol	210	99
Adipic acid	340	171	n-Butylamine	100	38
Allyl alcohol	340	171	Butyric acid	340	171
Allyl chloride	100	38	Calcium bisulfide	340	171
Alum	340	171	Calcium bisulfite	340	171
Aluminum chloride, aqueous	340	171	Calcium carbonate	340	171
Aluminum chloride, dry	340	171	Calcium chlorate 10%	140	60
Aluminum fluoride	340	171	Calcium chloride	340	171
Aluminum hydroxide	340	171	Calcium hydroxide 10%	200	93
Aluminum nitrate	340	171	Calcium hydroxide, sat.	250	121
Ammonia gas	340	171	Calcium hypochlorite	170	77
Ammonium bifluoride	390	199	Calcium nitrate	340	171
Ammonium carbonate	340	171	Calcium oxide	340	171
Ammonium chloride 10%	340	171	Calcium sulfate	340	171
Ammonium chloride 50%	340	171	Caprylic acid	340	171
Ammonium chloride, sat.	340	171	Carbon bisulfide	340	171
Ammonium fluoride 10%	330	166	Carbon dioxide, dry	340	171
Ammonium fluoride 25%	340	171	Carbon dioxide, wet	340	171
Ammonium hydroxide 25%	200	93	Carbon disulfide	340	171
Ammonium hydroxide, sat.	220	104	Carbon monoxide	340	171
Ammonium nitrate	340	171	Carbon tetrachloride	250	121
Ammonium persulfate	340	171	Carbonic acid	340	171
Ammonium phosphate	340	171	Cellosolve	200	93
Ammonium sulfate 10–40%	340	171	Chloroacetic acid, 50% water	340	171
Ammonium sulfide	340	171	Chloroacetic acid	340	171
Amyl acetate	340	171	Chlorine gas, dry	180	82
Amyl alcohol	200	93	Chlorine gas, wet	80	27
Amyl chloride	210	99	Chlorobenzene	340	171
Aniline	340	171	Chloroform	340	171
Barium carbonate	250	121	Chlorosulfonic acid	340	171
Barium chloride	250	121	Chromic acid 10%	x	x
Barium hydroxide	250	121	Chromic acid 50%	x	x
Barium sulfate	250	121	Citric acid 15%	340	171
Barium sulfide	250	121	Citric acid, concd	340	171
Benzaldeyde	340	171	Copper carbonate	340	171
Benzene	200	93	Copper chloride	340	171
				(Continued)	

TABLE 2.2 Continued

Chemical	Maximum temp. °F	Maximum temp. °C	Chemical	Maximum temp. °F	Maximum temp. °C
Copper cyanide	340	171	Nitric acid 5%	180	82
Copper sulfate	340	171	Nitric acid 20%	140	60
Cresol	400	204	Nitric acid 70%	x	x
Cupric chloride 5%	340	171	Nitric acid, anhydrous	x	x
Cupric chloride 50%	340	171	Nitrous acid, concd	x	x
Cyclohexane	340	171	Perchloric acid 10%	340	171
D-butyl phthalate	90	32	Perchloric acid 70%	340	171
Ethylene glycol	340	171	Phenol	340	171
Ferric chloride	340	171	Phosphoric acid 50–80%	200	93
Ferric chloride 50% in water	340	171	Picric acid	100	38
Ferric nitrate 10–50%	340	171	Potassium bromide 30%	340	171
Ferrous chloride	340	171	Salicylic acid	340	171
Ferrous nitrate	340	171	Sodium bromide	340	171
Fluorine gas, dry	x	x	Sodium carbonate	340	171
Hydrobromic acid, dilute	340	171	Sodium chloride	340	171
Hydrobromic acid 20%	340	171	Sodium hydroxide 10%	240	116
Hydrobromic acid 50%	340	171	Sodium hydroxide 50%	270	132
Hydrochloric acid 20%	340	171	Sodium hydroxide, concd	260	127
Hydrochloric acid 38%	340	171	Sodium hypochlorite 20%	x	x
Hydrocyanic acid 10%	340	171	Sodium hypochlorite, concd	x	x
Hydrofluoric acid 30%	340	171	Sodium sulfide to 50%	120	49
Hydrofluoric acid 70%	x	x	Stannic chloride	340	171
Hydrofluoric acid 100%	x	x	Sulfuric acid 10%	340	171
Hypochlorous acid	100	38	Sulfuric acid 50%	340	171
Ketones, general	340	171	Sulfuric acid 70%	340	171
Lactic acid 25%	340	171	Sulfuric acid 90%	180	82
Lactic acid, concd	340	171	Sulfuric acid 98%	x	x
Magnesium chloride	170	77	Sulfuric acid 100%	x	x
Malic acid	100	38	Sulfurous acid	340	171
Manganese chloride	400	204	Toluene	340	171
Methyl chloride	340	171	Trichloroacetic acid	340	171
Methyl ethyl ketone	340	171	White liquor	100	38
Methyl isobutyl ketone	340	171	Zinc chloride	340	171
Muriatic acid	340	171			

[a] The chemicals listed are in the pure state or in a saturated solution unless otherwise indicated. Compatibility is shown to the maximum allowable temperature for which data are available. Incompatibility is shown by an x.

Source: Extracted from Schweitzer, PA. Corrosion Resistance Tables, 4th ed., New York: Marcel Dekker, 1995.

TABLE 2.3 Compatibility of Impervious Graphite with Selected Corrodents[a]

Chemical	Resin	Maximum temp. °F	Maximum temp. °C	Chemical	Resin	Maximum temp. °F	Maximum temp. °C
Acetaldehyde	Phenolic	460	238	Butyric acid	Furan	460	238
Acetamide	Phenolic	460	238	Calcium bisulfite	Furan	460	238
Acetic acid 10%	Furan	400	204	Calcium carbonate	Phenolic	460	238
Acetic acid 50%	Furan	400	204	Calcium chloride	Furan	460	238
Acetic acid 80%	Furan	400	204	Calcium hydroxide 10%	Furan	250	121
Acetic acid, glacial	Furan	400	204	Calcium hydroxide, sat.	Furan	250	121
Acetic anhydride	Furan	400	204	Calcium hypochlorite	Furan	170	77
Acetone	Furan	400	204	Calcium nitrate	Furan	460	238
Acetyl chloride	Furan	460	238	Calcium oxide			
Acrylic acid				Calcium sulfate	Furan	460	238
Acrylonitrile	Furan	400	204	Carbon bisulfide	Furan	400	204
Adipic acid	Furan	460	238	Carbon dioxide, dry	Phenolic	460	238
Allyl alcohol		x	x	Carbon dioxide, wet	Phenolic	460	238
Aluminum chloride, aqueous	Phenolic	120	49	Carbon disulfide	Furan	400	204
Aluminum chloride, dry	Furan	460	238	Carbon monoxide	Phenolic	460	238
Aluminum fluoride	Furan	460	238	Carbon tetrachloride	Furan	400	204
Aluminum hydroxide	Furan	250	121	Carbonic acid	Phenolic	400	204
Aluminum nitrate	Furan	460	238	Cellosolve	Furan	460	238
Aluminum sulfate	Phenolic	460	238	Chloroacetic acid, 50% water	Furan	400	204
Ammonium bifluoride	Phenolic	390	199	Chloroacetic acid	Furan	400	204
Ammonium carbonate	Phenolic	460	238	Chlorine gas, dry	Furan	400	204
Ammonium chloride 10%	Phenolic	400	204	Chlorine, liquid	Phenolic	130	54
Ammonium hydroxide 25%	Phenolic	400	204	Chlorobenzene	Phenolic	400	204
Ammonium hydroxide, sat.	Phenolic	400	204	Chloroform	Furan	400	204
Ammonium nitrate	Furan	460	238	Chlorosulfonic acid		x	x
Ammonium persulfate	Furan	250	121	Chromic acid 10%		x	x
Ammonium phosphate	Furan	210	99	Chromic acid 50%		x	x
Ammonium sulfate 10–40%	Phenolic	400	204	Citric acid 15%	Furan	400	204
Amyl acetate	Furan	460	238	Citric acid, concd	Furan	400	204
Amyl alcohol	Furan	400	204	Copper chloride	Furan	400	204
Amyl chloride	Phenolic	210	99	Copper cyanide	Furan	460	238
Aniline	Furan	400	204	Copper sulfate	Phenolic	400	204
Aqua regia 3:1		x	x	Cresol	Furan	400	204
Barium chloride	Furan	250	121	Cupric chloride 5%	Furan	400	204
Barium hydroxide	Phenolic	250	121	Cupric chloride 50%	Furan	400	204
Barium sulfate	Phenolic	250	121	Cyclohexane	Furan	460	238
Barium sulfide	Phenolic	250	121	Ethylene glycol	Furan	330	166
Benzaldehyde	Furan	460	238	Ferric chloride 60%	Phenolic	210	99
Benzene	Furan	400	204	Ferric chloride 50% in water	Furan	260	127
Benzenesulfonic acid 10%	Phenolic	460	238	Ferric nitrate 10–50%	Furan	210	99
Borax	Furan	460	238	Ferrous chloride	Furan	400	204
Boric acid	Phenolic	460	238	Fluorine gas, dry	Phenolic	300	149
Bromine gas, dry		x	x	Fluorine gas, moist		x	x
Bromine gas, moist		x	x	Hydrobromic acid, dil	Furan	120	49
Butadiene	Furan	460	238	Hydrobromic acid 20%	Furan	250	121
Butyl acetate	Furan	460	238	Hydrobromic acid 50%	Furan	120	49
Butyl alcohol	Furan	400	204	Hydrochloric acid 20%	Phenolic	400	204
n-Butylamine	Phenolic	210	99	Hydrochloric acid 38%	Phenolic	400	204

(Continued)

TABLE 2.3 Continued

Chemical	Resin	Maximum emp. °F	°C	Chemical	Resin	Maximum emp. °F	°C
Hydrocyanic acid 10%	Furan	460	238	Potassium bromide 30%	Furan	460	238
Hydrofluoric acid 30%	Phenolic	460	238	Salicylic acid	Furan	340	171
Hydrdofluoric acid 70%		x	x	Sodium carbonate	Furan	400	204
Hydrofluoric acid 100%		x	x	Sodium chloride	Phenolic	400	204
Iodine solution 10%	Phenolic	120	49	Sodium hydroxide 10%	Furan	400	204
Ketones, general	Furan	400	204	Sodium hydroxide 50%	Furan	400	204
Lactic acid 25%	Furan	400	204	Sodium hydroxide, concd		x	x
Lactic acid, concd	Furan	400	204	Sodium hypochlorite 20%		x	x
Magnesium chloride	Furan	170	77	Sodium hypochlorite, concd		x	x
Manganese chloride	Furan	460	238	Stannic chloride	Furan	400	204
Methyl chloride	Phenolic	460	238	Sulfuric acid 10%	Phenolic	400	204
Methyl ethyl ketone	Furan	460	238	Sulfuric acid 50%	Phenolic	400	204
Methyl isobutyl ketone	Furan	460	238	Sulfuric acid 70%	Phenolic	400	204
Muriatic acid	Phenolic	400	204	Sulfuric acid 90%	Phenolic	400	204
Nitric acid 5%	Phenolic	220	104	Sulfuric acid 98%		x	x
Nitric acid 20%	Phenolic	220	104	Sulfuric acid 100%		x	x
Nitric acid 70%		x	x	Sulfuric acid, fuming		x	x
Nitric acid, anhydrous		x	x	Sulfurous acid	Phenolic	400	204
Nitrous acid, concd		x	x	Thionyl chloride	Phenolic	320	160
Oleum		x	x	Toluene	Furan	400	204
Phenol	Furan	400	204	Trichloroacetic acid	Furan	340	171
Phosphoric acid 50–80%	Phenolic	400	204	Zinc chloride	Phenolic	400	204

[a] The chemicals listed are in the pure state or in a saturated solution unless otherwise indicated. Compatibility is shown to the maximum allowable temperature for which data are available. Incompatibility is shown by an x. A blank space indicates that data are unavailable. *Source*: Schweitzer, PA. Corrosion Resistance Tables, 4th ed. Vols. 1–3. New York: Marcel Dekker, 1995.

dissimilar polymers are to be joined because of solvent and polymer compatibility problems.

A universal solvent should never be used. Solvent cements should be selected that have approximately the same solubility parameters as the polymer to be bonded. Table 2.7 lists the typical solvents used to bond the major polymers. In all cases manufacturers recommendations should be followed.

The three principal methods for joining thermoplasts by means of heat fusion are heated-tool welding, hot gas welding, and resistance-wire welding.

1. Heated Tool Welding

In this method, the surfaces to be joined are heated by holding them against a hot surface, then bringing them into contact and allowing them to harden under a slight pressure.

TABLE 2.4 Abbreviations Used for Thermoplasts

ABS	Acrylonitrile-butadiene-styrene
CPE	Chlorinated polyether
CPVC	Chlorinated polyvinyl chloride
CTFE	Chlorotrifluoroethylene
ECTFE	Ethylenechlorotrifluoroethylene
ETFE	Ethylenetetrafluoroethylene
FEP	Flouroethylene–propylene copolymer
HDPE	High-density polyethylene
LDPE	Low-density polyethylene
LLDPE	Linear low-density polyethylene
PA	Polyamide (Nylon)
PAI	Polyamide–imide
PAN	Polyacrylonitrile
PAS	Polyarylsulfone
PB	Polybutylene
PBT	Polybutylene terephthalate
PC	Polycarbonate
PE	Polyethylene
PEEK	Polyetheretherketone
PEI	Polyether–imide
PEK	Polyetherketone
PEKK	Polyetherketoneketone
PES	Polyethersulfone
PET	Polyethylene terephthalate
PFA	Perfluoralkoxy
PI	Polyimide
PP	Polypropylene
PPE	Polyphenylene ether
PPO	Polyphenylene oxide
PPS	Polyphenylene sulfide
PPSS	Polyphenylene sulfide sulfone
PSF	Polysulfone
PTFE	Polytetrafluoroethylene (Teflon) also TFE
PUR	Polyurethane
PVC	Polyvinyl chloride
PVDC	Polyvinylidene chloride
PVDF	Polyvinylidene fluoride
PVF	Polyvinyl fluoride
SAN	Styrene–acrylonitrile
UHMWPE	Ultra high molecular weight polyethylene

TABLE 2.5 Atmospheric Resistance of Thermoplastic Polymers

Polymer	UV degradation[a,b]	Moisture[c] absorption	Weathering	Ozone	SO_2	NO_x	H_2S
ABS	R	0.30	R	x[d]	R[g]		R[g]
CPVC	R	0.03	R	R	R		R[g]
ECTFE	R	<0.1	R	R	R	R	R
ETFE	R	<0.029	R	R	R	R	R
FEP	R	<0.01	R	R	R	R	R
HDPE	RS		R	R	R		R
PA	R	0.6–1.2	R	x[d]	R[f]		x[d]
PC	RS	0.15	L[e]	R			
PCTFE	R		R	R	R	R	R
PEEK	R	0.50	L[e]	R	R		R
PEI	R	0.25	R	R			
PES					x[d]		R
PF					R		R
PFA	R	<0.03	R	R	R	R	R
PI	R		R	R			
PP	RS	0.02	L[e]	L[e]	R	R	R
PPS	R	0.01	R	R			R
PSF	R	0.30	R	R			
PTFE	R	<0.01	R	R	R	R	R
PVC	R	0.05	R	R	R[h]	x[d]	R
PVDC	R		R	R	R	R	R
PVDF	R	<0.04	R	R	R	R	R
UHMWPE	RS	<0.01	R	R	R		R

[a] R = Resistant.
[b] RS = Resistant only if stabilized with a UV protector.
[c] Water absorption at 73°F/23°C(%).
[d] Not resistant.
[e] L = Limited resistance.
[f] Wet only.
[g] = Dry only.
[h] = Type 1 only.

2. Hot Gas Welding

In this method, the ends to be joined are beveled and positioned with a small gap between them. A welding rod made of the same polymer is laid in the joint under a steady pressure. An electric or gas-heated welding gun with an orifice temperature of 425 to 700°F is used. Heat from the gun is directed to the tip of the rod where it fills the gap.

TABLE 2.6 Allowable Temperature Range for Thermoplastic Polymers[a]

| Polymer | Allowable temperature (°F/°C) | |
	Minimum	Maximum
ABS	−40/−40	140/60
CPVC	0/−18	180/82
ECTFE	−105/−76	340/171
ETFE	−370/−223	300/149
FEP	−50/−45	400/205
HDPE	−60/−51	180/82
PA	−60/−51	300/149
PAI	−300/−190	500/260
PB		200/93
PC	−200/−129	250−275/120−135
PCTFE	−105/−76	380/190
PEEK	−85/−65	480/250
PEI	−310/−190	500/260
PES		340/170
PFA	−310/−190	500/260
PI	−310/−190	500−600/260−315
PP	32/0	215/102
PPS		450/230
PSF	−150/−101	300/149
PTFE	−20/−29	500/260
PVC	0/−18	140/60
PVDC	0/−18	175/80
PVDF	−50/−45	320/160
UHWMPE	40/4	200/93

[a] Temperature limits may have to be modified depending upon the corrodent if in direct contact.

When fusing polyolefins, the heated gas must be inert, since the air will oxidize the surface of the polymer. After welding, the joint should not be stressed for several hours.

3. Resistance-Wire Welding

This method employs an electrical resistance heating wire laid between mating surfaces to generate heat of fusion. After the bond has been made, the exterior wire is cut off.

TABLE 2.7 Typical Solvents for Solvent Cementing

Polymer	Solvent
PVC	Cyclohexane, tetrahydrofuran, dichlorobenzene
ABS	Methyl ethyl ketone, methyl isobutyl ketone, tetrahydrofuran, methylene chloride
Acetate	Methylene chloride, acetone, chloroform, methyl ethyl ketone, ethyl acetate
Acrylic	Methylene chloride, ethylene dichloride
PA (Nylon)	Aqueous phenol, solutions of resorcinol in alcohol, solutions of calcium chloride in alcohol
PPO	Trichloroethylene, ethylene dichloride, chloroform, methylene chloride
PC	Methylene chloride, ethylene dichloride
Polystyrene	Methylene chloride, ethylene dichloride, ethylene ketone, trichloroethylene, toluene, xylene
PSF	Methylene chloride

4. Use of Adhesive

The physical and chemical properties of both the solidified adhesive and the polymer affect the quality of the bonded joint. Major elements of concern are the corrosion resistance of the adhesive, its thermal expansion coefficient, and the glass transition temperature of the polymer relative to the adhesive.

Large differences in the thermal expansion coefficient between the polymer and adhesive can result in stress at the polymer's joint interface. These stresses are compounded by thermal cycling and low temperature service requirements.

A structural adhesive must have a glass transition temperature higher than the operating temperature to prevent a cohesively weak bond and possible creep problems. Engineering polymers such as polyamide or polyphenylene sulfide have very high glass transition temperatures while most adhesives have very low glass transition temperatures which means that the weakest thermal link in the joint may be the adhesive.

Low peel or cleavage strength may result when an adhesive is used that has a glass transition temperature too far below that of the polymer. Brittleness of the adhesive at very low temperatures could result in poor impact strength.

All polymers cannot be joined by all of the methods cited. Table 2.8 lists the methods by which individual polymers may be joined.

Preparation of the surface prior to "bonding" with an adhesive is critical. The quality of the surface will determine the initial bond strength and joint performance. Surface treatments prior to bonding are designed to remove weak boundary layers and provide easily wettable surfaces.

TABLE 2.8 Methods of Joining Thermoplasts

Polymer	Adhesives	Mechanical fastening	Solvent welding	Thermal welding
ABS	x	x	x	x
Acetals	x	x	x	x
Acrylics	x	x	x	
CPE	x	x		
Ethylene copolymers				x
Fluoroplastics	x			
PA	x	x		x
PPO		x	x	x
Polyesters	x	x	x	
PAI	x	x		
PAS	x	x		
PC	x	x	x	x
PA/ABS	x	x		x
PE	x	x		x
PI	x	x		
PPS	x	x		
PP	x	x		x
Polystyrene	x	x	x	
PSF	x	x		
PVC/Acrylic alloy	x	x		
PVC/ABS alloy	x	x		x
PVC	x	x	x	x
CPVC	x	x	x	

Surface preparations can vary depending upon the specific polymer. They may be as simple as solvent wiping and as involved as a combination of mechanical abrading, chemical cleaning, and acid etching. Extensive surface preparation may be unnecessary in many low to medium strength applications. However, when the application requires maximum bond strength, performance, and reliability, carefully controlled surface treating processes are necessary. When choosing a surface preparation process, the following factors should be considered:

1. The ultimate bond strength required
2. The degree of performance necessary and the service requirement
3. The degree and type of contamination on the polymer
4. The type of polymer and adhesive.

Surface preparations are designed to improve the quality of the bonded joint by performing one or more of the following functions:

1. Remove contaminants
2. Control absorbed water
3. Control oxide formation
4. Poison surface atoms which catalyze adhesive breakdown
5. Protect the polymer from the adhesive and vice versa
6. Match the adhesive molecular structure to the polymer crystal structure
7. Control surface roughness

Many polymer surfaces are contaminated with mold-release agents or processing additives. These contaminants must be removed before bonding. Polymers such as PTFE, PE, PP, and certain others, because of their low surface energy, are completely unsuitable for adhesive bonding in their natural state. It is necessary to alter their surfaces chemically or physically to improve wetting prior to bonding.

The type of polymer surface will determine the specific chemical etching treatment. Chemical etching treatment involves the use of corrosive and hazardous chemicals. Polyethylene and polypropylene require the use of sulfuric–dichromate, while fluorocarbons require the use of a sodium–naphthalene etch.

Plasma treatments physically and chemically change the nature of polymeric surfaces. This treatment produces a strong, wettable, cross-linked skin.

When the surface treatment has been completed, a primer is applied to the surface prior to bonding. These primers serve four primary functions either individually or in combination.

1. Protecting the surfaces after treatment. Primers can be used to extend the time between surface treatment and bonding.
2. Developing tack positioning or holding parts to be bonded.
3. Inhibiting corrosion during service.
4. Serving as an intermediate layer to improve the physical properties of the joint and bond stgrength.

Selection of the proper adhesive is important. Recommended adhesives for each specific polymer will be found in the chapter dealing with that polymer. However, which of the recommended adhesives is used will depend upon the specific operating conditions such as chemical environment and temperature. There is not a best adhesive for universal chemical environments. For example, an adhesive providing maximum resistance to acids in all probability will provide poor resistance to bases. It is difficult to select an adhesive that will not degrade in two widely differing chemical environments. In general, the adhesives that are most resistant to high temperatures usually exhibit the best resistance to chemicals and solvents.

The operating temperature is a significant factor in the aging process of the adhesive. As the temperature increases, the adhesive absorbs more fluid, and the degradation rate increases.

In general, chlorinated solvents and ketones are severe environments as are high boiling solvents such as dimethylformamide, dimethyl sulfoxide, and acetic acid. Amine curing agents for epoxies are poor in oxidizing acids while anhydride curing agents are poor in caustics. Table 2.9 provides the relative compatibilities of synthetic adhesives in selected environments.

Epoxy adhesives are generally limited to continuous applications below 300°F/149°C. However, there are epoxy formulations that can withstand short terms at 500°F/260°C and long-term service at 300–350°F/149–177°C. A combination epoxy–phenolic resin has been developed that will provide an adhesive capability at 700°F/371°C for short-term operation and continuous operation at 350°F/177°C.

Nitrile–phenolic adhesives have high shear strength up to 250–350°F/121–177°C, and the strength retention on aging at these temperatures is very good.

Silicone adhesives are used primarily in nonstructural applications. They have very good thermal stability but low strength.

TABLE 2.9 Chemical Resistance of Adhesives

Adhesive	Type[a]	A	B	C	D	E	F	G	H	J	K	L	M	N
								Environment[b,c]						
Cyanoacrylate	TS	x		x	x	x	x	3	3	x	x	x	4	4
Polyester + isocyanate	TS	3	2	1	3	3	2	2	2	3	2	2	x	2
Polyester + monomer	TS	x	3	3	x	3	x	2	2	2	x	x	x	x
Urea formaldehyde	TS	3	3	2	x	2	2	2	2	2	2	2	2	2
Melamine fomaldehyde	TS	2	2	2	x	2	2	2	2	2	2	2	2	2
Resorcinol fomaldehyde	TS	2	2	2	2	2	2	2	2	2	2	2	2	2
Epoxy + polyamine	TS	3	x	2	2	2	2	2	3	1	x	x	1	
Epoxy + polyamide	TS	x	2	2	x	3	x	2	2	1	x	x	3	
Polyimide	TS	1	1	2	4	2	2	2	2	2	2	2	2	2
Acrylic	TS	x	3	1	3	2	2	2	2	2	2	2	2	2
Cellulose acetate	TP	2	3	1	x	1	2		2	4	x	x	x	x
Cellulose nitrate	TP	3	3	3	3	3	x	2	2	x	x	x	x	x
Polyvinyl acetate	TP	x		3	x	3	3	2	2	x	x	x	x	x
Polyvinyl alcohol	TP	3		x	x	x	x	2	1	3	1	1	1	1
Polyamide	TP	x		x	x	x	2	2	2	x	2	2	2	x
Acrylic	TP	4	3	3	3			2			4	4		4
Phenoxy	TP	4	3	3	4	3	2	3	x	x			x	

[a] Type: TS = thermosetting adhesive; TP = thermoplastic adhesive.
[b] Environment: A = heat; B = cold; C = water; D = hot water; E = acid; F = alkali; G = oil, grease; H = fuels; J = alcohols; K = ketones; L = esters; M = aromatics; N = chlorinated solvents.
[c] Resistance: 1 = excellent; 2 = good; 3 = fair; 4 = poor; x = not recommended.

Polyimide adhesives have thermal endurance at temperatures greater than 500°F/260°C that is unmatched by any commercially available adhesive. Their short-term exposure at 1000°F/538°C is slightly better than the epoxy–phenolic alloy.

The successful application of an adhesive at low temperatures is dependent upon the difference in the coefficient of thermal expansion between adhesive and polymer, the elastic modulus, and the thermal conductivity of the adhesive Epoxy–polyimide adhesives can be made serviceable at very low temperatures by the addition of appropriate fillers to control thermal expansion.

Epoxy–phenolic adhesives exhibit good adhesive properties at both elevated and low temperatures. Polyurethane and epoxy–nylon systems exhibit outstanding cryogenic properties.

Other factors affecting the life of an adhesive bond are humidity, water immersion, and outdoor weathering. Moisture can affect adhesive strength in two ways. Some polymeric materials, notably ester-based polyurethanes will revert, i.e., lose hardness, strength, and in the worst case turn fluid during exposure to warm humid air. Water can also permeate the adhesive and displace the adhesive at the bond interface. Structural adhesives not susceptible to the reversion phenomenon are also likely to lose adhesive strength when exposed to moisture.

Adhesives exposed outdoors are affected primarily by heat and humidity. Thermal cycling, ultraviolet radiation and cold are relatively minor factors. Structural adhesives, when exposed to weather, rapidly lose strength during the first six months to a year. After 2 to 3 years, the rate of decline usually levels off, depending upon the climate zone, polymer, adhesive, and stress level. The following are important considerations when designing an adhesive joint for outdoor service:

1. The most severe locations are those with high humidity and warm temperatures.
2. Stressed panels deteriorate more rapidly than unstressed panels.
3. Heat-cured adhesive systems are generally more resistant than room-temperature cured systems.
4. With the better adhesives, unstressed bonds are relatively resistant to severe outdoor weathering, although all joints will eventually show some strength loss.

B. ABS (Acrylonitrile–Butadiene–Styrene)

ABS polymers are derived from acrylonitrile, butadiene, and styrene and have the following general chemical structure:

$$\left[\begin{array}{cc} H & H \\ | & | \\ -C-C- \\ | & | \\ H & C_n \end{array}\right]_X \left[\begin{array}{ccc} H & H & H \\ | & | & | \\ -C-C-\quad C- \\ | & & | \\ H & & H \end{array}\right]_Y \left[\begin{array}{cc} H & H \\ | & | \\ -C-C- \\ | & | \\ H & CH \end{array}\right]_Z$$

Acrylonitrile   Butadiene   Styrene

The properties of ABS polymers can be altered by the relative amounts of acrylonitrile, butadiene, and styrene present. Higher strength, better toughness, greater dimensional stability and other properties can be obtained at the expense of other characteristics.

1. Physical and Mechanical Properties

Impact resistance and toughness are the most outstanding mechanical properties. Impact resistance does not fall off rapidly at lower temperatures. Stability under limited load is excellent. When impact failure does take place, the failure is ductile rather than brittle.

Moisture has little effect on the physical properties of ABS which helps to maintain the dimensional stability of ABS products. The physical and mechanical properties of ABS, flame retardant grade of ABS and impact modified grade of ABS are shown in Table 2.10.

In order to increase the strength of ABS, it is reinforced with glass fibers. This reinforcement increases the tensile strength, tensile modulus, flexural strength, and compressive strength, while the thermal expansion is reduced. The degree of change is dependent upon the percentage of glass fiber reinforcement. Table 2.11 lists the physical and mechanical properties of 20% and 30% glass reinforced ABS.

When ABS is alloyed or blended with polycarbonate, some of the best qualities of both materials result. The resulting thermoplastic is easier to process, has high heat and impact resistance, and is more economical than polycarbonate alone.

The ABS/PC alloy has an impact strength of 10.7 ft-lb/in. notch which is well above the high impact strength of engineering thermoplastics but not as high as the impact strength of PC. Polycarbonate has a critical thickness with respect to impact strength which the ABS/PC alloy does not have. The notched impact value of the alloy drops by only 2–4 ft-lb/in. in the 1/8 to 1/4 inch range. One-eighth inch thick polycarbonate has a notched Izod value of approximately 16 ft-lb/in. but only 3–4 ft-lb/in. at thicknesses greater than 1/4 inch.

The ABS/PC alloy has a flexural strength approximately 15% greater than that of the polycarbonate alone. The alloy remains more rigid than polycarbonate

TABLE 2.10 Physical and Mechanical Properties of ABS, Flame Retardant Grade ABS, and Impact Modified ABS

Property	Unfilled	Flame-retardant grade	Impact-modified copolymer
Specific gravity	1.02–1.08	1.16–1.21	1.29–1.39
Water absorption (24 hr at 73°F/23°C) (%)	0.20–0.45	0.2–0.6	0.31–0.41
Dielectric strength, short-term (V/mil)	350–500	350–500	
Tensile strength at break (psi)	2500–8,000	3,300–8,000	
Tensile modulus ($\times 10^3$ psi)	130–420	270–400	187–319
Elongation at break (%)	20–100	1.5–80	60–300
Compressive strength (psi)	5,200–10,000	6,500–7,500	
Flexural strength (psi)	4,000–14,000	6,200–14,000	7,100
Compressive modulus ($\times 10^3$ psi)	150–390	130–310	
Flexural modulus ($\times 10^3$ psi) at 73°F/23°C 200°F/93°C 250°F/121°C	130–440	300–600	20–300
Izod impact (ft-lb/in. of notch)	1.5–12	1.4–12	1.7–4.7
Hardness, Rockwell	R75–115	R100–120	M35–70
Coefficient of thermal expansion 10^{-6} in./in./°F	60–130	65–95	130–150
Thermal conductivity (10^{-4}cal-cm/sec-cm^2°C or Btu/hr/ft^2/°F/in.)			
Deflection temperature at 264 psi (°F) at 66 psi (°F)	170–220 170–235	158–181 210–245	132–200 308–318
Max. operating temperature (°F/°C)	185/85		
Limiting oxygen index (%)			
Flame spread			
Underwriters Lab. rating (Sub. 94)			

TABLE 2.11 Physical and Mechanical Properties of Glass Fiber Reinforced ABS

Property	20% reinforcing	30% reinforcing
Specific gravity	1.18–1.22	1.29
Water absorption (24 hr at 73°F/23°C) (%)	0.18–0.20	0.3
Dielectric strength, short-term (V/mil)	450–460	
Tensile strength at break (psi)	10,500–13,000	13,000–16,000
Tensile modulus ($\times 10^3$ psi)	740–880	1,000–1,200
Elongation at break (%)	2–3	1.5–1.8
Compressive strength (psi)	13,000–14,000	15,000–17,000
Flexural strength (psi)	14,000–17,500	17,000–19,000
Compressive modulus ($\times 10^3$ psi)	800	
Flexural modulus ($\times 10^3$ psi) at		
73°F/23°C	650–800	1,000
200°F/93°C		
250°F/121°C		
Izod impact (ft-lb/in. of notch)	1.1–1.4	1.2–1.3
Hardness, Rockwell	R107, M85–98	M75–85
Coefficient of thermal expansion (10^{-6} in./in./°F)	20–21	
Thermal conductivity (10^{-4} cal-cm/sec-cm^2°C or Btu/hr/ft^2/°F/in.)	4.8	
Deflection temperature at 264 psi (°F)	210–220	215–230
at 66 psi (°F)	220–230	230–240
Max. operating temperature (°F/°C)		
Limiting oxygen index (%)		
Flame spread		
Underwriters Lab. rating (Sub. 94)		

alone up to approximately 200°F (93°C). Table 2.12 gives the physical and mechanical properties of a flame retardent ABS/PC alloy.

ABS can also be alloyed with other thermoplasts in order to improve certain properties. Table 2.13 lists the physical and mechanical properties of an ABS/Nylon alloy and an ABS/PVC flame retardant alloy.

In referring to the physical and mechanical property tables, it must be remembered that ABS polymer is very often altered by the percentage of acrylonitrile present and by the addition of other modifiers. Because of this, the data in the tables should be used only as guides, and the manufacturer should be checked for the properties of its specific material. The same applies to the ABS alloys.

ABS can be bonded to itself by either solvent cementing or any of the thermal fusion methods. For adhesive joining, the best adhesive materials to use

TABLE 2.12   Physical and Mechanical Properties of ABS/PC Flame-Retardant Alloy

Property	
Specific gravity	1.17–1.23
Water absorption (24 hr at 73°F/23°C) (%)	0.24
Dielectric strength, short-term (V/mil)	450–750
Tensile strength at break (psi)	5,800–9,300
Tensile modulus ($\times 10^3$ psi)	350–455
Elongation at break (%)	20–70
Compressive strength (psi)	11,000–11,300
Flexural strength (psi)	12,000–14,500
Compressive modulus ($\times 10^3$ psi)	230
Flexural modulus ($\times 10^3$ psi) at 73°F/23°C	350–400
200°F/93°C	
250°F/121°C	
Izod impact (ft-lb/in. of notch)	4.1–14.0
Hardness, Rockwell	R115–119
Coefficient of thermal expansion (10^{-6} in./in./°F)	67
Thermal conductivity (10^{-4} cal-cm/sec-cm^2°C or Btu/hr/ft^2/°F/in.)	
Deflection temperature at 264 psi (°F)	180–220
at 66 psi (°F)	195–244
Max. operating temperature (°F/°C)	
Limiting oxygen index (%)	
Flame spread	
Underwriters Lab. rating (Sub. 94)	

are epoxies, urethanes, thermosetting acrylics, nitrile–phenolics, and cyanoacrylates. These adhesives have greater strength than that of the ABS substrate. Only cleaning of the surface and removal of any contaminants are necessary. No other special surface preparation is required.

2.   Corrosion Resistance Properties

Pure ABS polymer will be attacked by oxidizing agents, strong acids, and will stress crack in the presence of certain organic compounds. It is resistant to aliphatic hydrocarbons but not to aromatic and chlorinated hydrocarbons.

ABS will be degraded by ultraviolet light unless protective additives are incorporated into the formulation.

When an ABS alloy or a reinforced ABS is used, all of the alloying ingredients and/or the reinforcing materials must be checked for chemical

TABLE 2.13 Physical and Mechanical Properties of an ABS/Nylon Alloy and an ABS/PVC Flame Retardant Alloy

Property	ABS/nylon	ABS/PVC flame-retardant
Specific gravity	1.06–1.07	1.13–1.25
Water absorption (24 hr at 73°F/23°C) (%)		
Dielectric strength, short-term (V/mil)		500
Tensile strength at break (psi)	4,000–6,000	5,800–6.500
Tensile modulus ($\times 10^3$ psi)	260–320	325–380
Elongation at break (%)	40–300	
Compressive strength (psi)		
Flexural strength (psi)	8,800–10,900	7,900–10,000
Compressive modulus ($\times 10^3$ psi)		
Flexural modulus ($\times 10^3$ psi) at		
73°F/23°C	250–310	320–400
200°F/93°C		
250°F/121°C		
Izod impact (ft-lb/in. of notch)	15–20	3.0–18.0
Hardness, Rockwell	R93–105	R100–106
Coefficient of thermal expansion (10^{-6} in./in./°F)	90–110	46–84
Thermal conductivity (10^{-4} cal-cm/ sec-cm^2°C or Btu/hr/ft^2/°F/in.)		
Deflection temperature at 264 psi (°F)	130–150	169–200
at 66 psi (°F)	180–195	
Max. operating temperature (°F/°C)		
Limiting oxygen index (%)		
Flame spread		
Underwriters Lab. rating (Sub. 94)		

compatibility. The manufacturer should also be checked. Table 2.14 provides the compatibility of ABS with selected corrodents. A more detailed listing may be found in Reference 1.

3. Typical Applications

ABS polymers are used to produce business machine and camera housings, blowers, bearings, gears, pump impellers, chemical tanks, fume hoods, ducts, piping, and electrical conduit.

TABLE 2.14 Compatibility of ABS with Selected Corrodents[a]

Chemical	Maximum temp.		Chemical	Maximum temp.	
	°F	°C		°F	°C
Acetaldehyde	x	x	Benzyl chloride	x	x
Acetic acid 10%	100	38	Borax	140	60
Acetic acid 50%	130	54	Boric acid	140	60
Acetic acid 80%	x	x	Bromine, liquid	x	x
Acetic acid, glacial	x	x	Butadiene	x	x
Acetic anhydride	x	x	Butyl acetate	x	x
Actone	x	x	Butyl alcohol	x	x
Acetyl chloride	x	x	Butyric acid	x	x
Adipic acid	140	60	Calcium bisulfite	140	60
Allyl alcohol	x	x	Calcium carbonate	100	38
Allyl chloride	x	x	Calcium chlorate	140	60
Alum	140	60	Calcium chloride	140	60
Aluminum chloride, aqueous	140	60	Calcium hydroxide, sat.	140	60
Aluminum fluoride	140	60	Calcium hypochlorite	140	60
Aluminum hydroxide	140	60	Calcium nitrate	140	60
Aluminum oxychloride	140	60	Calcium oxide	140	60
Aluminum sulfate	140	60	Calcium sulfate 25%	140	60
Ammonia gas, dry	140	60	Carbon bisulfide	x	x
Ammonium bifluoride	140	60	Carbon dioxide, dry	90	32
Ammonium carbonate	140	60	Carbon dioxide, wet	140	60
Ammonium chloride, sat.	140	60	Carbon disulfide	x	x
Ammonium fluoride 10%	x	x	Carbon monoxide	140	60
Ammonium fluoride 25%	x	x	Carbon tatrachloride	x	x
Ammonium hydroxide 25%	90	32	Carbonic acid	140	60
Ammonium hydroxide, sat.	80	27	Cellosolve	x	x
Ammonium nitrate	140	60	Chloroacetic acid	x	x
Ammonium persulfate	140	60	Chlorine gas, dry	140	60
Ammonium phosphate	140	60	Chlorine gas, wet	140	60
Ammonium sulfate 10–40%	140	60	Chlorine, liquid	x	x
Ammonium sulfide	140	60	Chlorobenzene	x	x
Amy acetate	x	x	Chloroform	x	x
Amyl alcohol	80	27	Chlorosulfonic acid	x	x
Amyl chloride	x	x	Chromic acid 10%	90	32
Aniline	x	x	Chromic acid 50%	x	x
Antimony trichloride	140	60	Citric acid 15%	140	60
Aqua regia 3 : 1	x	x	Citric acid 25%	140	60
Barium carbonate	140	60	Copper chloride	140	60
Barium chloride	140	60	Copper cyanide	140	60
Barium hydroxide	140	60	Copper sulfate	140	60
Barium sulfate	140	60	Cresol	x	x
Barium sulfide	140	60	Cyclohexane	80	27
Benzaldehyde	x	x	Cyclohexanol	80	27
Benzene	x	x	Dichloroacetic acid	x	x
Benzenesulfonic acid 10%	80	27	Dichloroethane (ethylene dichloride)	x	x
Benzoic acid	140	60	Ethylene glycol	140	60
Benzyl alcohol	x	x	Ferric chloride	140	60

TABLE 2.14 Continued

Chemical	Maximum temp. °F	Maximum temp. °C	Chemical	Maximum temp. °F	Maximum temp. °C
Ferric nitrate, 10–50%	140	60	Phosphoric acid 50–80%	130	54
Ferrous chloride	140	60	Picric acid	x	x
Fluorine gas, dry	90	32	Potassium bromide 30%	140	60
Hydrobromic acid 20%	140	60	Sodium carbonate	140	60
Hydrochloric acid 20%	90	32	Sodium chloride	140	60
Hydrochloric acid 38%	140	60	Sodium hydroxide 10%	140	60
Hydrofluoric acid 30%	x	x	Sodium hydroxide 50%	140	60
Hydrofluoric acid 70%	x	x	Sodium hydroxide, concd	140	60
Hydrofluoric acid 100%	x	x	Sodium hypochlorite 20%	140	60
Hypochlorous acid	140	60	Sodium hypochlorite, concd	140	60
Ketones, general	x	x	Sodium sulfide to 50%	140	60
Lactic acid 25%	140	60	Stannic chloride	140	60
Magnesium chloride	140	60	Stannous chloride	100	38
Malic acid	140	60	Sulfuric acid 10%	140	60
Methyl chloride	x	x	Sulfuric acid 50%	130	54
Methyl ethyl ketone	x	x	Sulfuric acid 70%	x	x
Methyl isobutyl ketone	x	x	Sulfuric acid 90%	x	x
Muriatic acid	140	60	Sulfuric acid 98%	x	x
Nitric acid 5%	140	60	Sulfuric acid 100%	x	x
Nitric acid 20%	130	54	Sulfuric acid, fuming	x	x
Nitric acid 70%	x	x	Sulfurous acid	140	60
Nitric acid, anhydrous	x	x	Thionyl chloride	x	x
Oleum	x	x	Toluene	x	x
Perchloric acid 10%	x	x	White liquor	140	60
Perchloric acid 70%	x	x	Zinc chloride	140	60
Phenol	x	x			

[a] The chemicals listed are in the pure state or in a saturated solution unless otherwise indicated. Compatibility is shown to the maximum allowable temperature for which data are available. Incompatibility is shown by an x. A blank space indicates that data are unavailable.
Source: Schweitzer, PA. Corrosion Resistance Tables, 4th ed. Vols. 1–3. New York: Marcel Dekker, 1995.

C. Acrylics

Acrylics are based on polymethyl methacrylate and have a chemical structure as follows:

$$\left[\begin{array}{cc} H & CH_3 \\ | & | \\ -C-C- \\ | & | \\ H & COOCH_3 \end{array} \right]_n$$

They are sold under the tradenames of Lucite by E. I. DuPont and Plexiglas by Rohm and Haas. Because of their chemical structure, acrylic resins are inherently resistant to discoloration and loss of light transmission.

1. Physical and Mechanical Properties

Parts molded from acrylyic powders in their pure state may be clear and nearly optically perfect. The total light transmission is as high as 92%, and haze measurements average only 1%. The index of refraction ranges from 1.486 to 1.496. Light transmittance and clarity can be modified by the addition of a wide variety of opaque and transparent colors, most of which are formulated for long outdoor service. The dimensional stability of acrylics is affected by temperature change and must be considered when the material is to be used for an exterior glazing unit. The amount of change for a given application can be calculated using the coefficient of linear thermal expansion.

In the case of an exterior glazing unit 1 ft by 1 ft (30.4 cm by 30.4 cm) which is subjected to a temperature variation of 32° to 100°F (0°C to 38°C), the space to be provided in the frame for expansion is calculated as follows:

$$\Delta L = e_t L T$$

where ΔL = change in length in inches or cm

$\quad\quad e_t$ = coefficient of thermal expansion

$\quad\quad$ = $(40 \times 10^{-6}$ in/in °F$)$

$\quad\quad L$ = initial length 12 in (30.4 cm)

$\quad\quad T$ = temperature variation

$\quad\quad$ = 68°F/38°C

Therefore

$$\Delta L = (40 \times 10^{-6})\,(12)\,(68) = 0.033 \text{ in}$$

Based on this, an expansion space of approximately 0.0165 in. is required at each end for the direction in which this calculation was made. This was based on the assumption of zero or negligible expansion of the frame.

The moisture level in acrylics is dependent upon the relative humidity of the environment. As the relative humidity of the air increases, acrylics will absorb moisture, which will result in a slight dimensional expansion. The equilibrium moisture level may be reached in a couple of weeks or months, depending upon the thickness of the part. For example, a 24 in. long lighting panel, 1/8 in. thick molded from a dry resin, installed in a 50% relative humidity environment, will expand approximately 1% or 0.024 in.

Acrylics can also be alloyed with polycarbonate. Table 2.15 lists the physical and mechanical properties of acrylics.

TABLE 2.15 Physical and Mechanical Properties of Acrylic and Acrylic/PC Alloy

Property	Acrylic	Acrylic/PC alloy
Specific gravity	1.17–1.20	1.15
Water absorption (24 hr at 73°F/23°C) (%)	0.2–0.4	0.3
Dielectric strength, short-term (V/mil)	450–550	
Tensile strength at break (psi)	66–11,000	8,000–9,000
Tensile modulus ($\times 10^3$ psi)	450–3,100	320–350
Elongation at break (%)	2–7	58
Compressive strength (psi)	11,000–19,000	
Flexural strength (psi)	12,000–17,000	11,300–12,500
Compressive modulus ($\times 10^3$ psi)	390–475	
Flexural modulus ($\times 10^3$ psi) at		
73°F/23°C	390–3,200	320–3,500
200°F/93°C		
250°F/121°C		
Izod impact (ft-lb/in. of notch)	0.3–0.4	26–30
Hardness, Rockwell	M80–102	M46–49
Coefficient of thermal expansion (10^{-6} in./in./°F)	50–90	52
Thermal conductivity (10^{-4}cal-cm/ sec-cm²°C or Btu/hr/ft²/°F/in.)	4.0–6.0	
Deflection temperature at 264 psi (°F)	98–215	101
at 66 psi (°F)	165–235	
Max. operating temperature (°F/°C)	140–190/60–90	
Limiting oxygen index (%)		
Flame spread		
Underwriters Lab. rating (Sub. 94)		

Acrylics can be joined by solvent cementing, thermal fusion, or adhesive bonding. Solvent cementing is accomplished by using methylene chloride or ethylene dichloride.

Preparing the surface for adhesive bonding only entails making sure that the surfaces are free of contamination. Adhesives recommended are epoxies, urethanes, cyanoacrylates, and thermosetting acrylics. The bond strengths will be greater than the strength of the acrylic.

2. Corrosion Resistance Properties

Acrylics exhibit outstanding weatherability. They are attacked by strong solvents, gasoline, acetone, and other similar fluids. Table 2.16 lists the compatibility of acrylics with selected corrodents. Reference 1 provides a more detailed listing.

TABLE 2.16 Compatibility of Acrylics with Selected Corrodents[a]

Chemical		Chemical	
Acetaldehyde	x	Carbon tetrachloride	x
Acetic acid 10%	x	Carbonic acid	R
Acetic acid 50%	x	Chloroacetic acid, 50% water	x
Acetic acid 80%	x	Chloroacetic acid	x
Acetic acid, glacial	x	Chlorobenzene	x
Acetic anhydride	x	Chloroform	x
Acetone	x	Chlorosulfonic acid	x
Alum	R	Chromic acid 10%	x
Aluminum chloride, aqueous	R	Chromic acid 50%	x
Aluminum hydroxide	R	Citric acid 15%	R
Aluminum sulfate	R	Citric acid, concd	R
Ammonia gas	R	Copper chloride	R
Ammonium carbonate	R	Copper cyanide	R
Ammonium chloride, sat.	R	Copper sulfate	x
Ammonium hydroxide 25%	R	Cresol	x
Ammonium hydroxide, sat.	R	Cupric chloride 5%	R
Ammonium nitrate	R	Cupric chloride 50%	R
Ammonium persulfate	R	Dichloroethane	x
Ammonium sulfate 10–40%	R	Ethylene chloride	R
Amyl acetate	x	Ferric chloride	R
Amyl alcohol	R	Ferric nitrate 10–50%	R
Aniline	x	Ferric chloride	R
Aqua regia 3 : 1	x	Hydrochloric acid 20%	R
Barium carbonate	R	Hydrochloric acid 38%	R
Barium chloride	R	Hydrofluoric acid 70%	x
Barium hydroxide	R	Hydrofluoric acid 100%	R
Barium sulfate	R	Ketones, general	x
Benzaldehyde	x	Lactic acid 25%	x
Benzene	x	Lactic acid, concd	x
Benzoic acid	x	Magnesium chloride	R
Borax	R	Methyl chloride	x
Boric acid	R	Methyl ethyl ketone	x
Bromine, liquid	x	Nitric acid 5%	R
Butyl acetate	x	Nitric acid 20%	x
Butyl alcohol	R	Nitric acid 70%	x
Butyric acid	x	Nitric acid anhydrous	x
Calcium bisulfite	R	Oleum	x
Calcium carbonate	R	Phenol	R
Calcium chlorate	R	Phosphoric acid 50–80%	R
Calcium chloride	R	Picric acid	R
Calcium hydroxide, sat.	R	Potassium bromide 30%	R
Calcium hypochlorite	R	Sodium carbonate	x
Calcium sulfate	R	Sodium chloride	R
Calcium bisulfide	R	Sodium hydroxide 10%	R
Carbon dioxide, wet	R	Sodium hydroxide 50%	R
Carbon disulfide	R	Sodium hydroxide, concd	R

TABLE 2.16 Continued

Chemical		Chemical	
Sodium hypochorite 20%	R	Sulfuric acid 90%	x
Sodium sulfide to 50%	R	Sulfuric acid 98%	x
Stanninc chloride	R	Sulfuric acid 100%	x
Stannous chloride	R	Sulfuric acid fuming	x
Sulfuric acid 10%	R	Sulfurous acid	R
Sulfuric acid 50%	R	Toluene	x
Sulfuric acid 70%	R	Zinc chloride	R

[a] The chemicals listed are in the pure state or in a saturated solution unless otherwise specified. Compatibility at 90°F/32°C is indicated by R. Incompatibility is shown by an x.
Source: Schweitzer, PA. Corrosion Resistance Tables, 4th ed. Vols. 1–3. New York: Marcel Dekker, 1995.

3. Typical Applications

Acrylics are used for lenses, aircraft and building glazing, lighting fixtures, coatings, textile fibers, fluorescent street lights, outdoor signs, and boat windshields.

D. CTFE (Chlorotrifluoroethylene)

Chlorotrifluoroethylene is sold under the tradename of Kel-F. It is a fluorocarbon with the following chemical structure:

$$\left[\begin{array}{cc} \overset{\displaystyle Cl}{\underset{\displaystyle F}{\overset{|}{\underset{|}{C}}}} & \overset{\displaystyle F}{\underset{\displaystyle F}{\overset{|}{\underset{|}{C}}}} \end{array}\right]_n$$

1. Physical and Mechanical Properties

CTFE has greater tensile and compressive strength than PTFE within its service temperature range. However, at the low and high end of its operating temperature range, it does not perform as well as PTFE for parts such as seals. At the low end of the range PTFE, has somewhat better physical properties, while at higher temperatures, CTFE is more prone to stress cracking and other difficulties.

The electrical properties of CTFE are generally excellent, although dielectric losses are higher than those of PTFE.

CTFE does not have the low-friction and bearing properties of PTFE. Table 2.17 lists the physical and mechanical properties of CTFE.

TABLE 2.17   Physical and Mechanical Properties of CTFE

Property	
Specific gravity	2.08–2.2
Water absorption (24 hr at 73°F/23°C) (%)	0
Dielectric strength, short-term (V/mil)	500–600
Tensile strength at break (psi)	4,500–6,000
Tensile modulus ($\times 10^3$ psi)	
Elongation at break (%)	80–250
Compressive strength (psi)	4,600–7,400
Flexural strength (psi)	7,400–11,000
Compressive modulus ($\times 10^3$ psi)	150–300
Flexural modulus ($\times 10^3$ psi) at 73°F/23°C	170–200
200°F/93°C	
250°F/121°C	
Izod impact (ft-lb/in. of notch)	2.5–5.0
Hardness, Rockwell	R75–112
Coefficient of thermal expansion (10^{-6} in./in./°F)	36–70
Thermal conductivity (10^{-4}cal-cm/sec-cm^2°C or Btu/hr/ft^2/°F/in.)	4.7–5.3
Deflection temperature at 264 psi (°F)	
at 66 psi (°F)	258
Max. operating temperature (°F/°C)	380/190
Limiting oxygen index (%)	
Flame spread	
Underwriters Lab. rating (Sub. 94)	

It is almost impossible to heat weld or solvent weld CTFE. For adhesive bonding, a sodium–naphthalene etch is necessary as a surface preparation. The sodium–naphthalene etched fluorocarbon surface is degraded by UV light and should be protected from direct exposure. The wettability of the fluorocarbon surface can also be increased by plasma treatment. Once treated, a conventional epoxy or urethane adhesive can be used.

2. Corrosion Resistance Properties

Although CTFE has a wide range of chemical resistance, it has less resistance than PTFE, FEP, and PFA. It is subject to swelling in some chlorinated solvents at elevated temperatures and is attacked by the same chemicals that attack PTFE. Refer to Table 2.18 for the compatibility of CTFE with selected corrodents. Reference 1 provides a more detailed listing.

TABLE 2.18 Compatibility of CTFE with Selected Corrodents[a]

Chemical	Maximum temp.		Chemical	Maximum temp.	
	°F	°C		°F	°C
Acetamide	200	93	Cyclohexanone	100	38
Acetaldehyde	122	50	Dimethyl aniline	200	93
Acetic acid 95–100%	122	50	Dimethyl phthalate	200	93
Acetic acid glacial	122	50	Ethanolamine	x	
Acetonitrile	100	38	Ethylene dichloride	100	38
Acetophenone	200	93	Ethylene glycol	200	92
Acetyl chloride, dry	100	38	Fluorine gas, dry	x	
Allyl alcohol	200	93	Fluorine gas, moist	70	23
Aluminum chloride 25%	200	93	Fuel oils	200	93
Ammonia anhydrous	200	93	Gasoline	200	93
Ammonia, wet	200	93	Glycerine	212	100
Ammonium chloride 25%	200	93	Heptane	200	93
Ammonium hydroxide, concd	200	93	Hexane	200	93
Amyl acetate	100	38	Hydrobromic acid 37%	300	149
Aniline	122	50	Hydrochloric acid 100%	300	149
Aqua regia	200	93	Hydrofluoric acid 50%	200	93
Barium sulfate	300	149	Magnesium chloride	300	149
Benzaldehyde	200	93	Magnesium sulfate	212	100
Benzene	200	93	Mineral oil	200	93
Benzontrile	200	93	Motor oil	200	93
Benzyl alcohol	200	93	Naphtha	200	93
Benzyl chloride	100	38	Naphthalene	200	93
Borax	300	149	Nitromethane	200	93
Bromine gas, dry	100	38	Oxalic acid 50%	212	100
Bromine gas, moist	212	100	Perchloric acid	200	93
Butadiene	200	93	Phosphorus trichloride	200	93
Butane	200	93	Potassium manganate	200	93
n-Butylamine	x		Seawater	250	121
Butyl ether	100	38	Sodium bicarbonate	300	149
Calcium chloride	300	149	Sodium bisulfate	140	60
Calcium sulfate	300	149	Sodium carbonate	300	149
Carbon dioxide	300	149	Sodium chromate 10%	250	121
Carbon disulfide	212	100	Sodium hypochlorite	250	121
Cellusolve	200	93	Sodium sulfate	300	149
Chlorine, dry	x		Sodium thiosulfate	300	149
Chloroacetic acid	125	52	Stannous chloride	212	100
Cupric chloride 5%	300	149	Sulfur	260	127

(Continued)

TABLE 2.18 Continued

Chemical	Maximum temp.		Chemical	Maximum temp.	
	°F	°C		°F	°C
Sulfur dioxide, wet	300	149	Turpentine	200	93
Sulfur dioxide, dry	300	149	Vinegar	200	93
Sulfuric acid, 0–100%	212	100	Water, fresh	250	121
Sulfuric acid, fuming	200	93	Zinc sulfate	212	100

[a] The chemicals listed are in the pure state or in a saturated solution, unless otherwise noted. Compatibility is shown to the maximum allowable temperature for which data are available. Incompatibility is shown by an x.

E. ECTFE (Ethylenechlorotrifluorethylene)

ECTFE is a 1 : 1 alternating copolymer of ethylene and chlorotrifluorethylene having a chemical structure as follows:

$$\begin{array}{c} \quad H \quad H \quad F \quad F \\ \quad | \quad\;\; | \quad\;\; | \quad\;\; | \\ -C-C-C-C- \\ \quad | \quad\;\; | \quad\;\; | \quad\;\; | \\ \quad H \quad H \quad F \quad Cl \end{array}$$

This chemical structure provides the polymer with a unique combination of properties. It possesses excellent chemical resistance, a broad use temperature range from cryogenic to 340°F (117°C) with continuous service to 300°F (149°C), and excellent abrasion resistance. It is sold under the tradename of Halar by Ausimont USA.

1. Physical and Mechanical Properties

ECTFE's resistance to permeation by oxygen, carbon dioxide, chlorine gas, or hydrochloric acid is 10–100 times better than that of PTFE and FEP. Water absorption is less than 0.01%. Other important physical properties include low coefficient of friction, excellent machineability, and the ability to be pigmented.

ECTFE is a strong, highly impact-resistant material that retains useful properties over a wide temperature range. Outstanding in this respect are properties related to impact at low temperatures. In addition to its excellent impact properties, ECTFE also possesses good tensile, flexural, and wear-resistant properties.

The resistance of ECTFE to degradation by heat is excellent. It is one of the most radiation-resistant polymers. Laboratory testing has determined that the following useful life can be expected at the temperature indicated:

Temperature		Useful life
°F	°C	(Years)
329	165	10
338	170	4.5
347	175	2
356	180	1.25

Refer to Table 2.19 for the physical and mechanical properties of ECTFE.

It is very difficult to bond ECTFE, and it is almost impossible to heat weld or solvent weld. For adhesive bonding, a sodium–naphthalene etch is necessary as a surface preparation. The sodium–naphthalene etched surface is subject to degradation by UV radiation and should be protected from direct exposure.

TABLE 2.19 Physical and Mechanical Properties of ECTFE

Property	
Specific gravity	1.68
Water absorption (24 hr at 73°F/23°C) (%)	<0.01
Dielectric strength, short-term (V/mil)	490
Tensile strength at break (psi)	4,500
Tensile modulus ($\times 10^3$ psi)	240
Elongation at break (%)	200–300
Compressive strength (psi)	
Flexural strength (psi)	7,000
Compressive modulus ($\times 10^3$ psi)	
Flexural modulus ($\times 10^3$ psi) at 73°F/23°C	240
200°F/93°C	
250°F/121°C	
Izod impact (ft-lb/in. of notch)	No break
Hardness, Shore D	75
Coefficient of thermal expansion (10^{-6} in./in./°F)	0.44–0.92
Thermal conductivity (10^{-4} cal-cm/sec-cm$^{2°}$C	
or Btu/hr/ft^2/°F/in.)	1.07
Deflection temperature at 264 psi (°F)	151
at 66 psi (°F)	195
Max. operating temperature (°F/°C)	340/171
Limiting oxygen index (%)	60
Flame spread	
Underwriters Lab. rating (Sub. 94)	V–O

The wettability of ECTFE may be increased by plasma treatment. Once the surface has been treated, conventional epoxy or urethane adhesives may be used to bond the surfaces.

2. Corrosion Resistance Properties

The chemical resistance of ECTFE is outstanding. It is resistant to most of the common corrosive chemicals encountered in industry. Included in this list of chemicals are strong mineral and oxidizing acids, alkalines, metal etchants, liquid oxygen, and practically all organic solvents except hot amines (aniline, dimethylamine, etc.) No known solvent dissolves or stress cracks ECTFE at temperatures up to 250°F (120°C).

Some halogenated solvents can cause ECTFE to become slightly plasticized when it comes into contact with them. Under normal circumstances, that does not impair the usefulness of the polymer. When the part is removed from contact with the solvent and allowed to dry, its mechanical properties return to their original values, indicating that no chemical attack has taken place.

As with other fluoropolymers, ECTFE will be attacked by metallic sodium and potassium.

The useful properties of ECTFE are maintained on exposure to cobalt-60 radiation of 200 Mrads.

ECTFE also exhibits excellent resistance to weathering and UV radiation. Table 2.20 shows the compatibility of ECTFE with selected corrodents. Reference 1 provides a more extensive listing.

3. Typical Applications

ECTFE has found application as a pipe and vessel liners and in piping systems, chemical process equipment, and high-temperature wire and cable insulation.

F. ETFE (Ethylenetetrafluorethylene)

ETFE is an alternating copolymer of ethylene and tetrafluorethylene sold by DuPont under the tradename of Tefzel and has the following structural formula:

$$
\begin{array}{cccc}
\text{H} & \text{H} & \text{H} & \text{H} \\
| & | & | & | \\
-\text{C} & -\text{C} & -\text{C} & -\text{C}- \\
| & | & | & | \\
\text{H} & \text{H} & \text{F} & \text{F}
\end{array}
$$

Tefzel is a high-temperature fluoropolymer with a maximum service temperature of 300°F.

1. Physical and Mechanical Properties

ETFE is a rugged thermoplastic with an outstanding balance of properties. Mechanically it is tough and has medium stiffness, excellent flex life, impact,

TABLE 2.20 Compatibility of ECTFE with Selected Corrodents[a]

Chemical	Maximum temp.		Chemical	Maximum temp.	
	°F	°C		°F	°C
Acetic acid 10%	250	121	Barium sulfide	300	149
Acetic acid 50%	250	121	Benzaldehyde	150	66
Acetic acid 80%	150	66	Benzene	150	66
Acetic acid, glacial	200	93	Benzenesulfonic acid 10%	150	66
Acetic anhydride	100	38	Benzoic acid	250	121
Acetone	150	66	Benzyl alcohol	300	149
Acetyl chloride	150	66	Benzyl chloride	300	149
Acrylonitrile	150	66	Borax	300	149
Adipic acid	150	66	Boric acid	300	149
Allyl chloride	300	149	Bromine gas, dry	x	x
Alum	300	149	Bromine, liquid	150	66
Aluminum chloride, aqueous	300	149	Butadiene	250	121
Aluminum chloride, dry			Butyl acetate	150	66
Aluminum fluoride	300	149	Butyl alcohol	300	149
Aluminum hydroxide	300	149	Butyric acid	250	121
Aluminum nitrate	300	149	Calcium bisulfide	300	149
Aluminum oxychloride	150	66	Calcium bisulfite	300	149
Aluminum sulfate	300	149	Calcium carbonate	300	149
Ammonia gas	300	149	Calcium chlorate	300	149
Ammonium bifluoride	300	149	Calcium chloride	300	149
Ammonium carbonate	300	149	Calcium hydroxide 10%	300	149
Ammonium chloride 10%	290	143	Calcium hydroxide, sat.	300	149
Ammonium chloride 50%	300	149	Calcium hypochlorite	300	149
Ammonium chloride, sat.	300	149	Calcium nitrate	300	149
Ammonium fluoride 10%	300	149	Calcium oxide	300	149
Ammonium fluoride 25%	300	149	Calcium sulfate	300	149
Ammonium hydroxide 25%	300	149	Caprylic acid	220	104
Ammonium hydroxide, sat.	300	149	Carbon bisulfide	80	27
Ammonium nitrate	300	149	Carbon dioxide, dry	300	149
Ammonium persulfate	150	66	Carbon dioxide, wet	300	149
Ammonium phosphate	300	149	Carbon disulfide	80	27
Ammonium sulfate 10–40%	300	149	Carbon monoxide	150	66
Ammonium sulfide	300	149	Carbon tetrachloride	300	149
Amyl acetate	160	71	Carbonic acid	300	149
Amyl alcohol	300	149	Cellosolve	300	149
Amyl chloride	300	149	Chloroacetic acid, 50% water	250	121
Aniline	90	32	Chloroacetic acid	250	121
Antimony trichloride	100	38	Chlorine gas, dry	150	66
Aqua regia 3 : 1	250	121	Chlorine gas, wet	250	121
Barium carbonate	300	149	Chlorine, liquid	250	121
Barium chloride	300	149	Chlorobenzene	150	66
Barium hydroxide	300	149	Chloroform	250	121
Barium sulfate	300	149	Chlorosulfonic acid	80	27

(*Continued*)

TABLE 2.20   Continued

Chemical	Maximum temp. °F	Maximum temp. °C	Chemical	Maximum temp. °F	Maximum temp. °C
Chromic acid 10%	250	121	Methyl isobutyl ketone	150	66
Chromic acid 50%	250	121	Muriatic acid	300	149
Citric acid 15%	300	149	Nitic acid 5%	300	149
Citric acid, concd	300	149	Nitric acid 20%	250	121
Copper carbonate	150	66	Nitic acid 70%	150	66
Copper chloride	300	149	Nitric acid, anhydrous	150	66
Copper cyanide	300	149	Nitrous acid, concd	250	121
Copper sulfate	300	149	Oleum	x	x
Cresol	300	149	Perchloric acid 10%	150	66
Cupric chloride 5%	300	149	Perchloric acid 70%	150	66
Cupric chloride 50%	300	149	Phenol	150	66
Cyclohexane	300	149	Phosphoric acid 50–80%	250	121
Cyclohexanol	300	149	Picric acid	80	27
Ethylene glycol	300	149	Potassium bromide 30%	300	149
Ferric chloride	300	149	Salicylic acid	250	121
Ferric chloride 50% in water	300	149	Sodium carbonate	300	149
Ferric nitrate 10–50%	300	149	Sodium chloride	300	149
Ferrous chloride	300	149	Sodium hydroxide 10%	300	149
Ferrous nitrate	300	149	Sodium hydroxide 50%	250	121
Fluorine gas, dry	x	x	Sodium hydroxide, concd	150	66
Fluorine gas, moist	80	27	Sodium hypochlorite 20%	300	149
Hydrobromic acid, dil	300	149	Sodium hypochlorite, concd	300	149
Hydrobromic acid 20%	300	149	Sodium sulfide to 50%	300	149
Hydrobromic acid 50%	300	149	Stannic chloride	300	149
Hydrochloric acid 20%	300	149	Stannous chloride	300	149
Hydrochloric acid 38%	300	149	Sulfuric acid 10%	250	121
Hydrocyanic acid 10%	300	149	Sulfuric acid 50%	250	121
Hydrofluoric acid 30%	250	121	Sulfuric acid 70%	250	121
Hydrofluoric acid 70%	240	116	Sufuric acid 90%	150	66
Hydrofluoric acid 100%	240	116	Sulfuric acid 98%	150	66
Hypochlorous acid	300	149	Sulfuric acid 100%	80	27
Iodine solution 10%	250	121	Sulfuric acid, fuming	300	149
Lactic acid 25%	150	66	Sulfurous acid	250	121
Lactic acid, concd	150	66	Thionyl chloride	150	66
Magnesium chloride	300	149	Toluene	150	66
Malic acid	250	121	Trichloroacetic acid	150	66
Methyl chloride	300	149	White liquor	250	121
Methyl ethyl ketone	150	66	Zinc chloride	300	149

[a] The chemicals listed are in the pure state or in a saturated solution unless otherwise indicated. Compatibility is shown to the maximum allowable temperature for which data are available. Incompatibility is shown by an x. A blank space indicates that data are unavailable.
Source: Schweitzer, PA. Corrosion Resistance Tables, 4th ed. Vols. 1–3. New York: Marcel Dekker, 1995.

cut through, and abrasion resistance. The carbon fiber reinforced compound has an even higher tensile strength (13,000 psi), stiffness, and creep resistance but is still tough and impact resistant.

ETFE can also be reinforced with glass fibers—being the first fluoroplastic that can be reinforced, not merely filled. Because the resin will bond to the fibers, strength, stiffness, creep resistance, heat distortion temperature, and dimensional stability are enhanced. Electrical and chemical properties approach those of the unreinforced resin while the coefficient of friction is actually lower. The service temperature limit is approximately 400°F (204°C). Table 2.21 lists the physical and mechanical properties of unfilled ETFE and ETFE with 20% carbon fiber.

ETFE being a fluorplastic cannot be solvent cemented or heat welded. For adhesive bonding, a sodium–naphthalene etch is necessary as a surface preparation. The sodium–naphthalene treated surface is degraded by UV light and should

TABLE 2.21 Physical and Mechanical Properties of Unfilled ETFE and ETFE with 20% Carbon Fiber

Property	Unfilled	20% carbon fiber
Specific gravity	1.7	1.72
Water absorption (24 hr at 73°F/23°C) (%)	<0.03	0.02
Dielectric strength, short-term (V/mil)	16	
Tensile strength at break (psi)	6,500	13,300
Tensile modulus ($\times 10^3$ psi)	217	
Elongation at break (%)	300	
Compressive strength (psi)		
Flexural strength (psi)		14,300
Compressive modulus ($\times 10^3$ psi)		
Flexural modulus ($\times 10^3$ psi) at 73°F/23°C	170	1,150
200°F/93°C		
250°F/121°C		
Izod impact (ft-lb/in. of notch)	No break	3.9
Hardness, Shore D	67	
Coefficient of thermal expansion (10^{-6} in./in./°F)		0.1
Thermal conductivity (10^{-4} cal-cm/sec-cm^2°C or Btu/hr/ft^2/°F/in.)	1.6	6.0
Deflection temperature at 264 psi (°F)	160	
at 66 psi (°F)	220	425
Max. operating temperature (°F/°C)	300/149	
Limiting oxygen index (%)	30	
Flame spread		
Underwriters Lab. rating (Sub. 94)	V–O	

be protected from direct exposure. Plasma treatment can also be used to increase the wetability of the fluorocarbon surfaces. Once treated, conventional epoxy or urethane adhesives may be used.

2. Corrosion Resistance Properties

ETFE is inert to strong mineral acids, halogens, inorganic bases, and metal salt solutions. Carboxylic acids, aldehydes, aromatic and aliphatic hydrocarbons, alcohols, ketones, esters, ethers, chlorocarbons, and classic polymer solvents have little effect on ETFE.

Tefzel is also weather resistant and ultraviolet light resistant.

Very strong oxidizing acids such as nitric, and organic bases such as amines and sulfuric acid at high concentrations and near their boiling points will affect ETFE to various degrees. Refer to Table 2.22 for the compatibility of ETFE with selected corrodents. Reference 1 provides a more extensive listing.

3. Typical Applications

ETFE finds application in process equipment, piping, chemical ware, wire insulation, tubing, and pump components.

G. FEP (Fluorinated Ethylene-Propylene)

Fluorinated ethylene-propylene is a fully fluorinated thermoplastic with some branching but consists mainly of linear chains having the following formula:

```
    F   F   F   F   F
    |   |   |   |   |
  —C—C—C—C—C—
    |   |   |   |   |
    F   F   |   F   F
            F—C—F
              |
              F
```

1. Physical and Mechanical Properties

FEP has a maximum operating temperature of 375°F/190°C. After prolonged exposure at 400°F (204°C), it exhibits changes in physical strength. It is a relatively soft plastic with lower tensile strength, wear resistance, and creep resistance than other plastics. It is insensitive to notched impact forces and has excellent permeation resistance except to some chlorinated hydrdocarbons. However, FEP may be subject to permeation by specific materials. Refer to Table 1.5.

TABLE 2.22 Compatibility of ETFE with Selected Corrodents[a]

Chemical	Maximum temp.		Chemical	Maximum temp.	
	°F	°C		°F	°C
Acetaldehyde	200	93	Barium hydroxide	300	149
Acetamide	250	121	Barium sulfate	300	149
Acetic acid 10%	250	121	Barium sulfide	300	149
Acetic acid 50%	250	121	Benzaldehyde	210	99
Acetic acid 80%	230	110	Benzene	210	99
Acetic acid, glacial	230	110	Benzenesulfonic acid 10%	210	99
Acetic anhydride	300	149	Benzoic acid	270	132
Acetone	150	66	Benzyl alcohol	300	149
Acetyl chloride	150	66	Benzyl chloride	300	149
Acrylonitrile	150	66	Borax	300	149
Adipic acid	280	138	Boric acid	300	149
Allyl alcohol	210	99	Bromine gas, dry	150	66
Allyl chloride	190	88	Bromine water 10%	230	110
Alum	300	149	Butadiene	250	121
Aluminum chloride, aqueous	300	149	Butyl acetate	230	110
Aluminum chloride, dry	300	149	Buty alcohol	300	149
Aluminum fluoride	300	149	n-Butylamine	120	49
Aluminum hydroxide	300	149	Butyric acid	250	121
Aluminum nitrate	300	149	Calcium bisulfide	300	149
Aluminum oxychloride	300	149	Calcium carbonate	300	149
Aluminum sulfate	300	149	Calcium chlorate	300	149
Ammonium bifluoride	300	149	Calcium chloride	300	149
Ammonium carbonate	300	149	Calcium hydroxide 10%	300	149
Ammonium chloride 10%	300	149	Calcium hydroxide, sat.	300	149
Ammonium chloride 50%	290	143	Calcium hypochlorite	300	149
Ammonium chloride, sat.	300	149	Calcium nitrate	300	149
Ammonium fluoride 10%	300	149	Calcium oxide	260	127
Ammonium fluoride 25%	300	149	Calcium sulfate	300	149
Ammonium hydroxide 25%	300	149	Caprylic acid	210	99
Ammonium hydroxide, sat.	300	149	Carbon bisulfide	150	66
Ammonium nitrate	230	110	Carbon dioxide, dry	300	149
Ammonium persulfate	300	149	Carbon dioxide, wet	300	149
Ammonium phosphate	300	149	Carbon disulfide	150	66
Ammonium sulfate 10–40%	300	149	Carbon monoxide	300	149
Ammonium sulfide	300	149	Carbon tetrachloride	270	132
Amyl acetate	250	121	Carbonic acid	300	149
Amyl alcohol	300	149	Cellosolve	300	149
Amyl chloride	300	149	Chloroacetic acid, 50% water	230	110
Aniline	230	110	Chloroacetic acid 50%	230	110
Antimony trichloride	210	99	Chlorine gas, dry	210	99
Aqua regia 3 : 1	210	99	Chlorine gas, wet	250	121
Barium carbonate	300	149	Chlorine, water	100	38
Barium chloride	300	149	Chlorobenzene	210	99

(*Continued*)

TABLE 2.22 Continued

Chemical	Maximum temp. °F	Maximum temp. °C	Chemical	Maximum temp. °F	Maximum temp. °C
Chloroform	230	110	Methyl ethyl ketone	230	110
Chlorosulfonic acid	80	27	Methyl isobutyl ketone	300	149
Chromic acid 10%	150	66	Muriatic acid	300	149
Chromic acid 50%	150	66	Nitric acid 5%	150	66
Chromyl chloride	210	99	Nitric acid 20%	150	66
Citric acid 15%	120	49	Nitric acid 70%	80	27
Copper chloride	300	149	Nitrous acid, anhydrous	x	x
Copper cyanide	300	149	Nitrous acid concd	210	99
Copper sulfate	300	149	Oleum	150	66
Cresol	270	132	Perchloric acid 10%	230	110
Cupric chloride 5%	300	149	Perchloric acid 70%	150	66
Cyclohexane	300	149	Phenol	210	99
Cyclohexanol	250	121	Phosphoric acid 50–80%	270	132
Dibutyl phthalate	150	66	Picric acid	130	54
Dichloroacetic acid	150	66	Potassium bromide 30%	300	149
Ethylene glycol	300	149	Salicylic acid	250	121
Ferric chloride 50% in water	300	149	Sodium carbonate	300	149
Ferric nitrate 10–50%	300	149	Sodium chloride	300	149
Ferrous chloride	300	149	Sodium hydroxide 10%	230	110
Ferrous nitrate	300	149	Sodium hydroxide 50%	230	110
Fluorine gas, dry	100	38	Sodium hypochlorite 20%	300	149
Fluorine gas, moist	100	38	Sodium hypochlorite, concd	300	149
Hydrobromic acid, dilute	300	149	Sodium sulfide to 50%	300	149
Hydrobromic acid 20%	300	149	Stannic chloride	300	149
Hydrobromic acid 50%	300	149	Stannous chloride	300	149
Hydrochloric acid 20%	300	149	Sulfuric acid 10%	300	149
Hydrochloric acid 38%	300	149	Sufuric acid 50%	300	149
Hydrocyanic acid 10%	300	149	Sulfuric acid 70%	300	149
Hydrofluoric acid 30%	270	132	Sufuric acid 90%	300	149
Hydrofluoric acid 70%	250	121	Sulfuric acid 98%	300	149
Hydrofluoric acid 100%	230	110	Sulfuric acid 100%	300	149
Hypochlorous acid	300	149	Sulfurous acid, fuming	120	49
Lactic acid 25%	250	121	Sulfurous acid	210	99
Lactic acid, concd	250	121	Thionly chloride	210	99
Magnesium chloride	300	149	Toluene	250	121
Malic acid	270	132	Trichloroacetic acid	210	99
Manganese chloride	120	49	Zinc chloride	300	149
Methyl chloride	300	149			

[a] The chemicals listed are in the pure state or in a saturated solution unless otherwise indicated. Compatibility is shown to the maximum allowable temperature for which data are available. Incompatibility is shown by an x. A blank space indicates that data are unavailable.
Source: Schweitzer, PA. Corrosion Resistance Tables, 4th ed. Vols. 1–3. New York: Marcel Dekker, 1995.

TABLE 2.23 Physical and Mechanical Properties of Unfilled FEP and 25% Glass-Filled FEP

Property	Unfilled	25% glass filled
Specific gravity	2.12–2.17	
Water absorption (24 hr at 73°F/23°C) (%)	<0.01	0.01
Dielectric strength, short-term (V/mil)	500–600	
Tensile strength at break (psi)	2,700–3,100	2,400
Tensile modulus ($\times 10^3$ psi)	50	
Elongation at break (%)	250–330	5
Compressive strength (psi)	16,000	
Flexural strength (psi)	3,000	4,000
Compressive modulus ($\times 10^3$ psi)		
Flexural modulus ($\times 10^3$ psi) at 73°F/23°C	80–95	250
200°F/93°C		
250°F/121°C		
Izod impact (ft-lb/in. of notch)	No break	3.2
Hardness, Shore D	D60–65	
Coefficient of thermal expansion (10^{-6} in./in./°F)	0.83–1.05	
Thermal conductivity (10^{-4} cal-cm/sec-cm^2°C	6.0	
or Btu/hr/ft^2/°F/in.)	0.11	
Deflection temperature at 264 psi (°F)	124	150
at 66 psi (°F)	158	
Max. operating temperature (°F/°C)	375	
Limiting oxygen index (%)	95	
Flame spread	nonflammable	
Underwriters Lab. rating (Sub. 94)	VO	

In order to improve some of the physical and mechanical properties of FEP, the polymer is filled with glass fibers. Table 2.23 lists the physical and mechanical properties of unfilled FEP and 25% glass filled FEP.

Because FEP is a fluoropolymer, it is not possible to join it by heat or solvent cementing. For adhesive bonding, a sodium–naphthalene etch is required as a surface preparation. The etched surface is subject to UV light degradation and should therefore be protected from direct exposure. Plasma treatment can also be used. After treatment, conventional epoxy or urethane adhesives may be used.

2. Corrosion Resistance Properties

FEP basically exhibits the same corrosion resistance as PTFE with a few exceptions but at lower operating temperatures. It is resistant to practically all chemicals, except for extremely potent oxidizers such as chlorine trifluoride and

TABLE 2.24 Compatibility of FEP with Selected Corrodents[a]

Chemical	Maximum temp. °F	°C	Chemical	Maximum temp. °F	°C
Acetaldehyde	200	93	Barium sulfate	400	204
Acetamide	400	204	Barium sulfide	400	204
Acetic acid 10%	400	204	Benzaldehyde[b]	400	204
Acetic acid 50%	400	204	Benzene[b,c]	400	204
Acetic acid 80%	400	204	Benzenesulfonic acid 10%	400	204
Acetic acid, glacial	400	204	Benzoic acid	400	204
Acetic anhydride	400	204	Benzyl alcohol	400	204
Acetone[b]	400	204	Benzyl chloride	400	204
Acetyl chloride	400	204	Borax	400	204
Acrylic acid	200	93	Boric acid	400	204
Acrylonitrile	400	204	Bromine gas, dry[c]	200	93
Adipic acid	400	204	Bromine gas, moist[c]	200	93
Allyl alcohol	400	204	Bromine, liquid[b,c]	400	204
Allyl chloride	400	204	Butadiene[b]	400	204
Alum	400	204	Butyl acetate	400	204
Aluminum acetate	400	204	Butyl alcohol	400	204
Aluminum chloride, aqueuos	400	204	n-Butylamine[b]	400	204
Aluminum chloride, dry	300	149	Butyric acid	400	204
Aluminum fluoride[c]	400	204	Calcium bisulfide	400	204
Aluminum hydroxide	400	204	Calcium bisulfite	400	204
Aluminum nitrate	400	204	Calcium carbonate	400	204
Aluminum oxychloride	400	204	Calcium chlorate	400	204
Aluminum sulfate	400	204	Calcium chloride	400	204
Ammonia gas[c]	400	204	Calcium hydroxide 10%	400	204
Ammonium bifluoride[b]	400	204	Calcium hydroxide, sat.	400	204
Ammonium carbonate	400	204	Calcium hypochlorite	400	204
Ammonium chloride 10%	400	204	Calcium nitrate	400	204
Ammonium chloride 50%	400	204	Calcium oxide	400	204
Ammonium chloride, sat.	400	204	Calcium sulfate	400	204
Ammonium fluoride 10%[c]	400	204	Caprylic acid	400	204
Ammonium fluoride 25%[c]	400	204	Carbon bisulfide[c]	400	204
Ammonium hydroxide 25%	400	204	Carbon dioxide, dry	400	204
Ammonium hydroxide, sat.	400	204	Carbon dioxide, wet	400	204
Ammonium nitrate	400	204	Carbon disulfide	400	204
Ammonium persulfate	400	204	Carbon monoxide	400	204
Ammonium phosphate	400	204	Carbon tetrachloride[b,c,d]	400	204
Ammonium sulfate 10–40%	400	204	Carbonic acid	400	204
Ammonium sulfide	400	204	Cellosolve	400	204
Ammonium sulfite	400	204	Chloracetic acid, 50% water	400	204
Amyl actate	400	204	Chloracetic acid	400	204
Amyl alcohol	400	204	Chlorine gas, dry	x	x
Amyl chloride	400	204	Chlorine gas, wet[c]	400	204
Aniline[b]	400	204	Chlorine, liquid[b]	400	204
Antimony trichloride	250	121	Chlorobenzene[c]	400	204
Aqua regia 3 : 1	400	204	Chloroform[c]	400	204
Barium carbonate	400	204	Chlorosulfonic acid[b]	400	204
Barium chloride	400	204	Chromic acid 10%	400	204
Barium hydroxide	400	204	Chromic acid 50%[b]	400	204

TABLE 2.24 Continued

Chemical	Maximum temp. °F	Maximum temp. °C	Chemical	Maximum temp. °F	Maximum temp. °C
Chromyl chloride	400	204	Methyl chloride[c]	400	204
Citric acid 15%	400	204	Methyl ethyl ketone[c]	400	204
Citric acid, concd	400	204	Methyl isobutyl ketone[c]	400	204
Copper acetate	400	204	Muriatic acid[c]	400	204
Copper carbonate	400	204	Nitric acid 5%[c]	400	204
Copper chloride	400	204	Nitric acid 20%[c]	400	204
Copper cyanide	400	204	Nitric acid 70%[c]	400	204
Copper sulfate	400	204	Nitric acid, anhydrous[c]	400	204
Cresol	400	204	Nitrous acid, concd	400	204
Cupric chloride 5%	400	204	Oleum	400	204
Cupric chloride 50%	400	204	Perchloric acid 10%	400	204
Cyclohexane	400	204	Perchloric acid 70%	400	204
Cyclohexanol	400	204	Phenol[c]	400	204
Dibutyl phthalate	400	204	Phosphoric acid 50–80%	400	204
Dichloracetic acid	400	204	Picric acid	400	204
Dichloroethane (ethylene dichloride)[c]	400	204	Potassium bromide 30%	400	204
Ethylene glycol	400	204	Salicylic acid	400	204
Ferric chloride	400	204	Silver bromide 10%	400	204
Ferric chloride 50% in water[b]	260	127	Sodium carbonate	400	204
Ferric nitrate 10–50%	260	127	Sodium chloride	400	204
Ferrous chloride	400	204	Sodium hydroxide 10%[b]	400	204
Ferrous nitrate	400	204	Sodium hydroxide 50%	400	204
Fluorine gas, dry	200	93	Sodium hydroxide, concd	400	204
Fluorine gas, moist	x	x	Sodium hypochlorite 20%	400	204
Hydrobromic acid, dil	400	204	Sodium hypochlorite, concd	400	204
Hydrobromic acid 20%[c,d]	400	204	Sodium sulfide to 50%	400	204
Hydrobromic acid 50%[c,d]	400	204	Stannic chloride	400	204
Hydrochloric acid 20%[c,d]	400	204	Stannous chloride	400	204
Hydrochloric acid 38%[c,d]	400	204	Sulfuric acid 10%	400	204
Hydrocyanic acid 10%	400	204	Sulfuric acid 50%	400	204
Hydrofluoric acid 30%[c]	400	204	Sulfuric acid 70%	400	204
Hydrofluoric acid 70%[c]	400	204	Sulfuric acid 90%	400	204
Hydrofluoric acid 100%[c]	400	204	Sulfuric acid 98%	400	204
Hypochlorous acid	400	204	Sulfuric acid 100%	400	204
Iodine solution 10%[c]	400	204	Sulfuric acid, fuming[c]	400	204
Ketones, general	400	204	Sulfurous acid	400	204
Lactic acid 25%	400	204	Thionyl chloride[c]	400	204
Lactic acid, concd	400	204	Toluene[c]	400	204
Magnesium chloride	400	204	Trichloroacetic acid	400	204
Malic acid	400	204	White liquor	400	204
Manganese chloride	300	149	Zinc Chloride[d]	400	204

[a] The chemicals listed are in the pure state or in a saturated solution unless otherwise indicated. Compatibility is shown to the maximum allowable temperature for which data are available. Incompatibility is shown by an x. A blank space indicates that data are unavailable.
[b] Material will be absorbed.
[c] Material will permeate.
[d] Material can cause stress cracking.
Source: Schweitzer, PA. Corrosion Resistance Tables, 4th ed. Vols. 1–3. New York: Marcel Dekker, 1995.

related compounds. Some chemicals will attack FEP when present in high concentrations at or near the service temperature limit. Refer to Table 2.24 for the compatibility of FEP with selected corrodents. Reference 1 provides a more detailed listing.

FEP is not degraded by UV light and has otherwise excellent weathering resistance.

3. Typical Applications

FEP finds extensive use as a lining material for process vessels and piping, laboratory ware, and use in other process equipment.

H. PA (Polyamides) (Nylon)

Polyamides are also known as Nylons. Polyamide polymers are available in several grades and are identified by the number of carbon atoms in the diamine and dibasic acid used to produce the particular grade. For example, Nylon 6/6 is the reaction product of hexamethylenediamine and adipic acid, both of which are compounds containing six carbon atoms. Some the commonly commercially available Nylons are 6, 6/6, 6/10, 11, and 12. Their structural formulas are as follows:

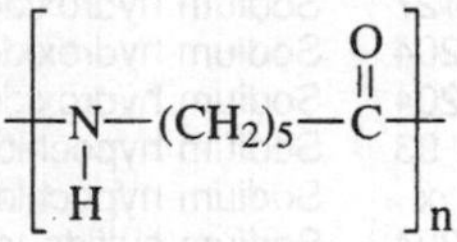

$$\left[-\underset{H}{\underset{|}{N}}-(CH_2)_5-\underset{\underset{O}{\|}}{C} - \right]_n$$

Nylon 6

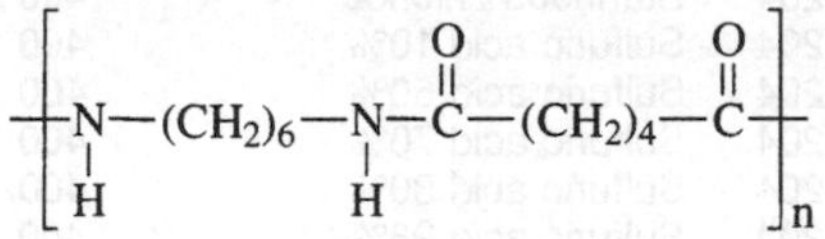

$$\left[-\underset{H}{\underset{|}{N}}-(CH_2)_6-\underset{H}{\underset{|}{N}}-\underset{\underset{O}{\|}}{C}-(CH_2)_4-\underset{\underset{O}{\|}}{C}- \right]_n$$

Nylon 6/6

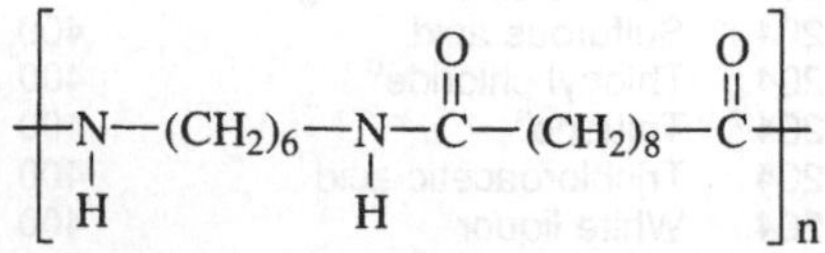

$$\left[-\underset{H}{\underset{|}{N}}-(CH_2)_6-\underset{H}{\underset{|}{N}}-\underset{\underset{O}{\|}}{C}-(CH_2)_8-\underset{\underset{O}{\|}}{C}- \right]_n$$

Nylon 6/10

$$\left[-\underset{H}{\underset{|}{N}}-(CH_2)_{10}-\underset{\underset{O}{\|}}{C}- \right]_n$$

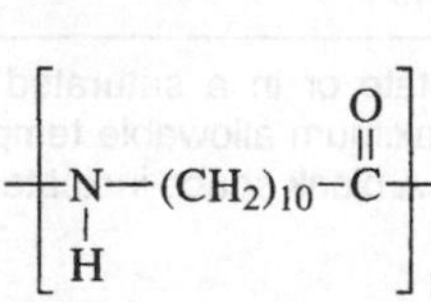

Nylon 11

$$\left[\begin{array}{c} N-(CH_2)_{11}-\overset{\overset{\displaystyle O}{\|}}{C} \\ | \\ H \end{array} \right]_n$$

Nylon 12

Grades 6 and 6/6 are the strongest structurally, grades 6/10 and 11 have the lowest moisture absorption, best electrical properties, and best dimensional stability, and grades 6, 6/6, and 6/10 are the most flexible. Nylon 12 has the same advantages as grades 6/10 and 11 but at a lower cost since it is more easily and economically processed.

1. Physical and Mechanical Properties

Polyamides are strong, tough thermoplastics with good impact, tensile, and flexural strengths, from freezing temperatures up to 300°F/149°C. They all exhibit good electrical resistivities and excellent low friction properties.

Moisture absorption properties must be considered when designing with polyamides since all Nylons absorb some moisture from environmental humidity. After a 24 hour immersion period, they will absorb from 0.5 to almost 2% moisture. However, this need not limit applications since there are low-moisture absorption grades available. The physical and mechanical properties of Nylon type 6 and type 6/6 are shown in Table 2.25.

Polyamide resins retain useful mechanical properties over a temperature range of −60 to +400°F (−51 to +204°C). Both short-term and long-term temperature effects must be considered. Such properties as stiffness and toughness are affected by short term exposure to elevated temperatures. There is also the possibility of stress relief and its effect on dimensions.

In long-term exposure at high temperatures, there is the possibility of gradual oxidative embrittlement. Such applications should make use of a heat-stabilized grade of resin. High-temperature oxidation reduces the impact strength more than the static strength properties. After several days at 250°F/121°C, the impact strength and elongation of Nylon are reduced considerably.

When Nylon is immersed in boiling water for 2500 hours, the tensile strength is slowly reduced until it levels off at 6000 psi. The elongation drops rapidly after 1500 hours of immersion. Hence this time has been taken as the limit for the use of basic polyamide. However, there are some compositions formulated especially to resist hot water exposure.

Polyamides, as do other thermoplasts, suffer from creep and dimensional-stability problems particularly under elevated temperature and load conditions. To offset these problems, glass fiber reinforcement is used. For example, 40% glass fiber reinforced nylon outperforms the unreinforced version by exhibiting two and one-half greater tensile and Izod impact strengths, four times greater flexural

TABLE 2.25 Physical and Mechanical Properties of Nylon Type 6 and Type 6/6

Property	Type 6	Type 6/6
Specific gravity	1.12–1.14	1.13–1.18
Water absorption (24 hr at 73°F/23°C) (%)	1.3–1.9	1.0–2.8
Dielectric strength, short-term (V/mil)	400	600
Tensile strength at break (psi)	6,000–24,000	13,700
Tensile modulus ($\times 10^3$ psi)	380–404	230–550
Elongation at break (%)	30–100	15–80
Compressive strength (psi)	13,000–16,000	12,500–15,000
Flexural strength (psi)	15,700	17,900
Compressive modulus ($\times 10^3$ psi)	250	
Flexural modulus ($\times 10^3$ psi) at		
73°F/23°C	390–410	410–470
200°F/93°C		
250°F/121°C		
Izod impact (ft-lb/in. of notch)	0.6–3.0	0.55–2.1
Hardness, Rockwell	R119	R120
Coefficient of thermal expansion (10^{-6} in./in./°F)	80–83	80
Thermal conductivity (10^{-4} cal-cm/ sec-cm²°C or Btu/hr/ft²/°F/in.)	5.8	5.8
Deflection temperature at 264 psi (°F)	155–185	158–212
at 66 psi (°F)	347–375	235–474
Max. operating temperature (°F/°C)	175–300/79–150	175–300/79–150
Limiting oxygen index (%)		
Flame spread		
Underwriters Lab. rating (Sub. 94)		

modulus, and only one-fifth of the tensile creep. Table 2.26 lists the physical and mechanical properties of glass reinforced type 6 polyamide while Table 2.27 does likewise for type 6/6 polyamide.

Polyamides can also be blended with other thermoplasts to produce polymers to meet specific applications. Since blends are only physical mixtures, the resulting polymer usually has physical and mechanical properties that lie somewhere between the values of its constituent materials. One such blend is PA and ABS. Table 2.28 lists the physical and mechanical properties of PA/ABS blend, while Table 2.29 shows the properties of a glass fiber reinforced PA/ceamic alloy.

Polyamides can be joined by solvent cementing using aqueous phenol, solutions of resorcinol in alcohol, or solutions of calcium chloride in alcohol. Adhesive bonding is also possible but is inferior to solvent cementing.

TABLE 2.26 Physical and Mechanical Properties of Glass Reinforced Type 6 PA

Property	15% glass-filled	25% glass-reinforced	30–35% glass-reinforced
Specific gravity	1.23	1.32	1.35–1.42
Water absorption (24 hr at 73°F/23°C) (%)	2.6	2.3	0.9–1.2
Dielectric strength, short-term (V/mil)			400–450
Tensile strength at break (psi)	18,900	23,200	18,900
Tensile modulus ($\times 10^3$ psi)	798	1,160	1,250
Elongation at break (%)	3.5	3.5	2.2–3.6
Compressive strength (psi)			19,000–29,000
Flexural strength (psi)			34,000–36,000
Compressive modulus ($\times 10^3$ psi)			
Flexural modulus ($\times 10^3$ psi) at			
73°F/23°C	700	910	1,250–1,400
200°F/93°C			
250°F/121°C			
Izod impact (ft-lb/in. of notch)	1.1	2.0	2.1–3.4
Hardness, Rockwell	M92	M95	M93–96
Coefficient of thermal expansion (10^{-6} in./in./°F)	52	40	16–80
Thermal conductivity (10^{-4}cal-cm/sec-cm^2°C or Btu/hr/ft^2/°F/in.)			5.8–11.4
Deflection temperature at			
264 psi (°F)	374	410	392–420
at 66 psi (°F)	419	428	420–430
Max. operating temperature (°F/°C)			
Limiting oxygen index (%)			
Flame spread			
Underwriters Lab. rating (Sub. 94)			

2. Corrosion Resistance Properties

The polyamides exhibit excellent resistance to a broad range of chemicals and harsh environments. They have good resistance to most inorganic alkalines, particularly ammonium hydroxide and ammonia, even at elevated temperatures, and to sodium and potassium hydroxides at ambient temperatures. They also display good resistance to almost all inorganic salts and to almost all hydrocarbons and petroleum-based fuels.

TABLE 2.27 Physical and Mechanical Properties of Glass Reinforced Type 6/6 PA

Property	13% reinforcing heat stabilized	15% reinforcing
Specific gravity	1.21–1.23	1.23
Water absorption (24 hr at 73°F/23°C) (%)	1.1	
Dielectric strength, short-term (V/mil)		
Tensile strength at break (psi)	15,000–17,000	18,900
Tensile modulus ($\times 10^3$ psi)		870
Elongation at break (%)	3.0–5	3
Compressive strength (psi)		
Flexural strength (psi)	27,500	25,600
Compressive modulus ($\times 10^3$ psi)		
Flexural modulus ($\times 10^3$ psi) at		
73°F/23°C	700–750	720
200°F/93°C		
250°F/121°C		
Izod impact (ft-lb/in. of notch)	0.95–1.1	1.1
Hardness, Rockwell	M95/R120	M97
Coefficient of thermal expansion (10^{-6} in./in./°F)		52
Thermal conductivity (10^{-4} cal-cm/sec-cm^2°C or Btu/hr/ft^2/°F/in.)		
Deflection temperature at		
264 psi (°F)	450–470	482
at 66 psi (°F)	494	482
Max. operating temperature (°F/°C)		
Limiting oxygen index (%)		
Flame spread		
Underwriters Lab. rating (Sub. 94)		

TABLE 2.28 Physical and Mechanical Properties of PA/ABS Blend

Property	
Specific gravity	1.06–1.07
Water absorption (24 hr at 73°F/23°C) (%)	1.1–1.5
Dielectric strength, short-term (V/mil)	425
Tensile strength at break (psi)	5,800–6,300
Tensile modulus ($\times 10^3$ psi)	245–295
Elongation at break (%)	140–290
Compressive strength (psi)	
Flexural strength (psi)	
Compressive modulus ($\times 10^3$ psi)	
Flexural modulus ($\times 10^3$ psi) at 73°F/23°C	0.17–0.3
200°F/93°C	
250°F/121°C	
Izod impact (ft-lb/in. of notch)	15–20
Hardness, Rockwell	R105
Coefficient of thermal expansion (10^{-6} in./in./°F)	
Thermal conductivity (10^{-4} cal-cm/sec-cm$^{2\circ}$C or Btu/hr/ft^2/°F/in.)	
Deflection temperature at 264 psi (°F)	149
at 66 psi (°F)	165
Max. operating temperature (°F/°C)	
Limiting oxygen index (%)	
Flame spread	
Underwriters Lab. rating (Sub. 94)	HB

Polyamides are also resistant to UV degradation, weathering, and ozone. Refer to Table 2.30 for the compatibility of PA with selected corrodents. A more detailed listing will be found in Reference 1.

3. Typical Applications

Polyamides are used to produce gears, cams, bearings, wire insulation, pipe fittings, and hose fittings. PA/ABS blends are used for appliances, lawn and garden equipment, power tools, and sporting goods. In the automotive industry, the blend is used to produce interior functional components, fasteners, housings, and shrouds.

I. PAI (Polyamide-Imide)

Polyamide–imides are heterocyclic polymers having an atom of nitrogen in one of the rings of the molecular chain. A typical chemical structure is shown below:

TABLE 2.29 Physical and Mechanical Properties of PA/Ceramic
Alloy, Glass Fiber Reinforced

Property	
Specific gravity	1.59–1.81
Water absorption (24 hr at 73°F/23°C) (%)	0.35–0.50
Dielectric strength, short-term (V/mil)	
Tensile strength at break (psi)	25,000–30,000
Tensile modulus (×10³ psi)	
Elongation at break (%)	3–4
Compressive strength (psi)	
Flexural strength (psi)	39,000–45,000
Compressive modulus (×10³ psi)	
Flexural modulus (×10³ psi) at 73°F/23°C	1,800–2,150
200°F/93°C	
250°F/121°C	
Izod impact (ft-lb/in. of notch)	1.6–2.0
Hardness, Rockwell	R120
Coefficient of thermal expansion (10⁻⁶ in./in./°F)	
Thermal conductivity (10⁻⁴ cal-cm/sec-cm²°C or Btu/hr/ft²/°F/in.)	
Deflection temperature at 264 psi (°F)	410–495
at 66 psi (°F)	420–515
Max. operating temperature (°F/°C)	
Limiting oxygen index (%)	
Flame spread	
Underwriters Lab. rating (Sub. 94)	

1. Physical and Mechanical Properties

This series of thermoplasts can be used at high and low temperatures and as such
finds applicaitons in the extreme environments of space. The temperature range is
−310°F to 500°F (−190°C to 260°C). The polyamide–imides possess excellent
electrical and mechanical properties that are relatively stable from low negative
temperatures to high positive temperatures, dimensional stability (low cold flow)
in most environments, excellent resistance to ionizing radiation, and very low
outgassing in high vacuum.

TABLE 2.30 Compatibility of Polyamides with Selected Corrodents[a]

Chemical	Maximum temp. °F	Maximum temp. °C	Chemical	Maximum temp. °F	Maximum temp. °C
Acetaldehyde	x	x	Benzenesulfonic acid 10%	x	x
Acetamide	250	121	Benzoic acid	80	27
Acetic acid 10%	200	93	Benzyl alcohol	200	93
Acetic acid 50%	x	x	Benzyl chloride	250	121
Acetic acid 80%	x	x	Borax	200	93
Acetic acid, glacial	x	x	Boric acid	x	x
Acetic anhydride	200	93	Bromine gas, dry	x	x
Acetone	80	27	Bromine gas, moist	x	x
Acetyl chloride	x	x	Bromine, liquid	x	x
Acrylonitrile	80	27	Butadiene	80	27
Allyl alcohol	80	27	Butyl acetate	250	121
Alum	x	x	Butyl alcohol	200	93
Aluminum chloride, aqueous	x	x	n-Butylamine	200	93
Aluminum chloride, dry	x	x	Butyric acid	x	x
Aluminum fluoride	80	27	Calcium bisulfite	140	60
Aluminum hydroxide	250	121	Calcium carbonate	200	93
Aluminum nitrate	80	27	Calcium chloride	250	121
Aluminum sulfate	140	60	Calcium hydroxide 10%	150	66
Ammonia gas	200	93	Calcium hydroxide, sat.	150	66
Ammonium carbonate	240	160	Calcium hypochlorite	x	x
Ammonium chloride 10%	200	93	Calcium nitrate	x	x
Ammonium chloride 50%	200	93	Calcium oxide	80	27
Ammonium fluoride 10%	80	27	Calcium sulfate	80	27
Ammonium fluoride 25%	80	27	Caprylic acid	230	110
Ammonium hydroxide 25%	250	121	Carbon bisulfide	80	27
Ammonium hydroxide, sat.	250	121	Carbon dioxide, dry	80	27
Ammonium nitrate	190	88	Carbon disulfide	80	27
Ammonium persulfate	x	x	Carbon monoxide	80	27
Ammonium phosphate	80	27	Carbon tetrachloride	250	121
Ammonium sulfite	80	27	Carbonic acid	100	38
Amyl acetate	150	66	Cellosolve	250	121
Amyl alcohol	200	93	Chloroacetic acid, 50% water	x	x
Amyl chloride	x	x	Chloroacetic acid	x	x
Aniline	x	x	Chlorine gas, dry	x	x
Antimony trichloride	x	x	Chlorine gas, wet	x	x
Aqua regia 3:1	x	x	Chlorine, liquid	x	x
Barium carbonate	80	27	Chlorobenzene	250	121
Barium chloride	250	121	Chloroform	130	54
Barium hydroxide	80	27	Chlorosulfonic acid	x	x
Barium sulfate	80	27	Chromic acid 10%	x	x
Barium sulfide	80	27	Chromic acid 50%	x	x
Benzaldehyde	150	66	Citric acid 15%	200	93
Benzene	250	121	Chloroacetic acid	x	x

(Continued)

TABLE 2.30 Continued

Chemical	Maximum temp. °F	Maximum temp. °C	Chemical	Maximum temp. °F	Maximum temp. °C
Citric acid, concd	210	99	Methyl isobutyl ketone	200	93
Copper acetate	100	38	Muriatic acid	x	x
Copper carbonate	80	27	Nitric acid 5%	90	32
Copper chloride	x	x	Nitric acid 20%	80	27
Copper cyanide	80	27	Nitric acid 70%	x	x
Copper sulfate	80	27	Nitric acid, anhydrous	x	x
Cresol	100	38	Nitrous acid, concd	x	x
Cupric chloride 5%	x	x	Oleum	x	x
Cupric chloride 50%	x	x	Phenol	570	299
Cyclohexanol	80	27	Phosphoric acid 50–80%	190	88
Dibutyl phthalate	240	116	Picric acid	x	x
Dichloroethane (ethylene dichloride)	200	93	Potassium bromide 30%	210	99
			Salicylic acid	80	27
Ethylene glycol	210	99	Sodium carbonate to 30%	210	99
Ferric chloride	x	x	Sodium chloride to 30%	210	99
Ferric chloride 50% in water	x	x	Sodium hydroxide 10%	300	149
Ferric nitrate 10–50%	x	x	Sodium hydroxide 50%[b]	300	149
Ferrous chloride	x	x	Sodium hydroxide, concd	80	27
Fluorine gas, dry	570	299	Sodium hypochlorite 20%	x	x
Fluorine gas, moist	60	16	Sodium hypochlorite, concd	x	x
Hydrobromic acid, dil	90	32	Sodium sulfide to 50%	210	99
Hydrobromic acid 20%	80	27	Stannic chloride	x	x
Hydrobromic acid 50%	x	x	Stannous chloride, dry	570	299
Hydrochloric acid 20%	80	27	Sulfuric acid 10%	x	x
Hydrochoric acid 38%	x	x	Sulfuric acid 50%	x	x
Hydrofluoric acid 30%	x	x	Sulfuric acid 70%	x	x
Hydrofluoric acid 70%	x	x	Sulfuric acid 90%	x	x
Hydrofluoric acid 100%	120	49	Sulfuric acid 98%	x	x
Lactic acid 25%	210	99	Sulfuric acid 100%	x	x
Lactic acid, concd	90	32	Sulfuric acid, fuming	x	x
Magnesium chloride 50%	130	54	Sulfurous acid	90	32
Malic acid	210	99	Toluene	210	99
Manganese chloride 37%	x	x	Trichloroacetic acid	80	27
Methyl chloride	210	99	Zinc chloride, dry	80	27
Methyl ethyl ketone	210	99			

[a] The chemicals listed are in the pure state or in a saturated solution unless otherwise indicated. Compatibility is shown to the maximum allowable temperature for which data are available. Incompatibility is shown by an x. A blank space indicates that data are unavailable.
[b] Material is subject to stress cracking.
[c] Material subject to pitting.
Source: Schweitzer, PA. Corrosion Resistance Tables, 4th ed. Vols. 1–3. New York: Marcel Dekker, 1995.

TABLE 2.31 Physical and Mechanical Properties of PAI

Property	Unfilled	Wear-resistant grade	30% glass fiber reinforced
Specific gravity	1.42	1.50	1.61
Water absorption (24 hr at 73°F/23°C) (%)			
Dielectric strength, short-term (V/mil)	580		840
Tensile strength at break (psi)			
Tensile modulus (×10³ psi)	700	870	1,560
Elongation at break (%)	15	9	7
Compressive strength (psi)	32,100	18,300	38,300
Flexural strength (psi)	34,900	27,000	48,300
Compressive modulus (×10³ psi)			1,150
Flexural modulus (×10³ psi) at 73°F/23°C 200°F/93°C 250°F/121°C	730	910	1,700
Izod impact (ft-lb/in. of notch)	2.7	1.3	1.5
Hardness, Rockwell	E86	E66	E94
Coefficient of thermal expansion (10⁻⁶ in./in./°F)	30.6	15	16.2
Thermal conductivity (10⁻⁴cal-cm/sec-cm²°C or Btu/hr/ft²/°F/in.)	6.2		8.8
Deflection temperature at 264 psi (°F) at 66 psi (°F)	532	532	539
Max. operating temperature (°F/°C)	500/260		
Limiting oxygen index (%)			
Flame spread			
Underwriters Lab. rating (Sub. 94)			

The polyamide–imides have very low coefficients of friction which can be further improved by the use of graphite or other fillers. Refer to Table 2.31 for the physical and mechanical properties.

The polyamide–imides can be solvent cemented or bonded with epoxy or urethane adhesives.

2. Corrosion Resistance Properties

PAI is resistant to acetic acid and phosphoric acid up to 35% and sulfuric acid to 30%. It is not resistant to sodium hydroxide. PAI also has resistance to UV light degradation. Refer to Table 2.32 for the compatibility of PAI with selected corrodents. Reference 1 provides a more detailed listing.

3. Typical Applications

PAI finds applications in under the hood applications and as bearings and pistons in compressors.

J. PB (Polybutylene)

Polybutylene is a semicrystalline polyolefin thermoplastic based on poly-1-butene and includes homopolymers and a series of copolymers (butene/ethylene). This thermoplast has the following structural formula:

$$
\begin{array}{c}
\quad \text{H} \quad \text{H} \quad \text{H} \\
\quad | \quad\; | \quad\; | \\
-\text{C}-\text{C}-\text{C}- \\
\quad | \quad\; | \\
\quad \text{H} \quad \text{H} \quad | \\
\quad\quad\quad\; \text{H}-\text{C}-\text{H} \\
\quad\quad\quad\quad\; | \\
\quad\quad\quad\quad\; \text{H}
\end{array}
$$

1. Physical and Mechanical Properties

Polybutylene maintains its mechanical properties at elevated temperatures. Its long-term strength is greater than that of high density polyethylene. Polybutylene has an upper temperature limit of 200°F (93°C). It possesses a combination of stress cracking resistance, chemical resistance, and abrasion resistance.

Refer to Table 2.33 for the physical and mechanical properties of polybutylene.

Polybutylene cannot be solvent cemented. Joining is by means of thermal welding or by adhesive bonding. Epoxy and nitrile–phenolic adhesives can be used after surface preparation. The surface must be etched with a sodium dichromate–sulfuric acid solution at elevated temperatures. However, the optimum surface process for polybutylene is by means of plasma treatment.

2. Corrosion Resistance Properties

Polybutylene is resistant to acids, bases, soaps, and detergents. It is partially soluble in aromatic and chlorinated hydrocarbons above 140°F (60°C) and is not completely resistant to aliphatic solvents at room temperatures. Chlorinated water will cause pitting attack. Refer to Table 2.34 for the compatibility of polybutylene with selected corrodents. PB is subject to degradation by UV light.

3. Typical Applications

Applications include piping, chemical process equipment, and also fly ash bottom ash lines containing abrasive slurries. PB is also used for molded appliance parts.

TABLE 2.32 Compatibility of PAI with Selected Corrodents[a]

Chemical	Maximum temp.		Chemical	Maximum temp.	
	°F	°C		°F	°C
Acetic acid 10%	200	93	Cellosolve	200	93
Acetic acid 50%	200	93	Chlorobenzene	200	93
Acetic acid 80%	200	93	Chloroform	120	49
Acetic acid glacial	200	93	Chromic acid 10%	200	93
Acetic anhydride	200	93	Cyclohexane	200	93
Acetone	80	27	Cyclohexanol	200	93
Acetyl chloride, dry	120	49	Dibutyl phthalate	200	93
Aluminum sulfate 10%	220	104	Ethylene glycol	200	93
Ammonium chloride 10%	200	93	Hydrochloric acid 20%	200	93
Ammonium hydroxide 25%	200	93	Hydrochloric acid 38%	200	93
Ammonium hydroxide, sat.	200	93	Lactic acid 25%	200	93
Ammonium nitrate 10%	200	93	Lactic acid concd	200	93
Ammonium sulfate 10%	200	93	Magnesium chloride, dry	200	93
Amyl acetate	200	93	Methyl ethyl ketone	200	93
Aniline	200	93	Oleum	120	49
Barium chloride 10%	200	93	Sodium carbonate 10%	200	93
Benzaldehyde	200	93	Sodium chloride 10%	200	93
Benzene	80	27	Sodium hydroxide 10%	x	
Benzenesulfonic acid 10%	x		Sodium hydroxide 50%	x	
Benzyl chloride	120	49	Sodium hydroxide concd.	x	
Bromine gas, moist	120	49	Sodium hypochlorite 10%	200	93
Butyl acetate	200	93	Sodium hypochlorite concd	x	
Butyl alcohol	200	93	Sodium sulfide to 50%	x	
n-Butylamine	200	93	Sulfuric acid 10%	200	93
Calcium chloride	200	93	Sulfuric acid, fuming	120	49
Calcium hypochlorite	x		Toluene	200	93

[a] The chemicals listed are in the pure state or in a saturated solution unless otherwise indicated. Compatibility is shown to the maximum allowable temperature for which data are available. Incompatibility is shown by an x.
Source: Schweitzer, PA. Corrosion Resistance Tables, 4th ed. Vols. 1–3. New York: Marcel Dekker, 1995.

K. Polycarbonate (PC)

This thermoplast is produced by General Electric under the tradename of Lexan. Polycarbonates are classified as engineering thermoplasts because of their high performance in engineering design. The generalized formula is

1. Physical and Mechanical Properties

Polycarbonates are noted for their impact strength which is several times higher than that of other engineering polymers. They are tough, rigid, and dimensionally stable and are available as transparent or colored parts.

They possess the useful properties of creep resistance, high heat resistance, dimensional stability, good electrical properties, product transparency, exceptional impact strength, and are self-extinguishing.

Immersion in water and exposure to high humidity at temperatures up to 212°F/100°C have little effect on dimensions. Polycarbonates are among the

TABLE 2.33 Physical and Mechanical Properties of PB

Property	
Specific gravity	0.91–0.925
Water absorption (24 hr at 73°F/23°C) (%)	0.01–0.02
Dielectric strength, short-term (V/mil)	>450
Tensile strength at break (psi)	3,800–4,400
Tensile modulus ($\times 10^3$ psi)	30–40
Elongation at break (%)	300–380
Compressive strength (psi)	
Flexural strength (psi)	2,000–2,300
Compressive modulus ($\times 10^3$ psi)	31
Flexural modulus ($\times 10^3$ psi) at 73°F/23°C	40–50
200°F/93°C	
250°F/121°C	
Izod impact (ft-lb/in. of notch)	No break
Hardness, Rockwell	
Coefficient of thermal expansion (10^{-6} in./in./°F)	128–150
Thermal conductivity (10^{-4} cal-cm/sec-cm^2°C or Btu/hr/ft^2/°F/in.)	5.2
Deflection temperature at 264 psi (°F)	130–140
at 66 psi (°F)	215–235
Max. operating temperature (°F/°C)	200–93
Limiting oxygen index (%)	
Flame spread	
Underwriters Lab. rating (Sub. 94)	

TABLE 2.34 Compatibility of PB with Selected Corrodents[a]

Acetic acid	R	Citric acid	R
Acetic anhydride	R	Cyclohexane	R
Allyl alcohol	x	Detergents	R
Aluminum chloride	x	Lactic acid	R
Ammonium chloride	x	Malic acid	R
Ammonium hydroxide	R	Methyl alcohol	x
Amyl alcohol	x	Phenol	R
Aniline	R	Picric acid	R
Benzaldehyde	R	Propyl alcohol	x
Benzene	R	Salicylic acid	R
Benzoic acid	R	Soaps	R
Boric acid	R	Sodium carbonate	R
Butyl alcohol	x	Sodium hydroxide 10%	R
Calcium carbonate	R	Sodium hydroxide 50%	R
Calcium hydroxide	R	Toluene	R
Calcium sulfate	R	Trichloracetic acid	R
Carbonic acid	R	Water (chlorine free)	R
Chloracetic acid	R	Xylene	R
Chlorobenzene	R		

[a]Materials in the pure state or saturated solution unless otherwise specified. R = PB is resistant at 70°F/23°C; x = PB is not resistant.

most stable polymers in a wet environment. The operating temperature range for polycarbonates is from −200°F to +250−275°F (−129°C to +120−135°C).

Refer to Table 2.35 for the physical and mechanical properties of polycarbonates.

To add additional strength, the polycarbonate can also be reinforced with glass fiber and/or blended or alloyed with other thermoplasts. Refer to Table 2.36.

Joining of polycarbonate can be accomplished either by solvent cementing or by thermal fusion methods. When joining by solvent cementing, methylene chloride or ethylene dichloride are the preferred solvents. Since PC can stress crack in the presence of certain solvents, care should be exercised as to which solvent cement is used. Cyanoacrylates are satisfactory.

2. Corrosion Resistance Properties

Polycarbonate is resistant to aliphatic hydrocarbons and weak acids and has limited resistance to weak alkalies. It is resistant to most oils and greases. PC will be attacked by strong alkalies and strong acids and is soluble in ketones, esters, and aromatic and chlorinated hydrocarbons. Polycarbonate is not affected by UV light and has excellent weatherability. Table 2.37 lists the compatibility of PC with selected corrodents. Reference 1 provides a more detailed listing.

TABLE 2.35 Physical and Mechanical Properties of PC

Property	Unfilled	30% glass-filled flame-retardant grade
Specific gravity	1.2	1.63
Water absorption (24 hr at 73°F/23°C) (%)	0.15	0.06–0.07
Dielectric strength, short-term (V/mil)	380–400	490
Tensile strength at break (psi)	9,100–10,500	17,400–20,000
Tensile modulus ($\times 10^3$ psi)	345	1,490–1,700
Elongation at break (%)	110–120	2.0–3.0
Compressive strength (psi)	10,000–12,500	18,000
Flexural strength (psi)	12,500–13,500	30,000
Compressive modulus ($\times 10^3$ psi)	350	
Flexural modulus ($\times 10^3$ psi) at 73°F/23°C 200°F/93°C 250°F/121°C	330–340	1,300–1,500
Izod impact (ft-lb/in. of notch)	2.3–12	1.3–1.6
Hardness, Rockwell	M70–75	M88–90
Coefficient of thermal expansion (10^{-6} in./in./°F)	68	1.5
Thermal conductivity (10^{-4}cal-cm/sec-cm^2°C or Btu/hr/ft^2/°F/in.)	4.7	
Deflection temperature at 264 psi (°F) at 66 psi (°F)	250–270 280–387	400–450 425–490
Max. operating temperature (°F/°C)	250–275/120–135	
Limiting oxygen index (%)	20	
Flame spread Underwriters Lab. rating (Sub. 94)	SEO–SEI	

3. Typical Applications

Because of its extremely high impact resistance and good clarity, it is widely used for windows in chemical equipment and glazing in chemical plants. It also finds wide use in outdoor energy management devices, network interfaces, electrical wiring blocks, telephone equipment, and lighting diffusers, globes, and housings.

TABLE 2.36 Physical and Mechanical Properties of Glass Fiber Reinforced PC

Property	10% reinforcing	30% reinforcing
Specific gravity	1.27–1.28	1.4–1.43
Water absorption (24 hr at 73°F/23°C) (%)	0.12–0.15	0.08–0.14
Dielectric strength, short-term (V/mil)	450–530	470–475
Tensile strength at break (psi)	7,000–10,000	19,000–20,000
Tensile modulus ($\times 10^3$ psi)	450–600	1,250–1,400
Elongation at break (%)	4–10	2–5
Compressive strength (psi)	12,000–16,000	18,000–20,000
Flexural strength (psi)	13,700–16,000	
Compressive modulus ($\times 10^3$ psi)	520	1,300
Flexural modulus ($\times 10^3$ psi) at 73°F/23°C 200°F/93°C 250°F/121°C	460–580	1,100
Izod impact (ft-lb/in. of notch)	2.4	1.7–3.0
Hardness, Rockwell	R118–122	R119
Coefficient of thermal expansion (10^{-6} in./in./°F)	32–38	22–23
Thermal conductivity (10^{-4} cal-cm/sec-cm^2°C or Btu/hr/ft^2/°F/in.)	4.6–5.2	5.2–7.6
Deflection temperature at 264 psi (°F) at 66 psi (°F)	280–288 295	295–300 300–305
Max. operating temperature (°F/°C)		
Limiting oxygen index (%)		
Flame spread		
Underwriters Lab. rating (Sub. 94)		

L. Polyetherether Ketone (PEEK)

PEEK is a linear polyaromatic thermoplast having the following chemical structure:

TABLE 2.37 Compatibility of PC with Selected Corrodents[a]

Acetic acid 5%	R	Hydrogen peroxide 0–5%	R
Acetic acid 10%	R	Ink	R
Acetone	x	Linseed oil	R
Ammonia 10%	x	Methanol	x
Benzene	x	Methyl ethyl ketone	x
Butyl acetate	x	Methylene chloride	x
Cadmium chloride	R	Milk	R
Carbon disulfide	x	Nitric acid 2%	R
Carbon tetrachloride	x	Paraffin oil	R
Chlorobenzene	x	Phosphoric acid 10%	x
Chloroform	x	Potassium hydroxide 50%	x
Citric acid 10%	R	Potassium dichromate	R
Diesel oil	x	Potassium permanganate 10%	R
Dioxane	x	Silicone oil	R
Edible oil	R	Soap solution	R
Ethanol	x	Sodium bisulfite	R
Ether	x	Sodium carbonate 10%	R
Ethyl acetate	x	Sodium chloride 10%	R
Ethylene chloride	x	Sodium hydroxide 50%	x
Ethylene glycol	R	Sodium hydroxide 5%	x
Formic acid	x	Sulfuric acid 98%	x
Fruit juice	R	Sulfuric acid 2%	R
Fuel oil	R	Toluene	x
Glycerine	x	Vaseline	R
Heptane/hexane	R	Water, cold	R
Hydrochloric acid 38%	R	Water, hot	x
Hydrochloric acid 2%	R	Wax, molten	R
Hydrogen peroxide 30%	R	Xylene	x

[a] R = material is resistant at 73°F/20°C; x = material is not resistant.

It is a proprietary product of ICI and is an engineering polymer suitable for applications that require mechanical strength with the need to resist difficult thermal and chemical environments.

1. Physical and Mechanical Properties

The tensile properties of PEEK exceed those of most engineering polymers. When reinforced with carbon fibers, tensile strengths of over 29,000 psi can be achieved with excellent properties retained up to 570°F/299°C.

PEEK also exhibits excellent creep properties, even at very high temperatures. Combined with good flexural and tensile characteristics of the material, the creep properties provide an excellent balance of properties for applications where the material is required to withstand high loadings for long periods at high temperatures without permanent deformation.

PEEK has exceptional stiffness which is reflected in its flexural modulus which is among the best of any thermoplastic. Its flexural modulus at very high temperatures can be improved by means of glass or carbon reinforcement. Such reinforcement also increases modulus, heat distortion temperature, creep and fatigue resistance, and thermal conductivity.

The physical and mechanical properties of PEEK are given in Table 2.38. PEEK can be bonded with epoxy or urethane adhesives without special surface treatment other than abrasion and solvent cleaning.

PEEK has an operating temperature range of −85 to +480°F/−65 to +250°C.

2. Corrosion Resistance Properties

PEEK is not chemically attacked by water. It has excellent long-term resistance to water at both ambient and elevated temperatures. It also has excellent rain erosion resistance. Because PEEK is not hydrolyzed by water at elevated temperatures in a continuous cycle environment, the material may be steam sterilized using conventional sterilization equipment.

PEEK is insoluble in all common solvents and has excellent resistance to a wide range of organic and inorganic solvents. It also exhibits excellent resistance to hard (gamma) radiation, absorbing over 1000 Mrads of radiation without suffering significant damage.

Like most polyaromatics, PEEK is subject to degradation by UV light during outdoor weathering. However, testing has shown that over a twelve-month period for both natural and pigmented moldings, the effect is minimal. In more extreme weathering conditions, painting or pigmenting of the polymer will protect it from excessive property degradation.

Refer to Table 2.39 for the compatibility of PEEK with selected corrodents. Reference 1 provides a more comprehensive listing.

3. Typical Applications

PEEK is used in many bearing applications because of the following advantages:

1. High strength and load-carrying capacity
2. Low friction
3. Good dimensional stability
4. Long life
5. A high continuous service temperature of 500°F/260°C
6. Excellent wear, abrasion, and fatigue resistance
7. Outstanding mechanical properties.

TABLE 2.38   Physical and Mechanical Properties of PEEK

Property	Unfilled	30% glass fiber reinforced	30% carbon fiber reinforced
Specific gravity	1.3–1.32	1.49–1.54	1.42–1.44
Water absorption (24 hr at 73°F/23°C) (%)	0.5	0.06–0.12	0.06–0.12
Dielectric strength, short-term (V/mil)	190		
Tensile strength break (psi)	14,500	22,500–28,500	29,800–33,000
Tensile modulus ($\times 10^3$ psi)	0.522	1,250–1,600	1,860–3,500
Elongation at break (%)	50	2.2	1.3
Compressive strength (psi)	17,110	21,300–22,400	34,800
Flexural strength (psi)	24,650	33,000–42,000	51,475
Compressive modulus ($\times 10^3$ psi)			
Flexural modulus ($\times 10^3$ psi) at			
73°F/23°C	594	1,260–1,600	2,900
248°F/120°C	580	1,334	2,697
482°F/250°C	43.5	739.5	739.5
Izod impact (ft-lb/in. of notch)	1.57	2.1–2.7	1.5–2.1
Hardness, Rockwell	R126	R124	R124
Coefficient of thermal expansion (10^{-6} in./in./°F)	2.6–6.0	1.2	15–22
Thermal conductivity (10^{-4} cal-cm/sec-cm^2°C or Btu/hr/ft^2/°F/in.)	1.75	4.9 0.43	4.9 0.43
Deflection temperature at 264 psi (°F) at 66 psi (°F)	320	550–599	550–610 615
Max. operating temperature (°F/°C)	482/250	482/250	482/250
Limiting oxygen index (%)	35		
Flame spread			
Underwriters Lab. rating (Sub. 94)	V-O	V-O	V-O

TABLE 2.39 Compatibility of PEEK with Selected Corrodents[a]

Chemical	Maximum temp. °F	Maximum temp. °C	Chemical	Maximum temp. °F	Maximum temp. °C
Acetaldehyde	80	27	Hydrobromic acid 20%	x	x
Acetic acid 10%	80	27	Hydrobromic acid 50%	x	x
Acetic acid 50%	140	60	Hydrochloric acid 20%	100	38
Acetic acid 80%	140	60	Hydrochloric acid 38%	100	38
Acetic acid, glacial	140	60	Hydrofluoric acid 30%	x	x
Acetone	210	99	Hydrofluoric acid 70%	x	x
Acrylic acid	80	27	Hydrofluoric acid 100%	x	x
Acrylonitrile	80	27	Hydrogen sulfide, wet	200	93
Aluminum sulfate	80	27	Ketones, general	80	27
Ammonia gas	210	99	Lactic acid 25%	80	27
Ammonium hydroxide, sat.	80	27	Lactic acid, concd	100	38
Aniline	200	93	Magnesium hydroxide	100	38
Aqua regia 3 : 1	x	x	Methyl alcohol	80	27
Benzaldehyde	80	27	Methyl ethyl ketone	370	188
Benzene	80	27	Naphtha	200	93
Benzoic acid	170	77	Nitric acid 5%	200	93
Boric acid	80	27	Nitric acid 20%	200	93
Bromine gas, dry	x	x	Nitrous acid 10%	80	27
Bromine gas, moist	x	x	Oxalic acid 5%	x	x
Calcium carbonate	80	27	Oxalic acid 10%	x	x
Calcium chloride	80	27	Oxalic acid, sat.	x	x
Calcium hydroxide, 10%	80	27	Phenol	140	60
Calcium hydroxide, sat.	100	38	Phosphoric acid 50–80%	200	93
Carbon dioxide, dry	80	27	Potassium bromide 30%	140	60
Carbon tetrachloride	80	27	Sodium carbonate	210	99
Carbonic acid	80	27	Sodium hydroxide 10%	220	104
Chlorine gas, dry	80	27	Sodium hydroxide 50%	180	82
Chlorine, liquid	x	x	Sodium hydroxide, concd	200	93
Chlorobenzene	200	93	Sodium hypochlorite 20%	80	27
Chloroform	80	27	Sodium hypochlorite, concd	80	27
Chlorosulfonic acid	80	27	Sulfuric acid 10%	80	27
Chromic acid 10%	80	27	Sulfuric acid 50%	200	93
Chromic acid 50%	200	93	Sulfuric acid 70%	x	x
Citric acid, concd	170	77	Sulfuric acid 90%	x	x
Cyclohexane	80	27	Sulfuric acid 98%	x	x
Ethylene glycol	160	71	Sulfuric acid 100%	x	x
Ferrous chloride	200	93	Sulfiric acid, fuming	x	x
Fluorine gas, dry	x	x	Toluene	80	27
Fluorine gas, moist	x	x	Zinc chloride	100	38
Hydbromic acid, dil	x	x			

[a] The chemicals listed are in the pure state or in a saturated solution unless otherwise indicated. Compatibility is shown to the maximum allowable temperature for which data are available. Incompatibility is shown by an x. A blank space indicates that data are unavailable.
Source: Schweitzer, PA. Corrosion Resistance Tables, 4th ed. Vols. 1–3. New York: Marcel Dekker, 1995.

PEEK has also found application in a number of aircraft exterior applications, such as radomes, and fairings as a result of its excellent erosion resistance.

M. Polyether–Imide (PEI)

PEI is a high-temperature engineering polymer with the structural formula shown below:

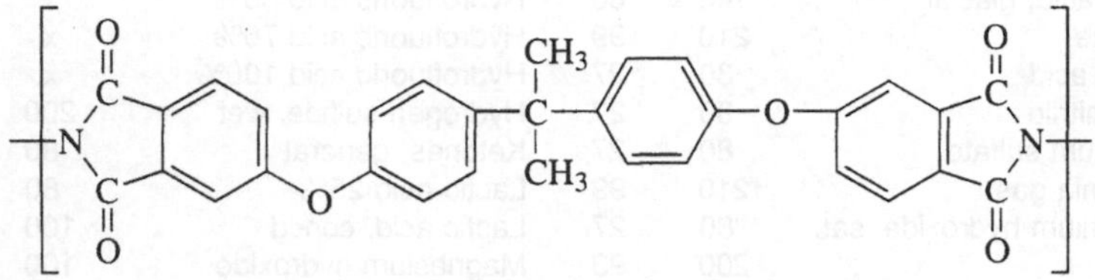

1. Physical and Mechanical Properties

Polyether–imides are noncrystalline polymers made up of alternating aromatic ether and imide units. This molecular structure has rigidity, strength, and impact resistance in fabricated parts over a wide range of temperature. PEI is one of the strongest thermoplasts even without reinforcement. Strength and rigidity of PEI will be increased by the addition of glass or carbon fiber reinforcement. PEI also exhibits outstanding resistance to creep, even at high stress levels and elevated temperatures. Refer to Table 2.40 for the physical and mechanical properties of PEI.

Polyether–imide can be joined by means of solvent cementing using methylene chloride or by means of adhesive bonding using polyurethane, silicone, or non-amine epoxy adhesives.

2. Corrosion Resistance Properties

PEI has a better chemical resistance than most noncrystalline polymers. It is resistant to acetic and hydrochloric acids, weak nitric and sulfuric acids, and alcohol. The unfilled polymer complies with FDA regulations and can be used in food and medical applications. Table 2.41 lists the compatibility of PEI with selected corrodents. Reference 1 provides a more comprehensive listing.

3. Typical Applications

PEI is useful in applications where high heat and flame resistance, low NBS smoke evolution, high tensile and flexural strength, stable electrical properties over a wide range of temperature and frequencies, chemical resistance, and superior finishing characteristics are required. One such application is under the hood of automobiles for connectors and MAP sensors.

TABLE 2.40 Physical and Mechanical Properties of PEI

Property	Unfilled	30% glass fiber reinforced
Specific gravity	1.27	1.49–1.51
Water absorption (24 hr at 73°F/23°C) (%)	0.25	0.16–0.20
Dielectric strength, short-term (V/mil)	500	495–630
Tensile strength at break (psi)	14,000	23,200–28,500
Tensile modulus ($\times 10^3$ psi)	430	1,300–1,600
Elongation at break (%)	60	2–5
Compressive strength (psi)	21,900	30,700
Flexural strength (psi)	22,000	33,000
Compressive modulus ($\times 10^3$ psi)	480	500–938
Flexural modulus ($\times 10^3$ psi) at 73°F/23°C	480	1,200–1,300
200°F/93°C	370	1,100
250°F/121°C	360	1,060
Izod impact (ft-lb/in. of notch)	1.0–1.2	1.7–2.0
Hardness, Rockwell	M109–110	M114, R123
Coefficient of thermal expansion (10^{-6} in./in./°F)	31	11
Thermal conductivity (10^{-4} cal-cm/sec-cm^2°C or Btu/hr/ft^2/°F/in.)	1.6 0.85	6.0–9.0 1.56
Deflection temperature at 264 psi (°F)	387–392	408–420
at 66 psi (°F)	440	412–415
Max. operating temperature (°F/°C)	338/170	
Limiting oxygen index (%)		
Flame spread		
Underwriters Lab. rating (Sub. 94)	V-O	V-O

N. Polyether Sulfone (PES)

Polyether sulfone is a high-temperature engineering thermoplastic with the combined characteristics of high thermal stability and mechanical strength. It is a linear polymer with the following structure:

1. Physical and Mechanical Properties

PES is a tough material with a dry weight impact strength similar to polycarbonate. However, it is sensitive to notches, and sharp corners should be avoided in design. The toughness is maintained at low temperatures, and even at cryogenic temperatures components fail in a ductile manner.

The long-term load bearing properties as measured by creep are outstanding. This resistance is maintained substantially even at a temperature of 300°F/149°C. Glass fiber reinforced materials are available for the most demanding loads.

Small dimensional changes may occur as a result of the polymer absorbing water from the atmosphere. At equilibrium water content for 65% relative humidity, these dimensional changes are of the order of magnitude of 0.15%, whereas in boiling water they are approximately 0.3%. The water may be removed by heating at 300–355°F/149–180°C. This predrying will also prevent outgassing at elevated temperatures.

PES is naturally flame retardant and when burnt emits very low levels of toxic gases and smoke.

TABLE 2.41 Compatibility of PEI with Selected Corrodents[a]

Acetic acid, glacial	R	Hydrochloric acid 30%	R
Acetic acid 80%	R	Iodine solution 10%	R
Acetic acid 10%	R	Isopropanol	R
Acetone	R	Methanol	R
Ammonia 10%	x	Methyl ethyl ketone	x
Ammonium hydroxide 25%	R	Methyl chloride	x
Ammonium hydroxide, sat.	R	Motor oil	R
Benzoic acid	R	Nitric acid 20%	R
Carbon tetrachloride	R	Nitric acid 2%	R
Chloroform	x	Perchloroethylene	R
Citric acid 10%	R	Phosphoric acid 50–80%	R
Citric acid, concd	R	Potassium hydroxide 10%	R
Cyclohexane	R	Potassium permanganate 10%	R
Diesel oil	R	Sodium chloride 10%	R
Ethanol	R	Sodium hydroxide 10%	R
Ether	R	Sodium hypochlorite 20%	R
Ethyl acetate	x	Sodium hypochlorite, concd	R
Ethylene glycol	R	Sulfuric acid 10%	R
Formic acid 10%	R	Sulfuric acid 50%	R
Fruit juice	R	Sulfuric acid, fuming	x
Fuel oil	R	Toluene	R
Gasoline	R	Trichloroethylene	x
Heptane/hexane	R	Water, cold	R
Hydrochloric acid 38%	R	Water, hot	R
Hydrochloric acid 2%	R		

[a] R = material resistant at 73°F/20°C; x = material not resistant.

Refer to Table 2.42 for the physical and mechanical properties of PES.

Polyether sulfone can be solvent cemented using methylene chloride or adhesive bonded with epoxy or urethane adhesives. No special surface treatments other than surface abrading and solvent cleaning are required.

2. Corrosion Resistance Properties

PES has excellent resistance to aliphatic hydrocarbons, some chlorinated hydrocarbons, and aromatics. It is also resistant to most inorganic chemicals. Hydrocarbons and mineral oils, greases and transmission fluids have no effect on PES.

Polyether sulfone will be attacked by strong oxidizing acids, but glass fiber reinforced grades are resistant to more dilute acids. PES is soluble in highly polar solvents and is subject to stress cracking in ketones and esters. Polyether sulfone does not have good outdoor weathering properties. Since it is susceptible to degradation by UV light, if used outdoors it must be stabilized by incorporating carbon black or by painting.

Refer to Table 2.43 for the compatibility of PES with selected corrodents. Reference 1 provides a more comprehensive listing.

3. Typical Applications

Polyether sulfone has found application in

> Medical appliances
> Chemical plants
> Fluid handling
> Aircraft and aerospace appliances
> Instrumentation housings
> Office equipment
> Photocopier parts
> Electrical components
> Automotive (carburetor parts, fuse boxes).

O. Perfluoralkoxy (PFA)

Perfluoralkoxy is a fully fluorinated polymer having the following formula:

$$
\begin{array}{ccccc}
 & F & F & F & F & F \\
 & | & | & | & | & | \\
-C & -C & -C & -C & -C- \\
 & | & | & | & | & | \\
 & F & F & O & F & F \\
 & & & | & & \\
 & & & R_f & &
\end{array}
\qquad R_f = C_nF_{2n} + 1
$$

TABLE 2.42   Physical and Mechanical Properties of PES

Property	Unfilled	10% glass fiber reinforced	30% carbon filled
Specific gravity	1.37	1.45	1.53
Water absorption (24 hr at 73°F/23°C) (%)			
Dielectric strength, short-term (V/mil)	16	20	20
Tensile strength at break (psi)	13,000	16,500	15,800
Tensile modulus ($\times 10^3$ psi)	410	740	1,740
Elongation at break (%)	15–40	4.3	1.4
Compressive strength (psi)	11,800–15,000	19,500–24,000	
Flexural strength (psi)	18,500	24,500	
Compressive modulus ($\times 10^3$ psi)			
Flexural modulus ($\times 10^3$ psi) at 73°F/23°C 200°F/93°C 250°F/121°C	370	650	
Izod impact (ft-lb/in. of notch)	1.5–1.6	1.3	
Hardness, Rockwell	M85	M94	
Coefficient of thermal expansion (10^{-6} in./in./°F)	31	19	12
Thermal conductivity (10^{-4} cal-cm/sec-cm^2°C or Btu/hr/ft^2/°F/in.)	1.25	1.4	1.5
Deflection temperature at 264 psi (°F)	383	414	433
at 66 psi (°F)	406	419	440
Max. operating temperature (°F/°C)	340–170		
Limiting oxygen index (%)	34		
Flame spread Underwriters Lab. rating (Sub. 94)	V-O		

TABLE 2.43 Compatibility of PES with Selected Corrodents[a]

Chemical	Maximum temp. °F	Maximum temp. °C	Chemical	Maximum temp. °F	Maximum temp. °C
Acetic acid 10%	80	27	Nitric acid 5%	80	27
Acetic acid 50%	140	60	Nitric acid 20%	x	x
Acetic acid 80%	200	93	Oxalic acid 5%	80	27
Acetic acid, glacial	200	93	Oxalic acid 29%	80	27
Acetone	x	x	Oxalic acid, sat.	80	27
Ammonia, gas	80	27	Phenol	x	x
Aniline	x	x	Phosphoric acid 50–80%	200	93
Benzene	80	27	Potassium bromide 30%	140	60
Benzenesulfonic acid 10%	100	38	Sodium carbonate	80	27
Benzoic acid	80	27	Sodium chloride	80	27
Carbon tetrachloride	80	27	Sodium hydroxide 20%	80	27
Chlorobenzene	x	x	Sodium hydroxide 50%	80	27
Chrolosulfonic acid	x	x	Sodium hypochlorite 20%	80	27
Chromic acid 10%	x	x	Sodium hypochlorite, concd	80	27
Chromic acid 50%	x	x	Sulfuric acid 10%	80	27
Citric acid, concd	80	27	Sulfuric acid 50%	x	x
Ethylene glycol	100	38	Sulfuric acid 70%	x	x
Ferrous chloride	100	38	Sulfuric acid 90%	x	x
Hydrochloric acid 20%	140	60	Sulfuric acid 98%	x	x
Hydrochloric acid 38%	140	60	Sulfuric acid 100%	x	x
Hydrogen sulfide, wet	80	27	Sulfuric acid, fuming	x	x
Methyl ethyl ketone	x	x	Toluene	80	27
Naphtha	80	27			

[a] The chemicals listed are in the pure state or in a saturated solution unless otherwise indicated. Compatibility is shown to the maximum allowable temperature for which data are available. Incompatibility is shown by an x.
Source: Schweitzer, PA. Corrosion Resistance Tables, 4th ed. Vols. 1–3. New York: Marcel Dekker, 1995.

1. Physical and Mechanical Properties

PFA lacks the physical strength of PTFE at elevated temperatures but has somewhat better physical and mechanical properties than FEP above 300°F/149°C and can be used up to 500°F/260°C. For example, PFA has reasonable tensile strength at 68°F/20°C, but its heat deflection temperature is the lowest of all the fluoroplastics. While PFA matches the hardness and impact strength of PTFE, it sustains only one-quarter of the life of PTFE in flexibility tests. Refer to Table 2.44 for the physical and mechanical properties of PFA.

Like PTFE, PFA is subject to permeation by certain gases and will absorb selected chemicals. Refer to Table 1.5 for the permeation of certain gases in PFA and Table 1.10 for the absorption of certain liquids by PFA.

Perfluoroalkoxy also performs well at cyrogenic temperatures. Table 2.45 compares the mechanical properties of PFA at room temperature and cryogenic temperatures.

Typical of the fluorocarbons, PFA is almost impossible to heat or solvent weld. In order to adhesive bond, the surfaces must be prepared with a sodium–naphthalene etch. The treated surface is degraded by UV light and should be protected from direct exposure. Once treated, conventional epoxy or urethane adhesives may be used.

2. Corrosion Resistance Properties

PFA is inert to strong mineral acids, organic bases, inorganic oxidizers, aromatics, some aliphatic hydrocarbons, alcohols, aldehydes, ketones, ethers, esters, chlorocarbons, fluorocarbons, and mixtures of these.

TABLE 2.44 Physical and Mechanical Properties of PFA

Property	
Specific gravity	2.12–2.17
Water absorption (24 hr at 73°F/23°C) (%)	0.03
Dielectric strength, short-term (V/mil)	500
Tensile strength at break (psi)	4,000–4,300
Tensile modulus ($\times 10^3$ psi)	70
Elongation at break (%)	300
Compressive strength (psi)	3,500
Flexural strength (psi)	
Compressive modulus ($\times 10^3$ psi)	
Flexural modulus ($\times 10^3$ psi) at 73°F/23°C	95–120
200°F/93°C	
250°F/121°C	
Izod impact (ft-lb/in. of notch)	No break
Hardness, Shore D	D64
Coefficient of thermal expansion (10^{-6} in./in./°F)	78–121
Thermal conductivity (10^{-4}cal-cm/sec-cm^2°C or Btu/hr/ft^2/°F/in.)	6.0
Deflection temperature at 264 psi (°F)	118
at 66 psi (°F)	164
Max. operating temperature (°F/°C)	500/260
Limiting oxygen index (%)	<95
Flame spread	10
Underwriters Lab. rating (Sub. 94)	V-O

TABLE 2.45 Comparison of Mechanical Properties of PFA at Room Temperature and Cryogenic Temperatures

Property	Temperature	
	73°F/23°C	−390°F/−234°C
Yield strength (psi)	2,100	No break
Ultimate tensile strength (psi)	2,600	18,700
Elongation (%)	260	8
Flexural modulus (psi)	81,000	840,000
Impact strength, notched (ft-lb/in.)	No break	12
Compressive strength (psi)	3,500	60,000
Compressive strain (%)	20	35
Modulus of elasticity (psi)	10,000	680,000

Perfluoroalkoxy will be attacked by certain halogenated complexes containing fluorine. This includes chlorine trifluoride, bromine trifluoride, iodine pentafluoride, and fluorine. It is also subject to attack by such metals as sodium or potassium, particularly in their molten state. Refer to Table 2.46 for the compatibility of PFA with selected corrodents. Reference 1 provides a more detailed listing.

PFA has excellent weatherability and is not subject to UV degradation.

3. Typical Applications

PFA finds many applications in the chemical process industry for corrosion resistance. Applications includes lining for pipes and vessels. Reference 2 details the use of PFA in lined process pipe and information on lining of vessels can be found in Reference 3.

P. Polytetrafluoroethylene (PTFE)

PTFE is marketed under the tradename Teflon by DuPont and the tradename Halon by Ausimont USA. It is a fully fluorinated thermoplastic having the following formula:

$$-\overset{\overset{\displaystyle F}{|}}{\underset{\underset{\displaystyle F}{|}}{C}}-\overset{\overset{\displaystyle F}{|}}{\underset{\underset{\displaystyle F}{|}}{C}}-$$

TABLE 2.46 Compatibility of PFA with Selected Corrodents[a]

Chemical	Maximum temp.		Chemical	Maximum temp.	
	°F	°C		°F	°C
Acetaldehyde	450	232	Benzene[b]	450	232
Acetamide	450	232	Benzene sulfonic acid 10%	450	232
Acetic acid 10%	450	232	Benzoic acid	450	232
Acetic acid 50%	450	232	Benzyl alcohol[c]	450	232
Acetic acid 80%	450	232	Benzyl chloride[b]	450	232
Acetic acid, glacial	450	232	Borax	450	232
Acetic anhydride	450	232	Boric acid	450	232
Acetone	450	232	Bromine gas, dry[b]	450	232
Acetyl chloride	450	232	Bromine, liquid[b,c]	450	232
Acrylonitrile	450	232	Butadiene[b]	450	232
Adipic acid	450	232	Butyl acetate	450	232
Allyl alcohol	450	232	Butyl alcohol	450	232
Allyl chloride	450	232	n-Butylamine[c]	450	232
Alum	450	232	Butyric acid	450	232
Aluminum chloride, aqueous	450	232	Calcium bisulfide	450	232
Aluminum fluoride	450	232	Calcium bisulfite	450	232
Aluminum hydroxide	450	232	Calcium carbonate	450	232
Aluminum nitrate	450	232	Calcium chlorate	450	232
Aluminum oxychloride	450	232	Calcium chloride	450	232
Aluminum sulfate	450	232	Calcium hydroxide 10%	450	232
Ammonia gas[b]	450	232	Calcium hydroxide, sat.	450	232
Ammonium bifluoride[b]	450	232	Calcium hypochlorite	450	232
Ammoinium carbonate	450	232	Calcium nitrate	450	232
Ammonium chloride 10%	450	232	Calcium oxide	450	232
Ammonium chloride 50%	450	232	Calcium sulfate	450	232
Ammonium chloride, sat.	450	232	Caprylic acid	450	232
Ammonium fluoride 10%[b]	450	232	Carbon bisulfide[b]	450	232
Ammonium fluoride 25%[b]	450	232	Carbon dioxide, dry	450	232
Ammonium hydroxide 25%	450	232	Carbon dioxide, wet	450	232
Ammonium hydroxide, sat.	450	232	Carbon disulfide[b]	450	232
Ammonium nitrate	450	232	Carbon monoxide	450	232
Ammonium persulfate	450	232	Carbon tetrachloride[b,c,d]	450	232
Ammonium phosphate	450	232	Carbonic acid	450	232
Ammonium sulfate 10–40%	450	232	Chloroacetic acid, 50% water	450	232
Ammonium sulfide	450	232	Chloroacetic acid	450	232
Amyl acetate	450	232	Chlorine gas, dry	x	x
Amyl alcohol	450	232	Chlorine gas, wet[b]	450	232
Amyl chloride	450	232	Chlorine, liquid[c]	x	x
Aniline[c]	450	232	Chlorobenzene[b]	450	232
Antimony trichloride	450	232	Chloroform[b]	450	232
Aqua regia 3 : 1	450	232	Chlorosulfonic acid[c]	450	232
Barium carbonate	450	232	Chromic acid 10%	450	232
Barium chloride	450	232	Chromic acid 50%[c]	450	232
Barium hydroxide	450	232	Chromyl chloride	450	232
Barium sulfate	450	232	Citric acid 15%	450	232
Barium sulfide	450	232	Citric acid, concd	450	232
Benzaldehyde[c]	450	232	Copper carbonate	450	232

TABLE 2.46 Continued

Chemical	Maximum temp. °F	Maximum temp. °C	Chemical	Maximum temp. °F	Maximum temp. °C
Copper carbonate	450	232	Methyl ethyl ketone[b]	450	232
Copper chloride	450	232	Methyl isobutyl ketone[b]	450	232
Copper cyanide	450	232	Muriatic acid[b]	450	232
Copper sulfate	450	232	Nitric acid 5%[b]	450	232
Cresol	450	232	Nitric acid 20%[b]	450	232
Cupric chloride 5%	450	232	Nitric acid 70%[b]	450	232
Cupric chloride 50%	450	232	Nitric acid, anhydrous[b]	450	232
Cyclohexane	450	232	Nitrous acid 10%	450	232
Cyclohexanol	450	232	Oleum	450	232
Dibutyl phthalate	450	232	Perchloric acid 10%	450	232
Dichloroacetic acid	450	232	Perchloric acid 70%	450	232
Dichloroethane	450	232	Phenol[b]	450	232
(ethylene dichloride)[b]			Phosphoric acid 50–80%[c]	450	232
Ethylene glycol	450	232	Picric acid	450	232
Ferric chloride	450	232	Potassum bromide 30%	450	232
Ferric chloride 50%	450	232	Salicylic acid	450	232
in water[c]			Sodium carbonate	450	232
Ferric nitrate 10–50%	450	232	Sodium chloride	450	232
Ferrous chloride	450	232	Sodium hydroxide 10%	450	232
Ferrous nitrate	450	232	Sodium hydroxide 50%	450	232
Fluorine gas, dry	x	x	Sodium hydroxide, concd	450	232
Fluorine gas, moist	x	x	Sodium hypochlorite 20%	450	232
Hydrobromic acid, dil[b,d]	450	232	Sodium hypochlorite, concd	450	232
Hydrobromic acid 20%[b,d]	450	232	Sodium sulfide to 50%	450	232
Hydrobromic acid 50%[b,d]	450	232	Stannic chloride	450	232
Hydrochloric acid 20%[b,d]	450	232	Stannous chloride	450	232
Hydrochloric acid 38%[b,d]	450	232	Sulfuric acid 10%	450	232
Hydrocyanic acid 10%	450	232	Sulfuric acid 50%	450	232
Hydrofluoric acid 30%[b]	450	232	Sulfuric acid 70%	450	232
Hydrofluoric acid 70%[b]	450	232	Sulfuric acid 90%	450	232
Hydrofluoric acid 100%[b]	450	232	Sulfuric acid 98%	450	232
Hydrochlorous acid	450	232	Sulfuric acid 100%	450	232
Iodine solution 10%[b]	450	232	Sulfuric acid, fuming[b]	450	232
Ketones, general	450	232	Sulfurous acid	450	232
Lactic acid 25%	450	232	Thionyl chloride[b]	450	232
Lactic acid, concd	450	232	Toluene[b]	450	232
Magnesium chloride	450	232	Trichloroacetic acid	450	232
Malic acid	450	232	White Liquor	450	232
Methyl chloride[b]	450	232	Zinc chloride[c]	450	232

[a] The chemicals listed are in the pure state or in a saturated solution unless otherwise indicated. Compatibility is shown to the maximum allowable temperature for which data are available. Incompatibility is shown by an x. A blank space indicates that data are unavailable.
[b] Material will permeate.
[c] Material will be absorbed.
[d] Material will cause stress cracking.
Source: Schweitzer, PA. Corrosion Resistance Tables, 4th ed. Vols. 1–3. New York: Marcel Dekker, 1995.

1. Physical and Mechanical Properties

PTFE has an operating temperature range of from $-20°$F to $430°$F$/ - 29°$C to $212°$C. This temperature range is based on the physical and mechanical properties of PTFE. When handling aggressive chemicals, it may be necessary to reduce the upper temperature limit.

PTFE is a relatively weak material and tends to creep under stress at elevated temperatures. It also has a low coefficient of friction. The hardness of PTFE is a function of temperature, decreasing with increasing temperature. It is also subject to permeation and absorption. Refer to chapter 1.2. Refer to Table 2.47 for its physical and mechanical properties.

The mechanical properties can be improved by the addition of glass filling. Glass filled material shows an increase in flexural modulus and tensile modulus.

By reinforcing PTFE with carbon fibers, the following characteristics are exhibited:

1. Virtually no compressive creep
2. Dimensional stability across a wide temperature range
3. High degree of toughness
4. Good wear resistance.

PTFE cannot be joined by solvent cementing or fusion welding. For adhesive bonding, a sodium–naphthalene etch is necessary as a minimum surface preparation. The treated surface is subject to degradation by UV light and should be protected from direct exposure.

2. Corrosion Resistance Properties

PTFE is unique in its corrosion resistant properties. It is chemically inert in the presence of most materials. There are very few chemicals that will attack PTFE at normal use temperatures. Among materials that will attack PTFE are the most violent oxidizing and reducing agents known. Elemental sodium removes fluorine from the polymer molecule. The other alkali metals (potassium, lithium, etc.) act in a similar manner.

Fluorine and related compounds (e.g., chlorine trifluoride) are absorbed into PTFE resin with to such a degree that the mixture becomes sensitive to a source of ignition, such as impact. These potent oxidizers should be handled only with great care and a recognition of the potential hazards.

The handling of 80% sodium hydroxide, aluminum chloride, amonia, and certain amines at high temperatures have the same effect as elemental sodium. Slow oxidation attack can be produced by 70% nitric acid under pressure at $480°$F$/250°$C.

PTFE also has excellent weathering properties and is not degraded by UV.

TABLE 2.47 Physical and Mechanical Properties of PTFE

Property	Unfilled	25% glass filled
Specific gravity	2.14–2.20	2.2–2.3
Water absorption (24 hr at 73°F/23°C) (%)	<0.01	
Dielectric strength, short-term (V/mil)	480	320
Tensile strength at break (psi)	3,000–5,000	2,000–2,700
Tensile modulus ($\times 10^3$ psi)	58–80	200–240
Elongation at break (%)	200–400	200–300
Compressive strength (psi)	1,700	1,000–1,400
Flexural strength (psi)	No break	2,000
Compressive modulus ($\times 10^3$ psi)	60	
Flexural modulus ($\times 10^3$ psi) at 73°F/23°C	80	190–235
200°F/93°C		
250°F/121°C		
Izod impact (ft-lb/in. of notch)	3	2.7
Hardness, Shore	D50–65	D60–70
Coefficient of thermal expansion (10^{-6} in./in./°F)	70–120	77–100
Thermal conductivity (10^{-4}cal-cm/sec-cm^2°C or Btu/hr/ft^2/°F/in.)	6.0 1.7	8–10
Deflection temperature at 264 psi (°F)	115	
at 66 psi (°F)	160–250	
Max. operating temperature (°F/°C)	500/260	
Limiting oxygen index (%)	95	
Flame spread Underwriters Lab. rating (Sub. 94)	V-O	

Refer to Table 2.48 for the compatibility of PTFE with selected corrodents. Reference 1 provides a more detailed listing.

3. Typical Applications

Applications for PTFE extend from exotic space-age usages to molded parts and wire and cable insulation to consumer use as a coating for cookware. One of the largest uses is for corrosion protection including linings for tanks and piping. Applications in the automotive industry take advantage of the low surface friction and chemical stability using it in seals and rings for transmission and power steering systems and in seals for shafts, compressors, and shock absorbers.

Reference 2 will provide details of the usage of PTFE in piping applications.

TABLE 2.48 Compatibility of PTFE with Selected Corrodents[a]

Chemical	Maximum temp.		Chemical	Maximum temp.	
	°F	°C		°F	°C
Acetaldehyde	450	232	Benzene[b]	450	232
Acetamide	450	232	Benzenesulfonic acid 10%	450	232
Acetic acid 10%	450	232	Benzoic acid	450	232
Acetic acid 50%	450	232	Benzyl alcohol	450	232
Acetic acid 80%	450	232	Benzyl chloride	450	232
Acetic acid, glacial	450	232	Borax	450	232
Acetic anhydride	450	232	Boric acid	450	232
Acetone	450	232	Bromine gas, dry[b]	450	232
Acetyl chloride	450	232	Bromine, liquid[b]	450	232
Acrylonitrile	450	232	Butadiene[b]	450	232
Adipic acid	450	232	Butyl acetate	450	232
Allyl alcohol	450	232	Butyl alcohol	450	232
Allyl chloride	450	232	n-Butylamine	450	232
Alum	450	232	Butyric acid	450	232
Aluminum chloride, aqueous	450	232	Calcium bisulfide	450	232
Aluminum fluoride	450	232	Calcium bisulfite	450	232
Aluminium hydroxide	450	232	Calcium carbonate	450	232
Aluminum nitrate	450	232	Calcium chlorate	450	232
Aluminum oxychloride	450	232	Calcium chloride	450	232
Aluminum sulfate	450	232	Calcium hydroxide 10%	450	232
Ammonia gas[b]	450	232	Calcium hydroxide, sat.	450	232
Ammonium bifluoride	450	232	Calcium hypochlorite	450	232
Ammonium carbonate	450	232	Calcium nitrate	450	232
Ammonium chloride 10%	450	232	Calcium oxide	450	232
Ammonium chloride 50%	450	232	Calcium sulfate	450	232
Ammonium chloride, sat.	450	232	Caprylic acid	450	232
Ammonium fluoride 10%	450	232	Carbon bisulfide[b]	450	232
Ammonium fluoride 25%	450	232	Carbon dioxide, dry	450	232
Ammonium hydroxide 25%	450	232	Carbon dioxide, wet	450	232
Ammonium hydroxide, sat.	450	232	Carbon disulfide	450	232
Ammonium nitrate	450	232	Carbon monoxide	450	232
Ammonium persulfate	450	232	Carbon tetrachloride[c]	450	232
Ammonium phosphate	450	232	Carbonic acid	450	232
Ammonium sulfate 10–40%	450	232	Chloroacetic acid, 50% water	450	232
Ammonium sulfide	450	232	Chloroacetic acid	450	232
Amyl acetate	450	232	Chlorine gas, dry	x	x
Amyl alcohol	450	232	Chlorine gas, wet[b]	450	232
Amyl chloride	450	232	Chlorine, liquid	x	x
Aniline	450	232	Chlorobenzene[b]	450	232
Antimony trichloride	450	232	Chloroform[b]	450	232
Aqua regia 3 : 1	450	232	Chlorosulfonic acid	450	232
Barium carbonate	450	232	Chromic acid 10%	450	232
Barium chloride	450	232	Chromic acid 50%	450	232
Barium hydroxide	450	232	Chromyl chloride	450	232
Barium sulfate	450	232	Citric acid 15%	450	232
Barium sulfide	450	232	Citric acid, concd	450	232
Benzaldehyde	450	232	Copper carbonate	450	232

TABLE 2.48 Continued

Chemical	Maximum temp.		Chemical	Maximum emp.	
	°F	°C		°F	°C
Copper chloride	450	232	Methyl isobutyl ketone[c]	450	232
Copper cyanide 10%	450	232	Muriatic acid[b]	450	232
Copper sulfate	450	232	Nitric acid 5%[b]	450	232
Cresol	450	232	Nitric acid 20%[b]	450	232
Cupric chloride 5%	450	232	Nitric acid 70%[b]	450	232
Cupric chloride 50%	450	232	Nitric acid, anhydrous[b]	450	232
Cyclohexane	450	232	Nitrous acid 10%	450	232
Cyclohexanol	450	232	Oleum	450	232
Dibutyl phthalate	450	232	Perchloric acid 10%	450	232
Dichloroacetic acid	450	232	Perchloric acid 70%	450	232
Dichloroethane	450	232	Phenol[b]	450	232
(ethylene dichloride)[b]			Phosphoric acid 50–80%	450	232
Ethylene glycol	450	232	Picric acid	450	232
Ferric chloride	450	232	Potassum bromide 30%	450	232
Ferric chloride 50% in water	450	232	Salicylic acid	450	232
Ferric nitrate 10–50%	450	232	Sodiium carbonate	450	232
Ferrous chloride	450	232	Sodium chloride	450	232
Ferrous nitrate	450	232	Sodium hydroxide 10%	450	232
Fluorine gas, dry	x	x	Sodium hydroxide 50%	450	232
Fluorine gas, moist	x	x	Sodium hydroxide, concd	450	232
Hydrobromic acid, dil[b,c]	450	232	Sodium hypochlorite 20%	450	232
Hydrobromic acid 20%[c]	450	232	Sodium hypochlorite, concd	450	232
Hydrobromic acid 50%[c]	450	232	Sodium sulfide to 50%	450	232
Hydrochloric acid 20%[c]	450	232	Stannic chloride	450	232
Hydrochloric acid 38%[c]	450	232	Stannous chloride	450	232
Hydrocyanic acid 10%	450	232	Sulfuric acid 10%	450	232
Hydrofluoric acid 30%[b]	450	232	Sulfuric acid 50%	450	232
Hydrofluoric acid 70%[b]	450	232	Sulfuric acid 70%	450	232
Hydrofluoric acid 100%[b]	450	232	Sulfuric acid 90%	450	232
Hypochlorous acid	450	232	Sulfuric acid 98%	450	232
Iodine solution 10%[b]	450	232	Sulfuric acid 100%	450	232
Ketones, general	450	232	Sulfuric acid, fuming[b]	450	232
Lactic acid 25%	450	232	Sulfurous acid	450	232
Lactic acid, concd	450	232	Thionyl chloride	450	232
Magnesium chloride	450	232	Toluene[b]	450	232
Malic acid	450	232	Trichloroacetic acid	450	232
Methyl chloride[b]	450	232	White liquor	450	232
Methyl ethyl ketone[b]	450	232	Zinc chloride[d]	450	232

[a] The chemicals listed are in the pure state or in a saturated solution unless otherwise indicated. Compatibility is shown to the maximum allowable temperature for which data are available. Incompatibility is shown by an x. A blank space indicates that data are unavailable.
[b] Material will permeate.
[c] Material will cause stress cracking.
[d] Material will be absorbed.
Source: Schweitzer, PA. Corrosion Resistance Tables, 4th ed. Vols. 1–3. New York: Marcel Dekker, 1995.

Q. Polyvinylidene Fluoride (PVDF)

Polyvinylidene fluoride is a crystalline, high molecular weight polymer containing 50% fluorine. It is similar in chemical structure to PTFE except that it is not fully fluorinated. The chemical structure is as follows:

$$
\begin{array}{cc}
\text{F} & \text{F} \\
| & | \\
-\text{C}-\text{C}- \\
| & | \\
\text{H} & \text{H}
\end{array}
$$

1. Physical and Mechanical Properties

Much of the strength and chemical resistance of PVDF is maintained through an operating range of −40 to 320°F/ − 40 to 160°C. It has high tensile strength and heat deflection temperature and is resistant to the permeation of gases. Approval has been granted by the Food and Drug Administration for repeated use in contact with food in food handling and processing equipment.

The abrasion resistance of PVDF is at a level comparable with that of polyamide (PA) and ultrahigh molecular weight polyethylene (UHMWPE), which places it among the best materials in this regard.

PVDF also shows a low surface tension, good flame resistance, V–O according to UL 94/1975, with a limiting oxygen index of 44. It belongs to the self-extinguishing (SE) class of materials defined by ASTM D 635.

Refer to Table 2.49 for the physical and mechanical properties of PVDF.

Polyvinylidene fluoride may be joined by heat fusion.

2. Corrosion Resistance Properties

PVDF is chemically resistant to most acids, bases, and organic solvents. It is also resistant to wet or dry chlorine, bromine, and other halogens.

It should not be used with strong alkalies, fuming acids, polar solvents, amines, ketones, and esters. When used with strong alkalies, it stress cracks. Refer to Table 2.50 for the compatibility of PVDF with selected corrodents. Reference 1 provides a more detailed listing.

Polyvinylidene fluoride also withstands UV light in the "visible" range and gamma radiations up to 100 MRad.

3. Typical Applications

Polyvinylidene fluoride finds many applications in the corrosion resistance field being used as lining material for vessels and piping, as solid piping, column packing, valving, pumps, and other processing equipment. Reference 2 provides details of the usage of PVDF piping.

TABLE 2.49 Physical and Mechanical Properties of PVDF

Property	
Specific gravity	1.77–1.78
Water absorption (24 hr at 73°F/23°C) (%)	0.03–0.08
Dielectric strength, short-term (V/mil)	260–280
Tensile strength at break (psi)	3,500–7,250
Tensile modulus ($\times 10^3$ psi)	200—80,000
Elongation at break (%)	12–600
Compressive strength (psi)	8,000–16,000
Flexural strength (psi)	9,700–13,600
Compressive modulus ($\times 10^3$ psi)	204–420
Flexural modulus ($\times 10^3$ psi) at 73°F/23°C	170–120,000
200°F/93°C	
250°F/121°C	
Izod impact (ft-lb/in. of notch)	2.5–8.0
Hardness, Rockwell	R79–83
Coefficient of thermal expansion (10^{-6} in./in./°F)	70–142
Thermal conductivity (10^{-4} cal-cm/sec-cm²°C	2.4–3.1
or Btu/hr/ft²/°F/in.)	0.79
Deflection temperature at 264 psi (°F)	183–244
at 66 psi (°F)	280–284
Max. operating temperature (°F/°C)	320/160
Limiting oxygen index (%)	44
Flame spread	0
Underwriters Lab. rating (Sub. 94)	V-O

TABLE 2.50 Compatibility of PVDF with Selected Corrodents[a]

Chemical	Maximum temp.		Chemical	Maximum temp.	
	°F	°C		°F	°C
Acetaldehyde	150	66	Aluminum hydroxide	260	127
Acetamide	90	32	Aluminum nitrate	300	149
Acetic acid 10%	300	149	Aluminum oxychloride	290	143
Acetic acid 50%	300	149	Aluminum sulfate	300	149
Acetic acid 80%	190	88	Ammonia gas	270	132
Acetic acid, glacial	190	88	Ammonium bifluoride	250	121
Acetic anhydride	100	38	Ammonium carbonate	280	138
Acetone	x	x	Ammonium chloride 10%	280	138
Acetyl chloride	120	49	Ammonium chloride 50%	280	138
Acrylic acid	150	66	Ammonium chloride, sat.	280	138
Acrylonitrile	130	54	Ammonium fluoride 10%	280	138
Adipic acid	280	138	Ammonium fluoride 25%	280	138
Allyl alcohol	200	93	Ammonium hydroxide 25%	280	138
Allyl chloride	200	93	Ammonium hydroxide, sat.	280	138
Alum	180	82	Ammonium nitrate	280	138
Aluminum acetate	250	121	Ammonium persulfate	280	138
Aluminum chloride, aqueous	300	149	Ammonium phosphate	280	138
Aluminum chloride, dry	270	132	Ammonium sulfate 10–40%	280	138
Aluminum fluoride	300	149	Ammonium sulfide	280	138

(*Continued*)

TABLE 2.50 Continued

Chemical	Maximum temp. °F	Maximum temp. °C	Chemical	Maximum temp. °F	Maximum temp. °C
Ammonium sulfite	280	138	Carbon tetrachloride	280	138
Amyl acetate	190	88	Carbonic acid	280	138
Amyl alcohol	280	138	Cellosolve	280	138
Amyl chloride	280	138	Chloroacetic acid,		
Aniline	200	93	50% water	210	99
Antimony trichloride	150	66	Chloroacetic acid	200	93
Aqua regia 3:1	130	54	Chlorine gas, dry	210	99
Barium carbonate	280	138	Chlorine gas, wet, 10%	210	99
Barium chloride	280	138	Chlorine, liquid	210	99
Barium hydroxide	280	138	Chlorobenzene	220	104
Barium sulfate	280	138	Chloroform	250	121
Barium sulfide	280	138	Chlorosulfonic acid	110	43
Benzaldehyde	120	49	Chromic acid 10%	220	104
Benzene	150	66	Chromic acid 50%	250	121
Benzenesulfonic acid 10%	100	38	Chromyl chloride	110	43
Benzoic acid	250	121	Citric acid 15%	250	121
Benzyl alcohol	280	138	Citric acid concd	250	121
Benzyl chloride	280	138	Copper acetate	250	121
Borax	280	138	Copper carbonate	250	121
Boric acid	280	138	Copper chloride	280	138
Bromine gas, dry	210	99	Copper cyanide	280	138
Bromine gas, moist	210	99	Copper sulfate	280	138
Bromine, liquid	140	60	Cresol	210	99
Butadiene	280	138	Cupric chlorides 5%	270	132
Butyl acetate	140	60	Cupric chloride 50%	270	132
Butyl alcohol	280	138	Cyclohexane	250	121
n-Butylamine	x	x	Cyclohexanol	210	99
Butyric acid	230	110	Dibutyl phthalate	80	27
Calcium bisulfide	280	138	Dichloroacetic acid	120	49
Calcium bisulfite	280	138	Dichloroethane		
Calcium carbonate	280	138	(ethylene dichloride)	280	138
Calcium chlorate	280	138	Ethylene glycol	280	138
Calcium chloride	280	138	Ferric chloride	280	138
Calcium hydroxide 10%	270	132	Ferric chloride 50% in water	280	138
Calcium hydroxide, sat.	280	138	Ferrous nitrate 10–50%	280	138
Calcium hypochlorite	280	138	Ferrous chloride	280	138
Calcium nitrate	280	138	Ferrous nitrate	280	138
Calcium oxide	250	121	Fluorine gas, dry	80	27
Calcium sulfate	280	138	Fluorine gas, moist	80	27
Caprylic acid	220	104	Hydrobromic acid, dilute	260	127
Carbon bisulfide	80	27	Hydrobromic acid 20%	280	138
Carbon dioxide, dry	280	138	Hydrobromic acid 50%	280	138
Carbon dioxide, wet	280	138	Hydrochloric acid 20%	280	138
Carbon disulfide	80	27	Hydrochloric acid 38%	280	138
Carbon monoxide	280	138	Hydrocyanic acid 10%	280	138

TABLE 2.50 Continued

Chemical	Maximum temp. °F	Maximum temp. °C	Chemical	Maximum temp. °F	Maximum temp. °C
Hydrofluoric acid 30%	260	127	Salicylic acid	220	104
Hydrofluoric acid 70%	200	93	Silver bromide 10%	250	121
Hydrofluoric acid 100%	200	93	Sodium carbonate	280	138
Hypochlorous acid	280	138	Sodium chloride	280	138
Iodine solution 10%	250	121	Sodium hydroxide 10%	230	110
Ketones, general	110	43	Sodium hydroxide 50%	220	104
Lactic acid 25%	130	54	Sodium hydroxide		
Lactic acid, concentrated	110	43	concentrated[b]	150	66
Magnesium chloride	280	138	Sodium hypochlorite 20%	280	138
Malic acid	250	121	Sodium hypochlorite,	280	138
Manganese chloride	280	138	concentrated		
Methyl chloride	x	x	Sodium sulfide to 50%	280	138
Methyl ethyl ketone	x	x	Stannic chloride	280	138
Methyl isobutyl ketone	110	43	Stannous chloride	280	138
Muriatic acid	280	138	Sulfuric acid 10%	250	121
Nitric acid 5%	200	93	Sulfuric acid 50%	220	104
Nitric acid 20%	180	82	Sulfuric acid 70%	220	104
Nitric acid 70%	120	49	Sulfuric acid 90%	210	99
Nitric acid, anhydrous	150	6	Sulfuric acid 98%	140	60
Nitrous acid, concentrated	210	99	Sulfuric acid 100%	x	x
Oleum	x	x	Sulfuric acid, fuming	x	x
Perchloric acid 10%	210	99	Sulfurous acid	220	104
Perchloric acid 70%	120	49	Thionyl chloride	x	x
Phenol	200	93	Toluene	x	x
Phosphoric acid 50–80%	220	104	Trichloroacetic acid	130	54
Picric acid	80	27	White liquor	80	27
Potassium bromide 30%	280	138	Zinc chloride	260	127

[a] The chemicals listed are in the pure state or in a saturated solution unless otherwise indicated. Compatibility is shown to the maximum allowable temperature for which data are available. Incompatibility is shown by an x. A blank space indicates that the data are unavailable.

[b] Source: Schweitzer, PA. Corrosion Resistance Tables, 4th ed. Vols 1–3, New York: Marcel Dekker, 1995.

Polyvinylidene fluoride is manufactured under the trade name of Kynar by Elf Atochem. Solef by Solvag, Hylar by Ausimont USA, and Super Pro 230 and ISO by Asahi/America.

R. Polyethylene (PE)

Polyethylenes are probably among the best known thermoplasts. Polyethylene is produced in various grades that differ in molecular structure, crystallinity, molecular weight, and molecular distribution. They are a member of the polyolefin family. The basic chemical structure is

$$\begin{array}{c} \quad\text{H}\quad\text{H} \\ \quad|\quad\quad| \\ -\text{C}-\text{C}- \\ \quad|\quad\quad| \\ \quad\text{H}\quad\text{H} \end{array}$$

PE is produced by polymerizing ethylene gas obtained from petroleum hydrocarbons. Changes in the polymerizing conditions are responsible for the various types of PE.

1. Physical and Mechanical Properties

Physical and mechanical properties differ in density and molecular weight. The three main classifications of density are low, medium, and high. These specific gravity ranges are 0.91 to 0.925, 0.925 to 0.940, and 0.940 to 0.965. These grades are sometimes referred to as types I, II, and III. All polyethylenes are relatively soft, and hardness increases as density increases. Generally, the higher the density, the better the dimensional stability and physical properties, particularly as a function of temperature. The thermal stability of polyethylenes ranges form 190°F/88°C for the low-density material up to 250°F/121°C for the high-density material. Toughness is maintained to low negative temperatures.

Industry practice breaks the molecular weight of polyethylenes into four distinct classifications which are

Medium molecular weight less than 100,000
High molecular weight 110,000 to 250,000
Extra high molecular weight 250,000 to 1,500,000
Ultrahigh molecular weight 1,500,000 and higher.

Usually the ultrahigh molecular weight material has a molecular weight of at least 3.1 million.

The molecular weights of the higher molecular weight polyethylene resins are usually determined by gel permeation chromatography (GPC), scanning electron chromatography (SEC), or relative specific viscosity (RVS) measurements. Reasonably accurate measurements of the extra high and ultrahigh molecular weight resins are usually obtained by GPC or SEC.

Every polyethylene resin consists of a mixture of large and small molecules, molecules of high and low molecular weight. The molecular weight distribution gives a general picture of the ratio of the large, medium, and small molecules in the resin. If the resin contains molecules close to the average weight, the distribution is called narrow. If the resin contains a wide variety of weights, the distribution is called broad. Figure 2.1 depicts this in graph form.

Linear low-density polyethylene (LLDPE) exhibits some of the same properties as high-density polyethylene (HDPE) with one main difference. LLDPE has a greater flexibility than HDPE. Its key properties are

Lightweight
Good impact resistance
Extreme flexibility
Easy cleanability.

Refer to Table 2.51 for physical and mechanical properties.
The key properties of HDPE are

Excellent impact resistance
Lightweight

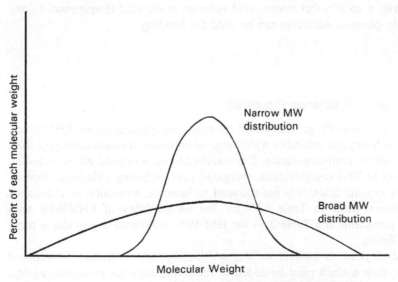

FIGURE 2.1 Schematic illustration of molecular weight distribution.

Low moisture absorption
High tensile strength
Nontoxicity
Nonstaining.

The key properties of extra high molecular weight polyethylene (EHMWPE) and
ultra high molecular weight polyethylene (UHMWPE) are

Good abrasion resistance
Excellent impact resistance
Lightweight
Easily heat fused
High tensile strength
Low moisture absorption
Nontoxicity
Non-staining
Corrosion resistance.

Polyethylene can also be filled with glass fiber to increase dimensional stability
and improve such mechanical properties as elongation, tensile strength, tensile
modulus, compressive strength, and impact strength. Table 2.51 lists the physical
and mechanical properties of HDPE and 30% glass filled HDPE while Table 2.52
lists the properties of EHMWPE and UHMWPE.

Polyethylenes can be joined by adhesive bonding or by thermal fusion.
They can not be solvent cemented. Prior to adhesive bonding, the surface must be
etched with a sodium dichromate–acid solution at elevated temperature. Epoxy
and nitrile–phenolic adhesives can be used for bonding.

2.   Corrosion Resistance Properties

The two varieties of PE generally used for corrosive applications are EHMW and
UHMW. Polyethylene exhibits a wide range of corrosion resistance, ranging from
potable water to corrosive wastes. It is resistant to most mineral acids, including
sulfuric up to 70% concentration, inorganic salts including chlorides, alkalies,
and many organic acids. It is not resistant to bromine, aromatics, or chlorinaed
hydrocarbons. Refer to Table 2.53 for the compatibility of EHMWPE with
selected corrodents and Table 2.54 for HMWPE. Reference 1 provides a more
detailed listing.

Polyethylene is subject to degradation by UV radiation. If exposed
outdoors, carbon black must be added to the formulation for protection against
UV degradation.

TABLE 2.51 Physical and Mechanical Properties of LLDPE and HDPE

Property	LLDPE	HDPE	30% Glass-filled HDPE
Specific gravity	0.918–0.940	0.952–0.965	1.18–1.28
Water absorption (24 hr at 73°F/23°C) (%)	<0.01	>0.01	0.02–0.06
Dielectric strength, short-term (V/mil)	18–39	450–500	500–550
Tensile strength at break (psi)	1,900–4,000	3,200–4,500	7,500–9,000
Tensile modulus ($\times 10^3$ psi)	38–75	155–158	700–900
Elongation at break (%)	100–965	10–1,200	1.5–2.5
Compressive strength (psi)		2,700–3,600	6,000–7,000
Flexural strength (psi)			11,000–12,000
Compressive modulus ($\times 10^3$ psi)			
Flexural modulus ($\times 10^3$ psi) at 73°F/23°C 200°F/93°C 250°F/121°C	40–105	145–225	700–800
Izod impact (ft-lb/in. of notch)	1.0–no break	0.4–4.0	1.1–1.5
Hardness, Rockwell	R10	R65	R75–90
Coefficient of thermal expansion (10^{-6} in./in./°F)	72–110	59–110	48
Thermal conductivity (10^{-4} cal-cm/sec-cm^2°C or Btu/hr/ft^2/°F/in.)		11–12	8.6–11
Deflection temperature at 264 psi (°F) at 66 psi (°F)	104 100–121	175–196	250 260–265
Max. operating temperature (°F/°C)	104/40	175/80	
Limiting oxygen index (%)			
Flame spread	Slow burning	Slow burning	Slow burning
Underwriters Lab. rating (Sub. 94)			

TABLE 2.52 Physical and Mechanical Properties of EHMWPE and UHMWPE

Property	EHMWPE	UHMWPE
Specific gravity	0.947–0.955	0.94
Water absorption (24 hr at 73°F/23°C) (%)		>0.01
Dielectric strength, short-term (V/mil)		710
Tensile strength at break (psi)	2,500–4,300	5,600–7,000
Tensile modulus ($\times 10^3$ psi)	136	
Elongation at break (%)	170–800	350–525
Compressive strength (psi)		
Flexural strength (psi)		
Compressive modulus ($\times 10^3$ psi)		
Flexural modulus ($\times 10^3$ psi) at 73°F/23°C	125–175	130–140
200°F/93°C		
250°F/121°C		
Izod impact (ft-lb/in. of notch)	3.2–4.5	No break
Hardness, Rockwell	D63–65	R50
Coefficient of thermal expansion	70–110	130–200
(10^{-6} in./in./°F)		
Thermal conductivity		
(10^{-4}cal-cm/sec-cm^2°C		
or Btu/hr/ft^2/°F/in.)	0.269	
Deflection temperature at 264 psi (°F)		110–120
at 66 psi (°F)	154–158	155–180
Max. operating temperature (°F/°C)		180/182
Limiting oxygen index (%)		
Flame spread	Slow burning	Slow burning
Underwriters Lab. rating (Sub. 94)		

3. Typical Applications

Applications for polyethylene vary depending upon the grade of resin. LLDPE, because of its flexibility, is used primarily for prosthetic devices and vacuum-formed parts. HDPE finds application in fabricated and machined parts, tanks, prosthetic devices, corrosion resistant wall coverings, vacuum-formed parts, and trays for the food industry. EHMW and UHMW polyethylenes find applications where corrosion resistance is required. Among such applications are structural tanks and covers, fabricated parts, and piping. The usage for piping is of major importance. PE pipe is used to transport natural gas, potable water systems, drainage piping, corrosive wastes, and underground fire main water. Reference 2 details usage of polyethylene piping.

Other miscellaneous applications include surgical implants, coating, wire and cable insulation, tubing (laboratory, instrumentation air), and irrigation pipe.

TABLE 2.53 Compatibility of EHMWPE with Selected Corrodents[a]

Chemical	Maximum temp. °F	°C	Chemical	Maximum temp. °F	°C
Acetaldehde 40%	90	32	Barium sulfate	140	60
Acetamide			Barium sulfide	140	60
Acetic acid 10%	140	60	Benzaldehyde	x	x
Acetic acid 50%	140	60	Benzene	x	x
Acetic acid 80%	80	27	Benzenesulfonic acid 10%	140	60
Acetic acid, glacial			Benzoic acid	140	60
Acetic anhydride	x	x	Benzyl alcohol	170	77
Acetone	120	49	Benzyl chloride		
Acetyl chloride			Borax	140	60
Acrylic acid			Boric acid	140	60
Acrylonitrile	150	66	Bromine gas, dry	x	x
Adipic acid	140	60	Bromine gas, moist	x	x
Allyl alcohol	140	60	Bromine, liquid	x	x
Allyl chloride	80	27	Butadiene	x	x
Alum	140	60	Butyl acetate	90	32
Aluminum acetate			Butyl alcohol	140	60
Aluminum chloride, aqueous	140	60	n-Butylamine	x	x
Aluminum chloride, dry	140	60	Butyric acid	130	54
Aluminum fluoride	140	60	Calcium bisulfide	140	60
Aluminum hydroxide	140	60	Calcium bisulfite	80	27
Aluminum nitrate			Calcium carbonate	140	60
Aluminum oxychloride			Calcium chlorate	140	60
Aluminum sulfate	140	60	Calcium chloride	140	60
Ammonia gas	140	60	Calcium hydroxide 10%	140	60
Ammonium bifluoride			Calcium hydroxide, sat.	140	60
Ammonium carbonate	140	60	Calcium hypochlorite	140	60
Ammonium chloride 10%	140	60	Calcium nitrate	140	60
Ammonium chloride 50%	140	60	Calcium oxide	140	60
Ammonium chloride, sat.	140	60	Calcium sulfate	140	60
Ammonium fluoride 10%	140	60	Caprylic acid		
Ammonium fluoride 25%	140	60	Carbon bisulfide	x	x
Ammonium hydroxide 25%	140	60	Carbon dioxide, dry	140	60
Ammonium hydroxide, sat.	140	60	Carbon dioxide, wet	140	60
Ammonium nitrate	140	60	Carbon disulfide	x	x
Ammonium persulfate	140	60	Carbon monoxide	140	60
Ammonium phosphate	80	27	Carbon tetrachloride	x	x
Ammonium sulfate 10–40%	140	60	Carbonic acid	140	60
Ammonium sulfide	140	60	Cellosolve		
Ammonium sulfite			Chloroacetic acid, 50% in		
Amyl acetate	140	60	water	x	x
Amyl alcohol	140	60	Choloracetic acid	x	x
Amyl chloride	x	x	Chlorine gas, dry	80	27
Aniline	130	54	Chlorine gas, wet, 10%	120	49
Antimony trichloride	140	60	Chlorine, liquid	x	x
Aqua regia 3 : 1	130	54	Chlorobenzene	x	x
Barium carbonate	140	60	Chloroform	80	27
Barium chloride	140	60	Chlorosulfonic acid	x	x
Barium hydroxide	140	60	Chromic acid 10%	140	60

(Continued)

TABLE 2.53 Continued

Chemical	Maximum temp. °F	°C	Chemical	Maximum temp. °F	°C
Chromic acid 50%	90	32	Manganese chloride	80	27
Chromyl chloride			Methyl chloride	x	x
Citric acid 15%	140	60	Methyl ethyl ketone	x	x
Citric acid, concd	140	60	Methyl isobutyl ketone	80	27
Copper acetate			Muriatic acid	140	60
Copper carbonate			Nitric acid 5%	140	60
Copper chloride	140	60	Nitric acid 20%	140	60
Copper cyanide	140	60	Nitric acid 70%	x	x
Copper sulfate	140	60	Nitric acid, anhydrous	x	x
Cresol	80	27	Nitrous acid, concd		
Cupric chloride 5%	80	27	Oleum		
Cupric chloride 50%			Perchloric acid 10%	140	60
Cyclohexane	130	54	Perchloric acid 70%	x	x
Cyclohexanol	170	77	Phenol	100	38
Dibutyl phthalate	80	27	Phosphoric acid 50–80%	100	38
Dichloroacetic acid	73	23	Picric acid	100	38
Dichloroethane			Potassium bromide 30%	140	60
(ethylene dichloride)	x	x	Salicylic acid		
Ethylene glycol	140	60	Silver bromide 10%		
Ferric chloride	140	60	Sodium carbonate	140	60
Ferric chloride 50% in water	140	60	Sodium chloride	140	60
Ferric nitrate 10–50%	140	60	Sodium hydroxide 10%	170	77
Ferrous chloride	140	60	Sodium hydroxide 50%	170	77
Ferrous nitrate	140	60	Sodium hydroxide, concd		
Fluorine gas, dry	x	x	Sodium hypochlorite 20%	140	60
Fluorine gas, moist	x	x	Sodium hypochlorite, concd	140	60
Hydrobromic acid, dilute	140	60	Sodium sulfide to 50%	140	60
Hydrobromic acid 20%	140	60	Stannic chloride	140	60
Hydrobromic acid 50%	140	60	Stannous chloride	140	60
Hydrochloric acid 20%	140	60	Sulfuric acid 10%	140	60
Hydrochloric acid 38%	140	60	Sulfuric acid 50%	140	60
Hydrocyanic acid 10%	140	60	Sulfuric acid 70%	80	27
Hydrofluoric acid 30%	80	27	Sulfuric acid 90%	x	x
Hydrofluoric acid 70%	x	x	Sulfuric acid 98%	x	x
Hydrofluoric acid 100%	x	x	Sulfuric acid 100%	x	x
Hypochlorous acid			Sulfuric acid, fuming	x	x
Iodine solution 10%	80	27	Sulfurous acid	140	60
Ketones, general	x	x	Thionyl chloride	x	x
Lactic acid 25%	140	60	Toluene	x	x
Lactic acid, concd	140	60	Trichloroacetic acid	140	60
Magnesium chloride	140	60	White liquor		
Malic acid	100	38	Zinc chloride	140	60

[a] The chemicals listed are in the pure state or in a saturated solution unless otherwise indicated. Compatibility is shown to the maximum allowable temperature for which data are available. Incompatibility is shown by an x. A blank space indicates that data are unavailable.
Source: Schweitzer, PA. Corrosion Resistance Tables, 4th ed. Vols. 1–3. New York: Marcel Dekker, 1995.

TABLE 2.54 Compatibility of HMWPE with Selected Corrodents[a]

Chemical	Maximum temp.		Chemical	Maximum temp.	
	°F	°C		°F	°C
Acetaldehyde	x		Benzaldehyde	x	
Acetamide	140	60	Benzene	x	
Acetic acid 10%	140	60	Benzoic acid	140	60
Acetic acid 50%	140	60	Benzyl alcohol	x	
Acetic acid 80%	80	27	Borax	140	60
Acetic anhydride	x		Boric acid	140	60
Acetone	80	27	Bromine gas, dry	x	
Acetyl chloride	x		Bromine gas, moist	x	
Acrylonitrile	150	66	Bromine, liquid	x	
Adipic acid	140	60	Butadiene	x	
Allyl alcohol	140	60	Butyl acetate	90	32
Allyl chloride	110	43	Butyl alcohol	140	60
Alum	140	60	n-Butylamine	x	
Aluminum chloride, aqueous	140	60	Butyric acid	x	
Aluminum chloride, dry	140	60	Calcium bisulfide	140	60
Aluminum fluoride	140	60	Calcium bisulfite	140	60
Aluminum hydroxide	140	60	Calcium carbonate	140	60
Aluminum nitrate	140	60	Calcium chlorate	140	60
Aluminum sulfate	140	60	Calcium chloride	140	60
Ammonium gas	140	60	Calcium hydroxide 10%	140	60
Ammonium bifluoride	140	60	Calcium hydroxide, sat.	140	60
Ammonium carbonate	140	60	Calcium hypochlorite	140	60
Ammonium chloride 10%	140	60	Calcium nitrate	140	60
Ammonium chloride 50%	140	60	Calcium oxide	140	60
Ammonium chloride, sat.	140	60	Calcium sulfate	140	60
Ammonium fluoride 10%	140	60	Carbon bisulfide	x	
Ammonium fluoride 25%	140	60	Carbonm dioxide, dry	140	60
Ammonium hydroxide 25%	140	60	Carbon dioxide, wet	140	60
Amonium hydroxide, sat.	140	60	Carbon disulfide	x	
Ammonium nitrate	140	60	Carbon monoxide	140	60
Ammonium persulfate	150	66	Carbon tetrachloride	x	
Ammonium phosphate	80	27	Carbonic acid	140	60
Ammonium sulfate to 40%	140	60	Cellosolve	x	
Ammonium sulfide	140	60	Chloroacetic acid	x	
Ammonium sulfite	140	60	Chlorine gas, dry	x	
Amyl acetate	140	60	Chlorine gas, wet	x	
Amyl alcohol	140	60	Chlorine, liquid	x	
Amyl chloride	x		Chlorobenzene	x	
Aniline	130	44	Chloroform	x	
Antimony trichloride	140	60	Chlorosulfonic acid	x	
Aqua regia 3:1	130	44	Chromic acid 10%	140	60
Barium carbonate	140	60	Chromic acid 50%	90	32
Barium chloride	140	60	Citric acid 15%	140	60
Barium hydroxide	140	60	Citric acid, concd	140	60
Barium sulfate	140	60	Copper chloride	140	60
Barium sulfide	140	60	Copper cyanide	140	60

(*Continued*)

TABLE 2.54 Continued

Chemical	Maximum temp. °F	°C	Chemical	Maximum temp. °F	°C
Copper sulfate	140	60	Nitric acid 20%	140	60
Cresol	x		Nitric acid 70%	x	
Cupric chloride 5%	140	60	Nitric acid, anhydrous	x	
Cupric chloride 50%	140	60	Nitrous acid, concd	120	49
Cyclohexane	80	27	Perchloric acid 10%	140	60
Cyclohexanol	80	27	Perchloric acid 70%	x	
Dibutyl phthalate	80	27	Phenol	100	38
Dichloroethane	80	27	Phosphoric acid 50–80%	100	38
Ethylene glycol	140	60	Picric acid	100	38
Ferric chloride	140	60	Potassium bromide 30%	140	60
Ferrous chloride	140	60	Salicylic acid	140	60
Ferrous nitrate	140	60	Sodium carbonate	140	60
Fluorine gas, dry	x		Sodium chloride	140	60
Fluorine gas, moist	x		Sodium hydroxide 10%	150	66
Hydrobromic acid, dil	140	60	Sodium hydroxide 50%	150	66
Hydrobromic acid 20%	140	60	Sodium hypochlorite 20%	140	60
Hydrobromic acid 50%	140	60	Sodium hypochlorite, concd	140	60
Hydrochloric acid 20%	140	60	Sodium sulfide to 50%	140	60
Hydrochloric acid 38%	140	60	Stannic chloride	140	60
Hydrocyanic acid 10%	140	60	Stannous chloride	140	60
Hydrofluoric acid 30%	140	60	Sulfuric acid 10%	140	60
Hydrofluoric acid 70%	x		Sulfuric acid 50%	140	60
Hydrochlorous acid	150	66	Sulfuric acid 70%	80	27
Iodine solution 10%	80	27	Sulfuric acid 90%	x	
Ketones, general	80	27	Sulfuric acid 98%	x	
Lactic acid 25%	150	66	Sulfuric acid 100%	x	
Magnesium chloride	140	60	Sulfuric acid, fuming	x	
Malic acid	140	60	Sufurous acid	140	60
Manganese chloride	80	27	Thionyl chloride	x	
Methyl chloride	x		Toluene	x	
Methyl ethyl ketone	x		Trichloroacetic acid	80	27
Methyl isobutyl ketone	80	27	Zinc chloride	140	60
Nitric acid 5%	140	60			

[a] The chemicals listed are in the pure state or in a saturated solution unless otherwise indicated. Compatibility is shown to the maximum allowable temperature for which data are available. Incompatibility is shown by an x.

Source: Schweitzer, PA. Corrosion Resistance Tables, 4th ed. Vols. 1–3. New York: Marcel Dekker, 1995.

S. Polyethylene Terephthalate (PET)

Polyethylene terephthalate is a semicrystalline engineering thermoplastic having the following chemical structure:

1. Physical and Mechanical Properties

PET exhibits excellent properties, low moisture absorption, and good mechanical properties. Its mechanical properties are in the same range as PA, but PET has better dimensional stability because of its low moisture absorption, while PA has superior impact strength. Refer to Table 2.55 for the physical and mechanical properties of PET.

TABLE 2.55 Physical and Mechanical Properties of PET

Property	
Specific gravity	1.38
Water absorption (24 hr at 73°F/23°C) (%)	0.10
Dielectric strength, short-term (V/mil)	400
Tensile strength at break (psi)	11,500
Tensile modulus ($\times 10^3$ psi)	400
Elongation at break (%)	70
Compressive strength (psi)	
Flexural strength (psi)	15,000
Compressive modulus ($\times 10^3$ psi)	
Flexural modulus ($\times 10^3$ psi) at 73°F/23°C	400
200°F/93°C	
250°F/121°C	
Izod impact (ft-lb/in. of notch)	0.7
Hardness, Rockwell	R117
Coefficient of thermal expansion (10^{-6} in./in./°F)	39
Thermal conductivity (10^{-4} cal-cm/sec-cm^2 °C or Btu/hr/ft^2/°F/in.)	2.01
Deflection temperature at 264 psi (°F)	175
at 66 psi (°F)	240
Max. operating temperature (°F/°C)	230/110
Limiting oxygen index (%)	
Flame spread	
Underwriters Lab. rating (Sub. 94)	HB

PET parts are generally joined by adhesives. Surfaces should be solvent cleaned before applying isocyanate-cured polyesters, epoxies, or urethane adhesives. Polyethylene terephthalate cannot be solvent cemented or thermally fused.

2. Corrosion Resistance Properties

PET is resistant to dilute mineral acids, aliphatic hydrocarbons, aromatic hydrocarbons, ketones, and esters, with limited resistance to hot water and washing soda. It is not resistant to alkalies and chloroinated hydrocarbons. PET has good resistance to UV degradation and weatherability. Refer to Table 2.56 for the compatibility of PET with selected corrodents.

TABLE 2.56 Compatibility of PET with Selected Corrodents[a]

Acetic acid 5%	R	Hydrogen peroxide 0–5%	R
Acetic acid 10%	R	Ink	R
Acetone	R	Linseed oil	R
Ammonia 10%	x	Methanol	R
Benzene	R	Methyl ethyl ketone	R
Butyl acetate	R	Methylene chloride	x
Cadmium chloride	R	Motor oil	R
Carbon disulfide	R	Nitric acid 2%	R
Carbon tetrachloride	R	Paraffin oil	R
Chlorobenzene	x	Phosphoric acid 10%	R
Chloroform	x	Potassium hydroxide 50%	R
Citric acid 10%	R	Potassium dichromate 10%	R
Diesel oil	R	Potassium permanganate 10%	R
Dioxane	R	Silicone oil	R
Edible oil	R	Soap solution	R
Ethanol	R	Sodium bisulfite	R
Ether	R	Sodium carbonate 10%	R
Ethyl acetate	R	Sodium chloride 10%	R
Ethylene chloride	x	Sodium hydroxide 50%	x
Ethylene glycol	R	Sodium nitrate	R
Formic acid	R	Sodium thiosulfate	R
Fruit juice	R	Sulfuric acid 98%	x
Fuel oil	R	Sufuric acid 2%	R
Gasoline	R	Toluene	R
Glycerine	R	Vaseline	R
Heptane/hexane	R	Water, cold	R
Hydrochloric acid 38%	x	Water, hot	R
Hydrochloric acid 2%	R	Wax, molten	R
Hydrogen peroxide 30%	R	Xylene	R

[a] R = material resistant at 73°F/20°C; x = material not resistant.

3. Typical Applications

PET is used in the automotive industry for housings, racks, minor parts, and latch mechanisms. It can be painted to match metal body panels and has been used as fenders.

PET is also used in water purification, food handling equipment, and for pump and valve components.

T. Polyimide (PI)

Polyimides are heterocyclic polymers having an atom of nitrogen in one of the rings in the molecular chain. The atom of nitrogen is in the inside ring as shown below:

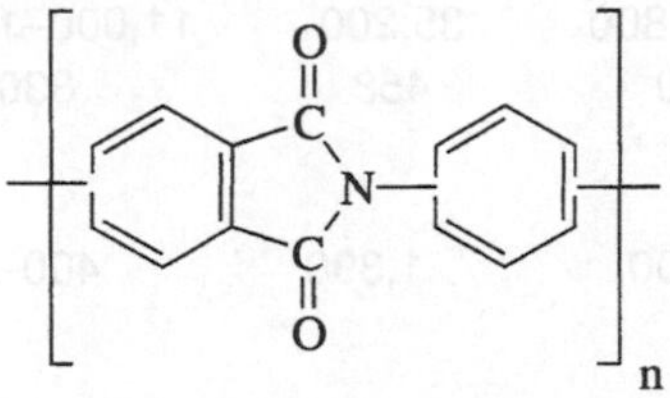

The fused rings provide chain stiffness essential to high-temperature strength retention. The low concentration of hydrogen provides oxidative resistance by preventing thermal degradation fracture of the chain.

1. Physical and Mechanical Properties

Polyimides exhibit outstanding properties resulting from their combination of high-temperature stability up to $500-600°F/260-315°C$ in continuous service and to $990°F/482°C$ for intermediate use. Polyimides also have a very low coefficient of friction, which can be further improved by use of graphite or other fillers, have excellent electrical and mechanical properties that are relatively stable from low negative temperatures $(-310°F/-190°C)$ to high positive temperatures, dimensional stability (low cold flow) in most environments, very low outgassing in high vacuum, and excellent resistance to ionizing radiation. Refer to Table 2.57 for physical and mechanical properties.

Polyimides can be adhesive bonded using epoxy adhesives after abrasion and solvent cleaning of the surfaces. The base polymer will usually have a higher temperature rating that the adhesive.

2. Corrosion Resistance Properties

The polyimides have excellent chemical and radiation inertness and are not subject to UV degradation. High oxidative resistance is another important property.

TABLE 2.57   Physical and Mechanical Properties of PI

Property	Unfilled	30% Glass fiber filled	15% Graphite filled
Specific gravity	1.33–1.43	1.56	1.41
Water absorption (24 hr at 73°F/23°C) (%)	0.24–0.34	0.23	0.19
Dielectric strength, short-term (V/mil)	415–550	528	250
Tensile strength at break (psi)	10,500–17,100	24,000	8,000–8,400
Tensile modulus ($\times 10^3$ psi)	300–400	1,720	
Elongation at break (%)	7.5–9.0	3	3.5
Compressive strength (psi)	17,500–40,000	27,500	25,000
Flexural strength (psi)	10,000–28,800	35,200	11,000–14,100
Compressive modulus ($\times 10^3$ psi)	315–350	458	330
Flexural modulus ($\times 10^3$ psi) at 73°F/23°C 200°F/93°C 250°F/121°C	360–500	1,390	400–500
Izod impact (ft-lb/in. of notch)	1.5–1.7	2.2	1.1
Hardness, Rockwell	R129	R128–M104	
Coefficient of thermal expansion (10^{-6} in./in./°F)	45–56	17–53	41
Thermal conductivity (10^{-4} cal-cm/sec-cm^2°C or Btu/hr/ft^2/°F/in.)	2.3–4.2	8.9	
Deflection temperature at 264 psi (°F) at 66 psi (°F)	460–680	489	680
Max. operating temperature (°F/°C)	500/260		
Limiting oxygen index (%)			
Flame spread			
Underwriters Lab. rating (Sub. 94)			

3. Typical Applications

Polyimides find applications that require high heat resistance and under the hood applications in the automobile industry. Other applications include bearings, compressors, valves, and piston rings.

U. Polyphenylene Oxide (PPO)

Noryls, patented by G.E. Plastics, are amorphous modified polyphenylene oxide resins. The basic phenylene oxide structure is as follows:

$$
\left[\begin{array}{c} \text{CH}_3 \\ \\ \text{---}\!\!\!\!\bigcirc\!\!\!\!\text{---O---} \\ \\ \text{CH}_3 \end{array}\right]_n
$$

Several grades of the resin are produced to provide a choice of performance characteristics to meet a wide range of engineering application requirements.

1. Physical and Mechanical Properties

PPO maintains excellent mechanical properties over a temperature range of from below $-40°F/-40°C$ to above $300°F/149°C$. It possesses excellent dimensional stability, is self-extinguishing with nonsagging characteristics, has low creep, high modulus, low water absorption, good electrical properties, and excellent impact strength. The physical and mechanical properties of PPO are shown in Table 2.58

2. Corrosion Resistance Properties

PPO has excellent resistance to aqueous environments, dilute mineral acids, and dilute alkalies. It is not resistant to aliphatic hydrocarbons, aromatic hydrocarbons, ketones, esters, or chlorinated hydrocarbons. Refer to Table 2.59 for the compatibility of PPO with selected corrodents.

3. Typical Applications

PPO finds application in business equipment, appliances, electronics, and electrical devices.

V. Polyphenylene Sulfide (PPS)

Polyphenylene sulfide is an engineering polymer capable of use at elevated temperatures. It has the following chemical structure:

$$
\left[\text{---}\!\!\!\!\bigcirc\!\!\!\!\text{---S---}\right]_n
$$

PPS has a symmetrical rigid backbone chain consisting of recurring para-substituted benzene rings and sulfur atoms. It is sold under the trade name of Ryton.

TABLE 2.58 Physical and Mechanical Properties of PPO Alloy with Polystyrenes

Property	Impact-modified	10% glass fiber reinforced
Specific gravity	1.27–1.36	1.14–1.31
Water absorption (24 hr at 73°F/23°C) (%)	0.01–0.07	0.06–0.07
Dielectric strength, short-term (V/mil)	530	420
Tensile strength at break (psi)	7,000–8,000	10,000–12,000
Tensile modulus ($\times 10^3$ psi)	345–360	
Elongation at break (%)	35	5–8
Compressive strength (psi)	10,000	
Flexural strength (psi)	8,200–11,000	20,000–23,000
Compressive modulus ($\times 10^3$ psi)		
Flexural modulus ($\times 10^3$ psi) at 73°F/23°C	325–345	760
200°F/93°C		
250°F/121°C		
Izod impact (ft-lb/in. of notch)	6.8	1.1–1.3
Hardness, Rockwell	R119	R121
Coefficient of thermal expansion	33	14
(10^{-6} in./in./°F)		
Thermal conductivity		
(10^{-4} cal-cm/sec-cm^2°C		
or Btu/hr/ft^2/°F/in.)	1.32	
Deflection temperature at 264 psi (°F)	190–275	252–260
at 66 psi (°F)	205–245	273–280
Max. operating temperature (°F/°C)	120–230/50–110	230/110
Limiting oxygen index (%)	22–39	26–36
Flame spread		
Underwriters Lab. rating (Sub. 94)	V-1	V-1

1. Physical and Mechanical Properties

PPS is noted for its high stiffness and good retention of mechanical properties at elevated temperatures. At normal ambient temperatures, unfilled PPS is a hard material with high tensile and flexural strengths. When the polymer is glass filled, appreciable increases in these properties are realized. Tensile strength and flexural modulus decrease as the temperature increases, leveling off at approximately 500°F/260°C. As the temperature increases, there is an increase in elongation and a corresponding increase in toughness. Long term exposure in air at 450°F/230°C has no effect on the mechanical properties of PPS. Refer to Table 2.60 for the physical and mechanical properties of PPS. The polymer is nonflammable.

TABLE 2.59 Compatibility of PPO with Selected Corrodents[a]

Acetic acid 5%	R	Hydrogen peroxide 0–5%	R
Acetic acid 10%	R	Hydrogen sulfide	R
Acetone	x	Linseed oil	R
Ammonia 10%	R	Methanol	R
Benzene	x	Methyl ethyl ketone	x
Carbon tetrachloride	x	Milk	R
Chlorobenzene	x	Motor oil	R
Chloroform	x	Nitric acid 2%	R
Citric acid 10%	R	Paraffin oil	R
Copper sulfate	R	Phosphoric acid 10%	R
Cyclohexane	x	Potassium hydroxide 50%	R
Cyclohexanone	x	Potassum dichromate	R
Diesel oil	R	Potassium permanganate 10%	R
Dioxane	x	Silicone oil	R
Edible oil	R	Soap solution	R
Ethanol	R	Sodium carbonate 10%	R
Ethyl acetate	R	Sodium chloride 10%	R
Ethylene chloride	x	Sodium hydroxide 50%	R
Ethylene glycol	R	Sodium hydroxide 5%	R
Formaldehyde 30%	R	Styrene	x
Formic acid	R	Sulfuric acid 98%	R
Fruit juice	R	Sulfuric acid 2%	R
Fuel oil	R	Trichloroethylene	R
Gasoline	R	Urea, aqueous	R
Glycerine	R	Water, cold	R
Hexane/heptane	R	Water, hot	R
Hydrochloric acid 38%	R	Wax, molten	R
Hydrochloric acid 2%	R	Xylene	x
Hydrogen peroxide 30%	R		

[a] R = material resistant at 73°F/20°C; x = material not resistant.

Polyphenylene sulfide cannot be thermally fused or solvent cemented. Joining must be by adhesive bonding. Surfaces to be joined must be solvent cleaned and abraded. Recommended adhesives include epoxies and urethanes.

2. Corrosion Resistance Properties

PPS has exceptional chemical resistance. It is resistant to aqueous inorganic salts and bases and many inorganic solvents. Relatively few materials react with PPS at high temperatures. It can also be used under highly oxidizing conditions.

Chlorinated solvents, some halogenated gases, and alkyl amines will attack PPS. It stress cracks in the presence of chlorinated solvents.

TABLE 2.60 Physical and Mechanical Properties of PPS

Property	Unfilled	30% glass fiber reinforced
Specific gravity	1.35	1.38–1.58
Water absorption (24 hr at 73°F/23°C) (%)	0.01–0.07	>0.03
Dielectric strength, short-term (V/mil)	380–450	
Tensile strength at break (psi)	7,000–12,500	22,000
Tensile modulus ($\times 10^3$ psi)	480	
Elongation at break (%)	1.6	1.5
Compressive strength (psi)	16,000	
Flexural strength (psi)	14,000–21,000	28,000
Compressive modulus ($\times 10^3$ psi)		
Flexural modulus ($\times 10^3$ psi) at 73°F/23°C	550–600	1,700
200°F/93°C		
250°F/121°C		
Izod impact (ft-lb/in. of notch)	>0.5	1.3
Hardness, Rockwell	R123–125	M102–103
Coefficient of thermal expansion	27–49	
(10^{-6} in./in./°F)		
Thermal conductivity		
(10^{-4}cal-cm/sec-cm^{2}°C	2.0–6.9	6.9–10.7
or Btu/hr/ft^2/°F/in.)	2.08	
Deflection temperature at 264 psi (°F)	212–275	507
at 66 psi (°F)	390	534
Max. operating temperature (°F/°C)	450/230	
Limiting oxygen index (%)	47	
Flame spread		
Underwriters Lab. rating (Sub. 94)	V-O	

Weak and strong alkalies have no effect. Refer to Table 2.61 for the compatibility of polyphenylene sulfide with selected corrodents. Reference 1 provides a more detailed listing.

PPS has good resistance to UV light degradation which can be increased by formulating it with carbon black.

3. Typical Applications

In the automotive industry, PPS molded components are used in the electrical, fuel handling, and emmision control systems. The chemical process industry makes use of PPS in valve and pump components as well as other industrial applications.

TABLE 2.61 Compatibility of PPS with Selected Corrodents[a]

Chemical	Maximum temp.		Chemical	Maximum temp.	
	°F	°C		°F	°C
Acetaldehyde	230	110	Bromine, liquid	x	x
Acetamide	250	121	Butadiene	100	38
Acetic acid 10%	250	121	Butyl acetate	250	121
Acetic acid 50%	250	121	Butyl alcohol	200	93
Acetic acid 80%	250	121	n-Butylamine	200	93
Acetic acid, glacial	190	88	Butyric acid	240	116
Acetic anhydride	280	138	Calcium bisulfite	200	93
Acetone	260	127	Calcium carbonate	300	149
Acrylic acid 25%	100	38	Calcium chloride	300	149
Acrylonitrile	130	54	Calcium hydroxide 10%	300	149
Adipic acid	300	149	Calcium hydroxide, sat.	300	149
Alum	300	149	Carbon bisulfide	200	93
Aluminum acetate	210	99	Carbon dioxide, dry	200	93
Aluminum chloride, aqueous	300	149	Carbon disulfide	200	93
Aluminum chloride, dry	270	132	Carbon tetrachloride	120	49
Aluminum hydroxide	250	121	Cellosolve	220	104
Aluminum nitrate	250	121	Choloroacetic acid	190	88
Aluminum oxychloride	460	238	Chlorine gas, dry	x	x
Ammonia gas	250	121	Chlorine gas, wet	x	x
Ammonium carbonate	460	238	Chlorine, liquid	200	93
Ammonium chloride 10%	300	149	Chloroform	150	66
Ammonium chloride 50%	300	149	Chlorosulfonic acid	x	x
Ammonium chloride, sat.	300	149	Chromic acid 10%	200	93
Ammonium hydroxide 25%	250	121	Chromic acid 50%	200	93
Ammonium hydroxide, sat.	250	121	Citric acid 15%	250	121
Ammonium nitrate	250	121	Citric acid, concd	250	121
Ammonium phosphate 65%	300	149	Copper acetate	300	149
Ammonium sulfate 10–40%	300	149	Copper chloride	220	104
Amyl acetate	300	149	Copper cyanide	210	99
Amyl alcohol	210	99	Copper sulfate	250	121
Amyl chloride	200	93	Cresol	200	93
Aniline	300	149	Cupric chloride 5%	300	149
Barium carbonate	200	93	Cyclohexane	190	88
Barium chloride	200	93	Cyclohexanol	250	121
Barium hydroxide	200	93	Dichloroethane		
Barium sulfate	220	104	(ethylene dichloride)	210	99
Barium sulfide	200	93	Ethylene glycol	300	149
Benzaldehyde	250	121	Ferric chloride	210	99
Benzene	300	149	Ferric chloride 50% in water	210	99
Benzenesulfonic acid 10%	250	121	Ferric nitrate 10–50%	210	99
Benzoic acid	230	110	Ferrous chloride	210	99
Benzyl alcohol	200	93	Ferrous nitrate	210	99
Benzyl chloride	300	149	Fluorine gas, dry	x	x
Borax	210	99	Hydrobromic acid, dil	200	93
Boric acid	210	99	Hydrobromic acid 20%	200	93
Bromine gas, dry	x	x	Hydrobromic acid 50%	200	93
Bromine gas, moist	x	x	Hydrochloric acid 20%	230	110

(Continued)

TABLE 2.61   Continued

Chemical	Maximum temp. °F	°C	Chemical	Maximum temp. °F	°C
Hydrochloric acid 38%	210	99	Sodium chloride	300	149
Hydrocyanic acid 10%	250	121	Sodium hydroxide 10%	210	99
Hydrofluoric acid 30%	200	93	Sodium hydroxide 50%	210	99
Lactic acid 25%	250	121	Sodium hypochlorite 5%	200	93
Lactic acid, concd	250	121	Sodium hypochlorite, concd	250	121
Magnesium chloride	300	149	Sodium sulfide to 50%	230	110
Methyl ethyl ketone	200	93	Stannic chloride	210	99
Methyl isobutyl ketone	250	121	Sulfuric acid 10%	250	121
Muriatic acid	210	99	Sulfuric acid 50%	250	121
Nitric acid 5%	150	66	Sulfuric acid 70%	250	121
Nitric acid 20%	100	38	Sulfuric acid 90%	220	104
Oleum	80	27	Sulfuric acid, fuming	80	27
Phenol 88%	300	149	Sulfurous acid 10%	200	93
Phosphoric acid 50–80%	220	104	Thionyl chloride	x	x
Potassium bromide 30%	200	93	Toluene	300	149
Sodium carbonate	300	149	Zinc chloride 70%	250	121

[a] The chemicals listed are in the pure state or in a saturated solution unless otherwise indicated. Compatibility is shown to the maximum allowable temperature for which data are available. Incompatibility is shown by an x. A blank space indicates that data are unavailable.
Source: Schweitzer, PA. Corrosion Resistance Tables, 4th ed. Vols. 1–3. New York: Marcel Dekker, 1995.

W.   Polypropylene (PP)

Polypropylene is one of the most common and versatile thermoplastics. It is closely related to polyethylene, both of which are members of a group known as polyolefins. The polyolefins are composed of only hydrogen and carbon. Within the chemical structure of PP, a distinction is made between isotactic PP and atactic PP; the isotactic form accounts for 97% of the polypropylene produced. This form is highly ordered having the structure shown below:

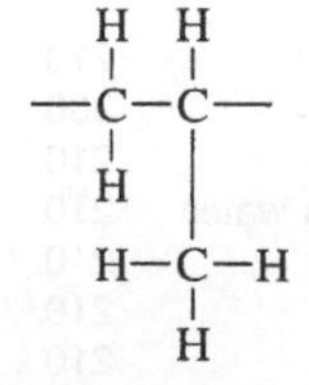

$$
\begin{array}{c}
\quad\ \ \text{H}\ \ \text{H} \\
\quad\ \ |\quad | \\
-\text{C}-\text{C}- \\
\quad\ \ |\quad | \\
\quad\ \ \text{H}\quad | \\
\quad\quad\ \ \text{H}-\text{C}-\text{H} \\
\quad\quad\quad\quad | \\
\quad\quad\quad\quad \text{H}
\end{array}
$$

Atactic PP is a viscous liquid type PP having a polypropylene matrix.

Polypropylene can be produced either as a homopolymer or as a copolymer with polyethylene. The copolymer has a structure as follows:

$$
\left[
\begin{array}{c}
\underset{\underset{\displaystyle H}{|}}{\overset{\overset{\displaystyle H}{|}}{C}} - \underset{\underset{\displaystyle \underset{\displaystyle H}{|}{C}{-}H}{|}}{\overset{\overset{\displaystyle H}{|}}{C}} - \underset{\underset{\displaystyle H}{|}}{\overset{\overset{\displaystyle H}{|}}{C}} - \underset{\underset{\displaystyle H}{|}}{\overset{\overset{\displaystyle H}{|}}{C}}
\end{array}
\right]_{n}
$$

1. Physical and Mechanical Properties

The specific gravity of the polypropylenes is among the lowest of all the polymers, ranging from 0.900 to 0.915. Polypropylene has good stiffness and tensile strength with excellent fatigue resistance, making PP an ideal material for living hinge designs, since they have an infinite life under flexing. Polypropylenes are perhaps the only thermoplastic surpassing all others in combined electrical properties, heat resistance, toughness, chemical resistance, dimensional stability, and surface gloss at a lower cost than most others. PP also has FDA approval for the handling of food products.

Polypropylene is available either as a homopolymer or as a copolymer with polyethylene. The homopolymers, being long chain high molecular weight molecules with a minimum of random orientation, have optimum chemical, thermal, and physical properties. For this reason, homopolymer material is preferred for difficult chemical, thermal, and physical conditions.

Copolymer PP is less brittle than the homopolymer and is able to withstand impact forces down to $-20°F/-29°C$, while the homopolymer is extremely brittle below $40°F$ ($4°C$).

Filling polypropylene with glass fibers greatly increases tensile strength, impact strength, flexural modulus, and deflection temperature under load with a corresponding reduction in elongation. Glass-filled PP can be used in load-bearing applications. Mineral-filled grades are used to reduce cost and prevent warpage in miscellaneous molded articles.

Table 2.62 lists the physical and mechanical properties of the polypropylene homopolymer, while Table 2.63 does likewise for the glass fiber reinforced homopolymer. Table 2.64 provides the physical and mechanical properties of the copolymer.

Polypropylene cannot be solvent cemented, but it may be thermally fused and joined by adhesive bonding. Surfaces to be bonded must be etched with a sodium dichromate–sulfuric acid solution at elevated temperatures. Epoxy adhesives should be used for bonding.

2. Corrosion Resistance Properties

PP is not affected by most inorganic chemicals except the halogens and severe oxidizing conditions. It can be used with sulfur-bearing compounds, caustics,

TABLE 2.62   Physical and Mechanical Properties of PP Homopolymers

Property	Unfilled	10–40% talc-filled
Specific gravity	0.9–0.91	0.97–1.27
Water absorption (24 hr at 73°F/23°C) (%)	0.01–0.03	0.01–0.03
Dielectric strength, short-term (V/mil)	600	500
Tensile strength at break (psi)	4,500–6,000	3,545–5,000
Tensile modulus ($\times 10^3$ psi)	165–225	450–575
Elongation at break (%)	10–600	30–60
Compressive strength (psi)	5,500–8,000	7,500
Flexural strength (psi)	6,000–8,000	7,000–9,200
Compressive modulus ($\times 10^3$ psi)	150–300	
Flexural modulus ($\times 10^3$ psi) at 73°F/23°C	170–250	210–670
200°F/93°C	50	400
250°F/121°C	35	
Izod impact (ft-lb/in. of notch)	0.4–1.4	0.4–1.4
Hardness, Rockwell	R80–102	R85–110
Coefficient of thermal expansion (10^{-6} in./in./°F)	81–100	42–80
Thermal conductivity (10^{-4} cal-cm/sec-cm^2°C or Btu/hr/ft^2/°F/in.)	2.8 1.2	7.6
Deflection temperature at 264 psi (°F)	120–140	132–180
at 66 psi (°F)	225–250	210–290
Max. operating temperature (°F/°C)	180/82	
Limiting oxygen index (%)	17	
Flame spread	Slow burning	
Underwriters Lab. rating (Sub. 94)	HB	

solvents, acids, and other organic chemicals. PP should not be used with oxidizing type acids, detergents, low-boiling hydrocarbons, alcohols, aromatics, and some organic materials.

If exposed to sunlight, an ultraviolet absorber or screening agent should be in the formulation to protect it from degradation. Thermal oxidative degradation, particularly where copper is involved, will pose a problem.

Refer to Table 2.65 for the compatibility of PP with selected corrodents. Reference 1 provides a more detailed listing.

3.   Typical Applications

Polypropylene is widely used in engineering fabrics such as bale wraps, filter cloths, bags, ropes, and strapping. Piping and small tanks of polypropylene are widely used, and 90% of all battery casings are made of PP. Ignition resistant

TABLE 2.63 Physical and Mechanical Properties of Glass Fiber Reinforced PP Homopolymers

Property	10–30% reinforcing	40% reinforcing
Specific gravity	0.97–1.14	1.22–1.23
Water absorption (24 hr at 73°F/23°C) (%)	0.01–0.05	0.05–0.06
Dielectric strength, short-term (V/mil)		500–510
Tensile strength at break (psi)	6,500–13,000	8,400–15,000
Tensile modulus ($\times 10^3$ psi)	700–1,000	1,100–1,500
Elongation at break (%)	1.8–7	1.5–4
Compressive strength (psi)	6,500–8,400	8,900–9,800
Flexural strength (psi)	7,000–20,000	10,500–22,000
Compressive modulus ($\times 10^3$ psi)		
Flexural modulus ($\times 10^3$ psi) at 73°F/23°C	310–780	950–1,000
200°F/93°C		
250°F/121°C		
Izod impact (ft-lb/in. of notch)	1.0–2.2	1.4–2.0
Hardness, Rockwell	R92–115	R102–111
Coefficient of thermal expansion (10^{-6} in./in./°F)	61–62	27–32
Thermal conductivity (10^{-4} cal-cm/sec-cm^2°C or Btu/hr/ft^2/°F/in.)	5.5–6.2	8.4–8.8
Deflection temperature at 264 psi (°F)	253–288	300–330
at 66 psi (°F)	290–320	330
Max. operating temperature (°F/°C)		
Limiting oxygen index (%)		
Flame spread		
Underwriters Lab. rating (Sub. 94)		

grades are available as a result of the addition of halogenated organic compounds. With this additive, polypropylene can be used in duct systems in the chemical industry. Because polypropylene exhibits good flex life, it is useful in the construction of integral hinges. PP also has many textule applications, such as carpet face and backing yarn, upholstery fabrics, and diaper cover stock. It is useful in outdoor clothing and sports clothes that are worn next to the body because its unique wicking qualities absorb body moisture and still leave the wearer dry. PP also finds application in the automotive industry for interior trim and under the hood components. In appliance areas, it is used in washer agitators and dishwasher components. Consumer goods such as straws, housewares, luggage, syringes, toys, and recreational items also make use of polypropylene.

TABLE 2.64 Physical and Mechanical Properties of PP Copolymers

Property	Unfilled	30–40% glass fiber reinforcing	10–40% talc filled
Specific gravity	0.89–0.905	1.11–1.21	0.97–1.24
Water absorption (24 hr at 73°F/23°C) (%)	0.03	0.01	0.02
Dielectric strength, short-term (V/mil)	600		
Tensile strength at break (psi)	4,000–6,500	6,000–10,000	3,000–3,775
Tensile modulus (×10³ psi)	130–180		
Elongation at break (%)	200–500	2.2–3.0	20–50
Compressive strength (psi)	3,500–8,000	5,400–5,700	
Flexural strength (psi)	5,000–7,000	9,000–15,000	4,500–5,100
Compressive modulus (×10³ psi)			
Flexural modulus (×10³ psi)			
at 73°F/23°C	130–200	600–960	160–400
200°F/93°C	40		
250°F/121°C	30		
Izod impact (ft-lb/in. of notch)	1.1–4.0	0.9–3.0	0.6–4.0
Hardness, Rockwell	R65–96	R104–105	R83–88
Coefficient of thermal expansion (10⁻⁶ in./in./°F)	68–95		
Thermal conductivity (10⁻⁴cal-cm/sec-cm²°C or Btu/hr/ft²/°F/in.)	3.5–4.0 1.3		
Deflection temperature			
at 264 psi (°F)	130–140	280	100–165
at 66 psi (°F)	185–220	310	195–260
Max. operating temperature (°F/°C)	200/93		
Limiting oxygen index (%)			
Flame spread	Slow burning		
Underwriters Lab. rating (Sub. 94)			

X. Styrene-Acrylonitrile (SAN)

SAN is a copolymer of styrene and acrylonitrile having the following chemical structure:

$$\left[\begin{array}{cc} H & H \\ | & | \\ C & C \\ | & | \\ H & CH \end{array}\right]\left[\begin{array}{cc} H & H \\ | & | \\ C & C \\ | & | \\ C & CN \end{array}\right]$$

TABLE 2.65 Compatibility of PP with Selected Corrodents[a]

Chemical	Maximum temp.		Chemical	Maximum temp.	
	°F	°C		°F	°C
Acetaldehyde	120	49	Benzaldehyde	80	27
Acetamide	110	43	Benzene	140	60
Acetic acid 10%	220	104	Benzenesulfonic acid 10%	180	82
Acetic acid 50%	200	93	Benzoic acid	190	88
Acetic acid 80%	200	93	Benzyl alcohol	140	60
Acetic acid, glacial	190	88	Benzyl chloride	80	27
Acetic anydride	100	38	Borax	210	99
Acetone	220	104	Boric acid	220	104
Acetyl chloride	x	x	Bromine gas, dry	x	x
Acrylic acid	x	x	Bromine gas, moist	x	x
Acrylonitrile	90	32	Bromine, liquid	x	x
Adipic acid	100	38	Butadiene	x	x
Allyl alcohol	140	60	Butyl acetate	x	x
Allyl chloride	140	60	Butyl alcohol	200	93
Alum	220	104	n-Butylamine	90	32
Aluminum acetate	100	38	Butyric acid	180	82
Aluminum chloride, aqueous	200	93	Calcium bisulfide	210	99
Aluminum chloride, dry	220	104	Calcium bisulfite	210	99
Aluminum fluoride	200	93	Calcium carbonate	210	99
Aluminum hydroxide	200	93	Calcium chlorate	220	104
Aluminum nitrate	200	93	Calcium chloride	220	104
Aluminum oxychloride	220	104	Calcium hydroxide 10%	200	93
Aluminum sulfate			Calcium hydroxide, sat.	220	104
Ammonia gas	150	66	Calcium hypochlorite	210	99
Ammonium bifluoride	200	93	Calcium nitrate	210	99
Ammonium carbonate	220	104	Calcium oxide	220	104
Ammonium chloride 10%	180	82	Calcium sulfate	220	104
Ammonium chloride 50%	180	82	Caprylic acid	140	60
Ammonium chloride, sat.	200	93	Carbon bisulfide	x	x
Ammonium fluoride 10%	210	99	Carbon dioxide, dry	220	104
Ammonium fluoride 25%	200	93	Carbon dioxide, wet	140	60
Ammonium hydroxide 25%	200	93	Carbon disulfide	x	x
Ammonium hydroxide, sat.	200	93	Carbon monoxide	220	104
Ammonium nitrate	200	93	Carbon tetrachloride	x	x
Ammonium persulfate	220	104	Carbonic acid	220	104
Ammonium phosphate	200	93	Cellosolve	200	93
Ammonium sulfate 10–40%	200	93	Chloroacetic acid, 50% water	80	27
Ammonium sulfide	220	104	Chloroacetic acid	180	82
Ammonium sulfite	220	104	Chlorine gas, dry	x	x
Amyl acetate	x	x	Chlorine gas, wet	x	x
Amyl alcohol	200	93	Chlorine, liquid	x	x
Amyl chloride	x	x	Chlorobenzene	x	x
Aniline	180	82	Chloroform	x	x
Antimony trichloride	180	82	Chlorosulfonic acid	x	x
Aqua regia 3 : 1	x	x	Chromic acid 10%	140	60
Bariium carbonate	200	93	Chromic acid 50%	150	66
Barium chloride	220	104	Chromyl chloride	140	60
Barium hydroxide	200	93	Citric acid 15%	220	104
Barium sulfate	200	93	Citric acid, concd	220	104
Barium sulfide	200	93	Copper acetate	80	27

(*Continued*)

TABLE 2.65 Continued

Chemical	Maximum temp. °F	Maximum temp. °C	Chemical	Maximum temp. °F	Maximum temp. °C
Copper carbonate	200	93	Methyl isobutyl ketone	80	27
Copper chloride	200	93	Muriatic acid	200	93
Copper cyanide	200	93	Nitric acid 5%	140	60
Copper sulfate	200	93	Nitric acid 20%	140	60
Cresol	x	x	Nitric acid 70%	x	x
Cupric chloride 5%	140	60	Nitric acid, anhydrous	x	x
Cupric chloride 50%	140	60	Nitrous acid, concd	x	x
Cyclohexane	x	x	Oleum	x	x
Cyclohexanol	150	66	Perchloric acid 10%	140	60
Dibutyl phthalate	180	82	Perchloric acid 70%	x	x
Dichloroacetic acid	100	38	Phenol	180	82
Dichloroethane			Phosphoric acid 50–80%	210	99
(ethylene dichloride)	80	27	Picric acid	140	60
Ethylene glycol	210	99	Potassium bromide 30%	210	99
Ferric chloride	210	99	Salicylic acid	130	54
Ferric chloride 50% in water	210	99	Silver bromide 10%	170	77
Ferrite nitrate 10–50%	210	99	Sodium carbonate	220	104
Ferrous chloride	210	99	Sodium chloride	200	93
Ferrous nitrate	210	99	Sodium hydroxide 10%	220	104
Fluorine gas, dry	x	x	Sodium hydroxide 50%	220	104
Fluorine gas, moist	x	x	Sodium hydroxide, concd	140	60
Hydrobromic acid, dilute	230	110	Sodium hypochlorite 20%	120	49
Hydrobromic acid 20%	200	93	Sodium hypochlorite, concd	110	43
Hydrobromic acid 50%	190	88	Sodium sulfide to 50%	190	88
Hydrochloric acid 20%	220	104	Stannic chloride	150	66
Hydrochloric acid 38%	200	93	Stannous chloride	200	93
Hydrocyanic acid 10%	150	66	Sulfuric acid 10%	200	93
Hydrofluoric acid 30%	180	82	Sulfuric acid 50%	200	93
Hydrofluoric acid 70%	200	93	Sulfuric acid 70%	180	82
Hydrofluoric acid 100%	200	93	Sulfuric acid 90%	180	82
Hypochlorous acid	140	60	Sulfuric acid 98%	120	49
Iodine solution 10%	x	x	Sulfuric acid 100%	x	x
Ketones, general	110	43	Sulfuric acid, fuming	x	x
Lactic acid 25%	150	66	Sulfurous acid	180	82
Lactic acid, concd	150	66	Thionyl chloride	100	38
Magnesium chloride	210	99	Toluene	x	x
Malic acid	130	54	Trichloroacetic acid	150	66
Manganese chloride	120	49	White liquor	220	104
Methyl chloride	x	x	Zinc chloride	200	93
Methyl ethyl ketone	x	x			

[a] The chemicals listed are in the pure state or in a saturated solution unless otherwise indicated. Compatibility is shown to the maximum allowable temperature for which data are available. Incompatibility is shown by an x. A blank space indicates that data are unavailable.
Source: Schweitzer, PA. Corrosion Resistance Tables, 4th ed. Vols. 1–3. New York: Marcel Dekker, 1995.

1. Physical and Mechanical Properties

The unfilled material has glass-like clarity with good strength and rigidity, heat, and abrasion resistance. It is also available glass filled. The physical and mechanical properties are given in Table 2.66.

SAN may be solvent cemented, thermally fused, or adhesive bonded. Solvent cementing can be accomplished using methyl ethyl ketone, methyl isobutyl ketone, tetrahydrofuran, or methylene chloride. The best adhesives for SAN are epoxies, urethanes, thermosetting acrylics, nitrile–phenolic, and cyanoacrylates. The surfaces to be joined do not require any special treatment other than simple cleaning and removal of possible contaminants.

TABLE 2.66 Physical and Mechanical Properties of SAN

Property	Unfilled	20% glass fiber reinforced
Specific gravity	1.07	1.22–1.40
Water absorption (24 hr at 73°F/23°C) (%)		0.1–0.2
Dielectric strength, short-term (V/mil)		500
Tensile strength at break (psi)	10,500	15,500–18,000
Tensile modulus ($\times 10^3$ psi)	475	1,200–1,710
Elongation at break (%)		1, 2–1.8
Compressive strength (psi)		17,000–21,000
Flexural strength (psi)	16,700	20,000–22,700
Compressive modulus ($\times 10^3$ psi)	500	1,000–1,280
Flexural modulus ($\times 10^3$ psi) at 73°F/23°C		
200°F/93°C		
250°F/121°C		
Izod impact (ft-lb/in. of notch)	0.4	1.0–3.0
Hardness, Rockwell	M83	M89–100
Coefficient of thermal expansion (10^{-6} in./in./°F)		23.4–41.4
Thermal conductivity (10^{-4} cal-cm/sec-cm^2°C or Btu/hr/ft^2/°F/in.)		6.6
Deflection temperature at 264 psi (°F)	205	210–230
at 66 psi (°F)		220
Max. operating temperature (°F/°C)		
Limiting oxygen index (%)		
Flame spread		
Underwriters Lab. rating (Sub. 94)	HB	

2. Corrosion Resistance Properties

SAN is resistant to aliphatic hydrocarbons but not to aromatic and chlorinated hydrocarbons. It will be attached by oxidizing agents, strong acids, and will stress crack in the presence of certain organic compounds.

SAN will be degraded by ultraviolet light unless protective additives are incorporated into the formulation.

3. Typical Applications

Styrene–acrylonitrile finds application as food and beverage containers, dinnerware, housewares, appliances, interior refrigerator components, and toys. Industrial applications include fan blades and filter housings. Medical applications include tubing connectors, and valves, labware, and urine bottles. The packaging industry makes use of SAN for cosmetic containers and displays.

Y. Polyvinylidine Chloride (PVDC)

Polyvinylidene chloride is manufactured under the tradename Saran by Dow Chemical. It is a variant of PVC having both chlorine atoms at the same end of the monomer instead of at the opposite ends as shown below:

$$
\begin{array}{c} \text{Cl} \quad \text{H} \\ | \quad\quad | \\ -\text{C}-\text{C}- \\ | \quad\quad | \\ \text{Cl} \quad \text{H} \end{array}
$$

1. Physical and Mechanical Properties

Saran has improved strength, hardness, and chemical resistance over that of PVC. The operating temperature range is from 0 to 175°F (−18 to 80°C).

PVDC can be joined by solvent cementing, thermal fusion, and adhesive bonding. For solvent welding, cyclohexane, tetrahydrofuran, and dichlorobenzene are used. Adhesive bonding can be accomplished using epoxies, urethane, cyanoacrylates, and thermosetting acrylics.

PVDC complies with FDA regulations for food processing and potable water and also with regulations prescribed by the Meat Inspection Division of the Department of Agriculture for transporting fluids used in meat production. Table 2.67 lists its physical and mechanical properties.

2. Corrosion Resistance Properties

Saran is resistant to oxidants, mineral acids and solvents. In applications such as plating solutions, chlorides and certain other chemicals polyvinylidene chloride is superior to polypropylene and finds many applications in the handling of

TABLE 2.67 Physical and Mechanical Properties of PVDC

Property	Unplasticized	Plasticized
Specific gravity	1.65–1.72	1.65–1,70
Water absorption (24 hr at 73°F/23°C) (%)	0.1	0.1
Dielectric strength, short-term (V/mil)		400–600
Tensile strength at break (psi)	2,800	3,500
Tensile modulus ($\times 10^3$ psi)	50–80	50–80
Elongation at break (%)	350–400	250–300
Compressive strength (psi)		
Flexural strength (psi)		
Compressive modulus ($\times 10^3$ psi)		
Flexural modulus ($\times 10^3$ psi) at 73°F/23°C		
200°F/93°C		
250°F/121°C		
Izod impact (ft-lb/in. of notch)	0.3–1.0	0.3–1.0
Hardness, Rockwell	R98–106	R98–106
Coefficient of thermal expansion (10^{-6} in./in./°F)	190	190
Thermal conductivity (10^{-4}cal-cm/sec-cm²°C or Btu/hr/ft²/°F/in.)	3	3
Deflection temperature at 264 psi (°F) at 66 psi (°F)	130–150	130–150
Max. operating temperature (°F/°C)	170/80	
Limiting oxygen index (%)		
Flame spread	Self-extinguishing	
Underwriters Lab. rating (Sub. 94)		

municipal water supplies and waste waters. Saran is also resistant to weathering and UV degradation. Refer to Table 2.68 for the compatibility of PVDC with selected corrodents. Refer to Reference 1 for a more detailed listing.

3. Typical Applications

Saran has found wide application in the plating industry and for handling deionized water, pharmaceuticals, food processing, and other applications where stream purity protection is critical. It also finds application as a lined piping system. Refer to Reference 2.

Saran is also used to manufacture auto seat covers, film, bristles, and paperboard coatings.

TABLE 2.68 Compatibility of PVDC with Selected Corrodents[a]

Chemical	Maximum temp. °F	Maximum temp. °C	Chemical	Maximum temp. °F	Maximum temp. °C
Acetaldehyde	150	66	Benzene	x	x
Acetic acid 10%	150	66	Benzenesulfonic acid 10%	120	49
Acetic acid 50%	130	54	Benzoic acid	120	49
Acetic acid 80%	130	54	Benzyl chloride	80	27
Acetic acid, glacial	140	60	Boris acid	170	77
Acetic anhydride	90	32	Bromine, liquid	x	x
Acetone	90	32	Butadiene	x	x
Acetyl chloride	130	54	Butyl acetate	120	49
Acrylonitrile	90	32	Butyl alcohol	150	66
Adipic acid	150	66	Butyric acid	80	27
Allyl alcohol	80	27	Calcium bisulfite	80	27
Alum	180	82	Calcium carbonate	180	82
Aluminum chloride, aqueous	150	66	Calcium chlorate	160	71
Aluminum fluoride	150	66	Calcium chloride	180	82
Aluminum hydroxide	170	77	Calcium hydroxide 10%	160	71
Aluminum nitrate	180	82	Calcium hydroxide, sat.	180	82
Aluminum oxychloride	140	60	Calcium hypochlorite	120	49
Aluminum sulfate	180	82	Calcium nitrate	150	66
Ammonia gas	x	x	Calcium oxide	180	82
Ammonium bifluoride	140	60	Calcium sulfate	180	82
Ammonium carbonate	180	82	Caprylic acid	90	32
Ammonium chloride, sat.	160	71	Carbon bisulfide	90	32
Ammonium fluoride 10%	90	32	Carbon dioxide, dry	180	82
Ammonium fluoride 25%	90	32	Carbon dioxide, wet	80	27
Ammonium hydroxide 25%	x	x	Carbon disulfide	80	27
Ammonium hydroxide, sat.	x	x	Carbon monoxide	180	82
Ammonium nitrate	120	49	Carbon tetrachloride	140	60
Ammonium persulfate	90	32	Carbonic acid	180	82
Ammonium phosphate	150	66	Cellosolve	80	27
Ammonium sulfate 10–40%	120	49	Chloroacetic acid, 50% water	120	49
Ammonium sulfide	80	27	Chloroacetic acid	120	49
Amyl acetate	120	49	Chlorine gas, dry	80	27
Amyl alcohol	150	66	Chlorine gas, wet	80	27
Amyl chloride	80	27	Chlorine, liquid	x	x
Aniline	x	x	Chlorobenzene	80	27
Antimony trichloride	150	66	Chloroform	x	x
Aqua regia 3 : 1	120	49	Chlorosulfonic acid	x	x
Barium carbonate	180	82	Chromic acid 10%	180	82
Barium chloride	180	82	Chromic acid 50%	180	82
Barium hydroxide	180	82	Citric acid 15%	180	82
Barium sulfate	180	82	Citric acid, concd	180	82
Barium sulfide	150	66	Copper carbonate	180	82
Benzaldehyde	x	x	Cooper chloride	180	82

TABLE 2.68 Continued

Chemical	Maximum temp. °F	Maximum temp. °C	Chemical	Maximum temp. °F	Maximum temp. °C
Copper cyanide	130	54	Muriatic acid	180	82
Copper sulfate	180	82	Nitric acid 5%	90	32
Cresol	150	66	Nitric acid 20%	150	66
Cupric chloride 5%	160	71	Nitric acid 70%	x	x
Cupric chloride 50%	170	77	Nitric acid, anhydrous	x	x
Cyclohexane	120	49	Oleum	x	x
Cyclohexanol	90	32	Perchloric acid 10%	130	54
Dibutyl phthalate	180	82	Perchloric acid 70%	120	49
Dichloroacetic acid	120	49	Phenol	x	x
Dichloroethane			Phosphoric acid 50–80%	130	54
(ethylene dichloride)	80	27	Picric acid	120	49
Ethylene glycol	180	82	Potassium bromide 30%	110	43
Ferric chloride	140	60	Salicylic acid	130	54
Ferric chloride 50% in water	140	60	Sodium carbonate	180	82
Ferric nitrate 10–50%	130	54	Sodium chloride	180	82
Ferrus chloride	130	54	Sodium hydroxide 0%	90	32
Ferrous nitrate	80	27	Sodium hydroxide 50%	150	66
Fluorine gas, dry	x	x	Sodium hydroxide, concd	x	x
Fluorine gas, moist	x	x	Sodium hypochlorite 10%	130	54
Hydrobromic acid, dil	120	49	Sodium hypochlorite, concd	120	49
Hydrobromic acid 20%	120	49	Sodium sulfide to 50%	140	60
Hydrobromic acid 50%	130	54	Stannic chloride	180	82
Hydrochloric acid 20%	180	82	Stannous chloride	180	82
Hydrochloric acid 38%	180	82	Sulfuric acid 10%	120	49
Hydrocyanic acid 10%	120	49	Sulfuric acid 50%	x	x
Hydrofluoric acid 30%	160	71	Sulfuric acid 70%	x	x
Hydrofluoric acid 100%	x	x	Sulfuric acid 90%	x	x
Hypochlorous acid	120	49	Sulfuric acid 98%	x	x
Ketones, general	90	32	Sulfuric acid 100%	x	x
Lactic acid, concd	80	27	Sulfuric acid, fuming	x	x
Magnesium chloride	180	82	Sulfurous acid	80	27
Malic acid	80	27	Thionyl chloride	x	x
Methyl chloride	80	27	Toluene	80	27
Methyl ethyl ketone	x	x	Trichloroacetic acid	80	27
Methyl isobutyl ketone	80	27	Zinc chloride	170	77

[a] The chemicals listed are in the pure state or in a saturated solution unless otherwise indicated. Compatibility is shown to the maximum allowable temperature for which data are available. Incompatibility is shown by an x. A blank space indicates that data are unavailable.
Source: Schweitzer, PA. Corrosion Resistance Tables, 4th ed. Vols. 1–3. New York: Marcel Dekker, 1995.

Z. Polysulfone (PSF)

Polysulfone is an engineering polymer which can be used at elevated temperatures. It has the following chemical structure:

The linkages connecting the benzene rings are hydrolytically stable.

1. Physical and Mechanical Properties

PSF has high tensile strength and stress-strain behavior which is typical of that found in a ductile material. In addition, as temperatures increase, flexural modulus remains high. These resins also remain stable at elevated temperatures resisting creep and deformation under continuous load. PSF has an operating temperature range of from $-150°F$ to $300°F$ ($-101°C$ to $149°C$). Thermal gravimetric analysis shows that PSF is stable in air up to $932°F$ ($500°C$) .

PSF also exhibits excellent electrical properties that remain stable over a wide temperature range up to $350°F/177°C$. Refer to Table 2.69 for its physical and mechanical properties.

Polysulfone can be reinforced with glass fiber to improve its mechanical properties, as shown in Table 2.70.

PSF can be joined by solvent cementing, thermal fusion, and adhesive bonding. Methylene chloride is used to solvent cement PSF while epoxy adhesives are used for adhesive bonding. No special surface treatment is required.

2. Corrosion Resistance Properties

PSF is resistant to repeated sterilization by several techniques including steam autoclave, dry heat, ethylene oxide, certain chemicals, and radiation. It will withstand exposure to soap, detergent solutions, and hydrocarbon oils, even at elevated temperatures and under moderate stress levels. Polysulfone is unaffected by hydrolysis and has a very high resistance to mineral acids, alkali, and salt solutions.

PSF is not resistant to polar organic solvents such as ketones, chlorinated hydrocarbons, and aromatic hydrocarbons.

Polysulfone has good weatherability and is not degraded by UV radiation. Refer to Table 2.71 for the compatibility of PSF with selected corrodents. Reference 1 provides a more detailed listing.

TABLE 2.69 Physical and Mechanical Properties of Unfilled PSF

Property	Flame retardant	Unfilled
Specific gravity	1.24–1.25	1.24
Water absorption (24 hr at 73°F/23°C) (%)	0.3	0.62
Dielectric strength, short-term (V/mil)	425	20
Tensile strength at break (psi)		10,200
Tensile modulus ($\times 10^3$ psi)	360–390	390
Elongation at break (%)	50–100	40–80
Compressive strength (psi)	40,000	
Flexural strength (psi)	15,400–17,500	17,500
Compressive modulus ($\times 10^3$ psi)	374	
Flexural modulus ($\times 10^3$ psi) at 73°F/23°C	390	370
200°F/93°C	370	
250°F/121°C	350	
Izod impact (ft-lb/in. of notch)	1.0–1.3	1.0–1.2
Hardness, Rockwell	M69	R120
Coefficient of thermal expansion (10^{-6} in./in./°F)	56	31
Thermal conductivity (10^{-4}cal-cm/sec-cm^2°C	6.2	
or Btu/hr/ft^2/°F/in.)		1.8
Deflection temperature at 264 psi (°F)	345	340
at 66 psi (°F)	358	360
Max. operating temperature (°F/°C)		300/149
Limiting oxygen index (%)		
Flame spread		
Underwriters Lab. rating (Sub. 94)	V-O	

3. Typical Applications

Polysulfone finds application as hot-water piping, lenses, iron handles, switches, and circuit breakers. Its rigidity and high temperature performance make it ideal for medical, microwave, and electronic application.

AA. Polyvinyl Chloride (PVC)

Polyvinyl chloride is the most widely used of any of the thermoplasts. PVC is polymerized vinyl chloride, which is produced from ethylene and anhydrous hydrochloric acid. The structure is

```
    H  Cl  H  Cl
    |  |   |  |
  —C—C—C—C—
    |  |   |  |
    H  H   H  H
```

TABLE 2.70 Physical and Mechanical Properties of Glass Fiber Reinforced PSF

Property	10% reinforcing	30% reinforcing
Specific gravity	1.31	1.46–1.49
Water absorption (24 hr at 73°F/23°C) (%)		0.3
Dielectric strength, short-term (V/mil)		
Tensile strength at break (psi)	14,500	14,500–18,100
Tensile modulus ($\times 10^3$ psi)	667–670	1,360–1,450
Elongation at break (%)	4.2	1.5–1.8
Compressive strength (psi)		19,000
Flexural strength (psi)	20,000	20,000–23,500
Compressive modulus ($\times 10^3$ psi)		
Flexural modulus ($\times 10^3$ psi) at 73°F/23°C	600	1,050–1,250
200°F/93°C		
250°F/121°C		
Izod impact (ft-lb/in. of notch)	1.3	1.1–1.5
Hardness, Rockwell	M79	M87–100
Coefficient of thermal expansion	18–32	20–25
(10^{-6} in./in./°F)		
Thermal conductivity		
(10^{-4} cal-cm/sec-cm^2°C		
or Btu/hr/ft^2/°F/in.)		
Deflection temperature at 264 psi (°F)	361	350–365
at 66 psi (°F)	367	360–372
Max. operating temperature (°F/°C)		
Limiting oxygen index (%)		
Flame spread		
Underwriters Lab. rating (Sub. 94)		

1. Physical and Mechanical Properties

PVC is stronger and more rigid than other general purpose thermoplastic materials. It has a high tensile strength and modulus of elasticity. Additives are used to further specific end uses, such as thermal stabilizers, lubricity, impact modifiers, and pigmentation.

Two types of PVC are produced, normal impact (type 1) and high impact (type 2). Type 1 is a rigid unplasticized PVC having normal impact with optimum chemical resistance. Type 2 has optimum impact resistance and reduced chemical resistance. It is modified by the addition of styrene–butadiene rubber which improves notch toughness and impact strength. PVCs are basically tough and strong, resist water and abrasion, and are excellent electrical insulators. Special tougher types are available to provide high wear resistance.

TABLE 2.71 Compatibility of PSF with Selected Corrodents[a]

Chemical	Maximum temp.		Chemical	Maximum temp.	
	°F	°C		°F	°C
Acetaldehyde	x		Calcium sulfate	200	93
Acetic acid 10%	200	93	Carbon bisulfide	x	
Acetic acid 50%	200	93	Carbon disulphide	x	
Acetic acid 80%	200	93	Carbon tetrachloride	x	
Acetic acid, glacial	200	93	Carbonic acid	200	93
Acetone	x		Cellosolve	x	
Acetyl chloride	x		Chloroacetic acid	x	
Aluminum chloride, aqueous	200	93	Chlorine, liquid	x	
Aluminum chloride, dry	200	93	Chlorobenzene	x	
Aluminum fluoride	200	93	Chloroform	x	
Aluminum oxychloride	150	66	Chlorosulfonic acid	x	
Aluminum sulfate	200	93	Chromic acid 10%	140	60
Ammonia gass	x		Chromic acid 50%	x	
Ammonium carbonate	200	93	Citric acid 15%	100	38
Ammonium chloride 10%	200	93	Citric acid 40%	80	27
Ammonium chloride 50%	200	93	Copper cyanide	200	93
Ammonium chloride, sat.	200	93	Copper sulfate	200	93
Ammonium hydroxide 25%	200	93	Cresol	x	
Ammonium hydroxide, sat.	200	93	Cupric chloride 5%	200	93
Ammonium nitrate	200	93	Cupric chloride 50%	200	93
Ammonium phosphate	200	93	Cyclohexane	200	93
Ammonium sulfate to 40%	200	93	Cyclohexanol	200	93
Amyl acetate	x		Dibutyl phthalate	180	82
Amyl alcohol	200	93	Ethylene glycol	200	93
Aniline	x		Ferric chloride	200	93
Aqua regia 3 : 1	x		Ferric nitrate	200	93
Barium carbonate	200	93	Ferrous chloride	200	93
Barium chloride 10%	200	93	Hydrobromic acid, dil	300	149
Barium hydroxide	200	93	Hydrobromic acid 20%	200	93
Barium sulfate	200	93	Hydrochloric acid 20%	140	60
Benzaldehyde	x		Hydrochloric acid 38%	140	60
Benzene	x		Hydrofluoric acid 30%	80	27
Benzoic acid	x		Ketones, general	x	
Benzyl chloride	x		Lactic acid 25%	200	93
Borax	200	93	Lactic acid, concd	200	93
Boric acid	200	93	Methyl chloride	x	
Bromine gas, moist	200	93	Methyl ethyl ketone	x	
Butyl acetate	x		Nitric acid 5%	x	
Butyl alcohol	200	93	Nitric acid 20%	x	
n-Butylamine	x		Nitric acid 70%	x	
Calcium bisulfite	200	93	Nitric acid, anhydrous	x	
Calcium chloride	200	93	Phosphoric acid 50–80%	80	27
Calcium hypochlorite	200	93	Potassium bromide 30%	200	93
Calcium nitrate	200	93	Sodium carbonate	200	93

(*Continued*)

TABLE 2.71 Continued

Chemical	Maximum temp. °F	°C	Chemical	Maximum temp. °F	°C
Sodium chloride	200	93	Sulfuric acid 50%	300	149
Sodium hydroxide 10%	200	93	Sulfuric acid 70%	x	
Sodium hydroxide 50%	200	93	Sulfuric acid 90%	x	
Sodium hypochlorate 20%	300	149	Sulfuric acid 98%	x	
Sodium hypochlorite concd.	300	149	Sulfuric acid 100%	x	
Sodium sulfite to 50%	200	93	Sulfuric acid, fuming	x	
Stannic chloride	200	93	Sulfurous acid	200	93
Sulfuric acid 10%	300	149	Toluene	x	

[a] The chemicals listed are in the pure state or in a saturated solution unless otherwise indicated. Compatibility is shown to the maximum allowable temperature for which data are available. Incompatibility is shown by an x.
Source: Schweitzer, PA. Corrosion Resistance Tables, 4th ed. Vols. 1–3. New York: Marcel Dekker, 1995.

Generally, PVC will withstand continuous exposure to temperatures ranging up to 103°F/54°C. Flexible types, filaments, and some rigids are unaffected by even higher temperatures. In certain operations, some of the PVCs may be health hazards. These materials are slow burning and certain types are self-extinguishing, but direct contact with an open flame or extreme heat must be avoided.

PVC has a wide range of flexibility. One of its advantages is the way it accepts compounding ingredients. For example, PVC can be plasticized with a variety of plasticizers to produce soft yielding materials with almost any degree of flexibility. Without plasticizers, it is a strong, rigid material, that can be machined, heat formed, solvent cemented, or thermally fused. Cyclohexane, tetrahydrofuran, or dichlorbenzene are solvents used for solvent cementing.

PVC can be glass filled to improve its mechanical properties. It may also be blended with other thermoplasts.

Table 2.72 provided the physical and mechanical properties of PVC while Table 2.73 does likewise for a PVC/acrylic blend.

2. Corrosion Resistance Properties

Type 1 PVC (unplasticized) resists attack by most acids and strong alkalies, gasoline, kerosene, aliphatic alcohols, and hydrocarbons. It is particularly useful in the handling of hydrocholoric acid.

The chemical resistance of type 2 PVC to oxidizing and highly alkaline material is reduced.

TABLE 2.72 Physical and Mechanical Properties of PVC

Property	Type 1	Type 2	20% glass fiber filled
Specific gravity	1.45	1.38	1.43–1.50
Water absorption (24 hr at 73°F/23°C) (%)	0.04	0.05	0.01
Dielectric strength, short-term (V/mil)			
Tensile strength at break (psi)	6,800	5,500	8,600–12,800
Tensile modulus ($\times 10^3$ psi)	500	420	680–970
Elongation at break (%)			2–5
Compressive strength (psi)	10,000	7,900	
Flexural strength (psi)	14,000	11,000	14,200–22,500
Compressive modulus ($\times 10^3$ psi)			
Flexural modulus ($\times 10^3$ psi) at 73°F/23°C			680–970
200°F/93°C			
250°F/121°C			
Izod impact (ft-lb/in. of notch)	0.88	12.15	1.0–1.9
Hardness, Rockwell			R108–119
Coefficient of thermal expansion (10^{-6} in./in./°F)	0.4	0.6	24–36
Thermal conductivity (10^{-4} cal-cm/sec-cm^2°C or Btu/hr/ft^2/°F/in.)	1.33	1.62	
Deflection temperature at 264 psi (°F)			165–174
at 66 psi (°F)	135		
Max. operating temperature (°F/°C)	150/66	140/60	
Limiting oxygen index (%)		43	
Flame spread		15–20	
Underwriters Lab. rating (Sub. 94)		V-O	

PVC may be attacked by aromatics, chlorinated organic compounds, and lacquer solvents.

PVC is resistant to all normal atmospheric polutants including weather and UV degradation.

Refer to Table 2-74 for the compatibility of type 2 PVC with selected corrodents and Table 2.75 for type 1 PVC. Reference 1 provies more detailed listing.

3. Typical Applications

The primary applications for PVC include water, gas, vent, drain, and corrosive chemical piping, electrical conduit, and wire insulation. It is also used as a liner. Reference 2 provides details of PVC piping.

TABLE 2.73 Physical and Mechanical Properties of a PVC/Acrylic Blend

Property	
Specific gravity	1.26–1.35
Water absorption (24 hr at 73°F/23°C) (%)	0.09–0.16
Dielectric strength, short-term (V/mil)	480
Tensile strength at break (psi)	6,400–7,000
Tensile modulus ($\times 10^3$ psi)	340–370
Elongation at break (%)	35–100
Compressive strength (psi)	6,800–8,500
Flexural strength (psi)	10,300–11,000
Compressive modulus ($\times 10^3$ psi)	350–380
Flexural modulus ($\times 10^3$ psi) at 73°F/23°C	
200°F/93°C	
250°F/121°C	
Izod impact (ft-lb/in. of notch)	1–12
Hardness, Rockwell	R106–110
Coefficient of thermal expansion (10^{-6} in./in./°F)	44–79
Thermal conductivity (10^{-4} cal-cm/sec-cm^2°C or Btu/hr/ft^2/°F/in.)	
Deflection temperature at 264 psi (°F)	167–185
at 66 psi (°F)	172–189
Max. operating temperature (°F/°C)	
Limiting oxygen index (%)	
Flame spread	
Underwriters Lab. rating (Sub. 94)	

The automotive industry makes use of PVC for exterior applications as body side moldings. Its ability to be pigmented and its excellent weathering ability to retain color makes it ideal for this application.

BB. Chlorinated Polyvinyl Chloride (CPVC)

When acetylene and hydrochloric acid are reacted to produce polyvinyl chloride, the chlorination is approximately 56.8%. Further chlorination of the PVC to approximately 67% produces CPVC whose chemical structure is

$$
\begin{array}{c}
\;\;\;\; \text{H}\;\;\; \text{H}\;\;\; \text{Cl}\;\; \text{H} \\
\;\;\;\; |\;\;\;\;\; |\;\;\;\;\; |\;\;\;\;\; | \\
-\text{C}-\text{C}-\text{C}-\text{C}- \\
\;\;\;\; |\;\;\;\;\; |\;\;\;\;\; |\;\;\;\;\; | \\
\;\;\;\; \text{H}\;\;\; \text{Cl}\;\;\; \text{H}\;\;\; \text{Cl}
\end{array}
$$

TABLE 2.74 Compatibility of Type 2 PVC with Selected Corrodents[a]

Chemical	Maximum temp.		Chemical	Maximum temp.	
	°F	°C		°F	°C
Acetaldehyde	x	x	Barium chloride	140	60
Acetamide	x	x	Barium hydroxide	140	60
Acetic acid 10%	100	38	Barium sulfate	140	60
Acetic acid 50%	90	32	Barium sulfide	140	60
Acetic acid 80%	x	x	Benzaldehyde	x	x
Acetic acid, glacial	x	x	Benzene	x	x
Acetic anhydride	x	x	Benzenesulfonic acid 10%	140	60
Acetone	x	x	Benzoic acid	140	60
Acetyl chloride	x	x	Benzyl alcohol	x	x
Acrylic acid	x	x	Borax	140	60
Acrylonitrile	x	x	Boric acid	140	60
Adipic acid	140	60	Bromine gas, dry	x	x
Allyl alcohol	90	32	Bromine gas, moist	x	x
Allyl chloride	x	x	Bromine, liquid	x	x
Alum	140	60	Butadiene	60	16
Aluminum acetate	100	38	Butyl acetate	x	x
Aluminum chloride, aqueous	140	60	Butyl alcohol	x	x
Aluminum fluoride	140	60	n-Butylamine	x	x
Aluminum hydroxide	140	60	Butyric acid	x	x
Aluminum nitrate	140	60	Calcium bisulfide	140	60
Aluminum oxychloride	140	60	Calcium bisulfite	140	60
Aluminum sulfate	140	60	Calcium carbonate	140	60
Ammonia gas	140	60	Calcium chlorate	140	60
Ammonium bifluoride	90	32	Calcium chloride	140	60
Ammonium carbonate	140	60	Calcium hydroxide 10%	140	60
Ammonium chloride 10%	140	60	Calcium hydroxide, sat.	140	60
Ammonium chloride 50%	140	60	Calcium hypochlorite	140	60
Ammonium chloride, sat.	140	60	Calcium nitrate	140	60
Ammonium fluoride 10%	90	32	Calcium oxide	140	60
Ammonium fluoride 25%	90	32	Calcium sulfate	140	60
Ammonium hydroxide 25%	140	60	Calcium bisulfide	x	x
Ammonium hydroxide, sat.	140	60	Carbon dioxide, dry	140	60
Ammonium nitrate	140	60	Carbon dioxide, wet	140	60
Ammonium persulfate	140	60	Carbon disulfide	x	x
Ammonium phosphate	140	60	Carbon monoxide	140	60
Ammonium sulfate 10–40%	140	60	Carbon tetrachloride	x	x
Ammonium sulfide	140	60	Carbonic acid	140	60
Amyl acetate	x	x	Cellosolve	x	x
Amyl alcohol	x	x	Chloroacetic acid	105	40
Amyl chloride	x	x	Chlorine gas, dry	140	60
Aniline	x	x	Chlorine gas, wet	x	x
Antimony trichloride	140	60	Chlorine, liquid	x	x
Aqua regia 3 : 1	x	x	Chlorobenzene	x	x
Barium carbonate	140	60	Chloroform	x	x

(Continued)

TABLE 2.74 Continued

Chemical	Maximum temp. °F	Maximum temp. °C	Chemical	Maximum temp. °F	Maximum temp. °C
Chlorosulfonic acid	60	16	Muriatic acid	140	60
Chromic acid 10%	140	60	Nitric acid 5%	100	38
Chromic acid 50%	x	x	Nitric acid 20%	140	60
Citric acid 15%	140	60	Nitric acid 70%	70	140
Citric acid, concd	140	60	Nitric acid, anhydrous	x	x
Copper carbonate	140	60	Nitirc acid, concd	60	16
Copper chloride	140	60	Oleum	x	x
Copper cyanide	140	60	Perchloric acid 10%	60	16
Copper sulfate	140	60	Perchloric acid 70%	60	16
Cresol	x	x	Phenol	x	x
Cyclohexanol	x	x	Phosphoric acid 50–80%	140	60
Dichloroacetic acid	120	49	Picric acid	x	x
Dichloroethane			Potassium bromide 30%	140	60
(ethylene dichloride)	x	x	Salicylic acid	x	x
Ethylene glycol	140	60	Silver bromide 10%	105	40
Ferric chloride	140	60	Sodium carbonate	140	60
Ferric nitrate 10–50%	140	60	Sodium chloride	140	60
Ferrous chloride	140	60	Sodium hydroxide 10%	140	60
Ferrous nitrate	140	60	Sodium hydroxide 50%	140	60
Fluorine gas, dry	x	x	Sodium hydroxide, concd	140	60
Fluorine gas, moist	x	x	Sodium hypochlorite 20%	140	60
Hydrobromic acid, dilute	140	60	Sodium hypochlorite concd	140	60
Hydrobromic acid 20%	140	60	Sodium sulfide to 50%	140	60
Hydrobromic acid 50%	140	60	Stannic chloride	140	60
Hydrochloric acid 20%	140	60	Stannous chloride	140	60
Hydrochloric acid 38%	140	60	Sulfuric acid 10%	140	60
Hydrocyanic acid 10%	140	60	Sulfuric acid 50%	140	60
Hydrofluoric acid 30%	120	49	Sulfuric acid 70%	140	60
Hydrofluoric acid 70%	68	20	Sulfuric acid 90%	x	x
Hypochlorous acid	140	60	Sulfuric acid 98%	x	x
Ketones, general	x	x	Sulfuric acid 100%	x	x
Lactic acid 25%	140	60	Sulfuric acid, fuming	x	x
Lactic acid, concd	80	27	Sulfurous acid	140	60
Magnesium chloride	140	60	Thionyl chloride	x	x
Malic acid	140	60	Toluene	x	x
Methyl chloride	x	x	Trichloroacetic acid	x	x
Methyl ethyl ketone	x	x	White liquor	140	60
Methyl isobutyl ketone	x	x	Zinc chloride	140	60

[a] The chemicals listed are in the pure state or in a saturated solution unless otherwise indicated. Compatibility is shown to the maximum allowable temperature for which data are available. Incompatibility is shown by an x. A blank space indicates that data are unavailable.
Source: Schweitzer, PA. Corrosion Resistance Tables, 4th ed. Vols. 1–3. New York: Marcel Dekker, 1995.

TABLE 2.75 Compatibility of Type 1 PVC with Selected Corrodents[a]

Chemical	Maximum temp. °F	Maximum temp. °C	Chemical	Maximum temp. °F	Maximum temp. °C
Acetaldehyde	x		Aqua regia 3:1	x	
Acetamide	x		Barium carbonate	140	60
Acetic acid 10%	140	60	Barium chloride	140	60
Acetic acid 50%	140	60	Barium hydroxide	140	60
Acetic acid 80%	140	60	Barium sulfate	140	60
Acetic acid, glacial	130	54	Barium sulfide	140	60
Acetic anhydride	x		Bensaldehyde	x	
Acetone	x		Benzene	x	
Acetyl chloride	x		Benzenesulfonic acid	140	60
Acrylic acid	x		Benzoic acid	140	60
Acrylonitrile	x		Benzyl alcohol	x	
Adipic acid	140	60	Benzyl chloride	x	
Allyl alcohol	90	32	Borax	140	60
Allyl chloride	x		Boric acid	140	60
Alum	140	60	Bromine gas, dry	x	
Aluminum acetate	100	38	Bromine gas, moist	x	
Aluminum chloride, aqueous	140	60	Bromine, liquid	x	
Aluminum chloride, dry	140	60	Butadiene	140	60
Aluminum fluoride	140	60	Butyl acetate	x	
Aluminum hydroxide	140	60	Butyl alcohol	x	
Aluminum nitrate	140	60	n-Butylamine	x	
Aluminum oxychloride	140	60	Butyric acid	x	
Aluminum sulfate	140	60	Calcium bisulfide	140	60
Ammonia gas	140	60	Calcium bisulfite	140	60
Ammonium bifluoride	90	32	Calcium carbonate	140	60
Ammonium carbonate	140	60	Calcium chlorate	140	60
Ammonium chloride 10%	140	60	Calcium chloride	140	60
Ammonium chloride 50%	140	60	Calcium hydroxide, sat.	140	60
Ammonium chloride, sat.	140	60	Calcium hypochlorite	140	60
Ammonium fluoride 1%	140	60	Calcium nitrate	140	60
Ammonium fluoride 25%	140	60	Calcium oxide	140	60
Ammonium hydroxide 25%	140	60	Calcium sulfate	140	60
Ammonium hydroxide, sat.	140	60	Caprylic acid	120	49
Ammonium nitrate	140	60	Carbonm bisulfide	x	
Ammonium persulfate	140	60	Carbon dioxide, dry	140	60
Ammonium phosphate	140	60	Carbon dioxide, wet	140	60
Ammonium sulfate 10–40%	140	60	Carbon disulfide	x	
Ammonium sulfide	140	60	Carbon monoxide	140	60
Ammonium sulfite	120	49	Carbon tetrachloride	x	
Amyl acetate	x		Carbonic acid	140	60
Amyl alcohol	140	60	Cellosolve	x	
Amyl chloride	x		Chloroacetic acid, 50% water	x	
Aniline	x		Chloroacetic acid	140	60
Antimony trichloride	140	60	Chlorine gas, dry	140	60

(Continued)

TABLE 2.75   Continued

Chemical	Maximum temp.		Chemical	Maximum temp.	
	°F	°C		°F	°C
Chlorine gas, wet	x		Malic acid	140	60
Chlorine, liquid	x		Manganese chloride	90	32
Chlorobenzene	x		Methyl chloride	x	
Chloroform	x		Methyl ethyl ketone	x	
Chlorosulfonic acid	x		Methyl isobutyl ketone	x	
Chromic acid 10%	140	60	Muriatic acid	160	71
Chromic acid 50%	x		Nitric acid 5%	140	60
Chromyl chloride	120	49	Nitric acid 20%	140	60
Citric acid 15%	140	60	Nitric acid 70%	140	60
Citric acid, concd	140	60	Nitric acid, anhydrous	x	
Copper acetate	80	27	Nitrous acid, concd	80	27
Copper carbonate	140	60	Oleum	x	
Copper chloride	140	60	Perchloric acid 10%	140	60
Copper cyanide	140	60	Perchloric acid 70%	x	
Copper sulfate	140	60	Phenol	x	
Cresol	120	49	Phosphoric acid 50–80%	140	60
Cupric chloride	140	60	Picric acid	x	
Cupric chloride 50%	150	66	Potassium bromide 30%	140	60
Cyclohexane	80	27	Salicylic acid	140	60
Cyclohexanol	x		Silver bromide 10%	140	60
Dibutyl phthalate	80	27	Sodium carbonate	140	60
Dichloroacetic acid	100	38	Sodium chloride	140	60
Dichloroethane	x		Sodium hydroxide 10%	140	60
Ethylene glycol	140	60	Sodium hydroxide 50%	140	60
Ferric chloride	140	60	Sodium hydroxide, concd	140	60
Ferric chloride 50% water	140	60	Sodiium hypochlorite 20%	140	60
Ferric nitrate 10–50%	140	60	Sodium hypochlorite, concd	140	60
Fluorine gas, dry	x		Sodium sulfide to 50%	140	60
Fluorine gas, moist	x		Stannic chloride	140	60
Hydrobromic acid dil	140	60	Stannous chloride	140	60
Hydrobromic acid 20%	140	60	Sulfuric acid 10%	140	60
Hydrobromic acid 50%	140	60	Sulfuric acid 50%	140	60
Hydrochloric acid 20%	140	60	Sulfuric acid 70%	140	60
Hydrochloric acid 38%	140	60	Sulfuric acid 90%	140	60
Hydrocyanic acid	140	60	Sulfuric acid 98%	x	
Hydrofluoric acid 30%	130	54	Sulfuric acid 100%	x	
Hydrofluoric acid 70%	x		Sulfuric acid, fuming	x	
Hypochlorous acid	140	60	Sulfurous acid	140	60
Iodine solution 10%	100	38	Thionyl chloride	x	
Ketones, general	x		Toluene	x	
Lactic acid 25%	140	60	Trichloracetic acid	90	32
Lactic acid, concd	80	27	White liquor	140	60
Magnesium chloride	140	60	Zinc chloride	140	60

[a] The chemicals listed are in the pure state or in a saturated solution unless otherwise indicated. Compatibility is shown to the maximum allowable temperature for which data are available. Incompatibility is shown by an x.
Source: Schweitzer, PA. Corrosion Resistance Tables, 4th ed. Vols. 1–3. New York: Marcel Dekker, 1995.

1. Physical and Mechanical Properties

The additional chlorine increases the heat deflection temperature and permits a higher allowable operating temperature. While PVC is limited to a maximum operating temperature of 140°F/60°C, CPVC has a maximum operating temperature of 190°F/82°C. There are special formulations of CPVC that permit operating temperatures of 200°F/93°C.

Because of the greater quantity of chlorine in CPVC versus PVC, its flame resistance is improved. While many ordinary combustibles such as wood have a flash ignition temperature of 500°F/260°C, chlorinated polyvinyl chloride has a flash ignition temperature of 900°F/482°C. CPVC will not sustain burning, and because of its very high limiting oxygen index (60%) it must be forced to burn. It will not burn unless a flame is constantly applied.

As with PVC, CPVC is a strong rigid material that can be machined, heat formed, solvent cemented, or thermally fused. Cyclohexane, tetrahydrofuran, or dichlorobenzene are solvents used for solvent cementing.

Refer to Table 2.76 for the physical and mechanical properties of CPVC.

2. Corrosion Resistance Properties

There are many similarities in the chemical resistance of CPVC and PVC. However, care must be exercised since there are differences. Overall, the corrosion resistance of CPVC is somewhat inferior to that of PVC.

In general, CPVC is inert to most mineral acids, bases, salts, and paraffinic hydrocarbons.

CPVC is not recommended for use with most polar organic materials including various solvents, chlorinated or aromatic hydrocarbons, esters, and ketones.

It is ideally suited for handling hot water and/or steam condensate.

Refer to Table 2.77 for the compatibility of CPVC with selected corrodents. Reference 1 provides a more detailed listing.

3. Typical Applications

The primary applications of CPVC include hot water and/or steam condensate piping, corrosive chemical piping at elevated temperatures, valves, fume ducts, internal column packings, and miscellaneous fabricated items. Reference 2 provides detailed information on CPVC piping.

CC. Chlorinated Polyether (CPE)

Chlorinated Polyether is sold under the trade name of Penton.

1. Physical and Mechanical Properties

At room temperature its physical properties are average. However, the values of these physical properties decrease only slightly as the temperature increases. It

TABLE 2.76 Physical and Mechanical Properties of CPVC

Property	
Specific gravity	1.49–1.58
Water absorption (24 hr at 73°F/23°C) (%)	0.02–0.15
Dielectric strength, short-term (V/mil)	600–625
Tensile strength at break (psi)	6,800–9,000
Tensile modulus ($\times 10^3$ psi)	341–475
Elongation at break (%)	4–100
Compressive strength (psi)	9,000–22,000
Flexural strength (psi)	14,500–17,000
Compressive modulus ($\times 10^3$ psi)	335–600
Flexural modulus ($\times 10^3$ psi) at 73°F/23°C	380–450
200°F/93°C	
250°F/121°C	
Izod impact (ft-lb/in. of notch)	1.0–5.6
Hardness, Rockwell	
Coefficient of thermal expansion (10^{-6} in./in./°F)	62–78
Thermal conductivity (10^{-4} cal-cm/sec-cm^2°C	3.3
or Btu/hr/ft^2/°F/in.)	0.95
Deflection temperature at 264 psi (°F)	202–234
at 66 psi (°F)	215–247
Max. operating temperature (°F/°C)	200/93
Limiting oxygen index (%)	60
Flame spread	15
Underwriters Lab. rating (Sub. 94)	V-O, 5VA, 5VB

has a maximum operating temperature of 225°F/107°C. Penton can be joined by thermal fusion. Refer to Table 2.78 for its physical and mechanical properties.

2. Corrosion Resistant Properties

CPE is resistant to most acids and bases, oxidizing agents, and common solvents. It has the advantage of being able to resist acids at elevated temperatures. It is not resistant to nitric acid above 10% concentration. Refer to Table 2.79 for the compatibility of CPE with selected corrodents. Reference 1 provides a more detailed listing.

3. Typical Applications

CPE finds application as bearing retainers, tanks, tank linings, and in process equipment.

TABLE 2.77 Compatibility of CPVC with Selected Corrodents[a]

Chemical	Maximum temp. °F	Maximum temp. °C	Chemical	Maximum temp. °F	Maximum temp. °C
Acetaldehyde	x	x	Barium hydroxide	180	82
Acetic acid 10%	90	32	Barium sulfate	180	82
Acetic acid 50%	x	x	Barium sulfide	180	82
Acetic acid 80%	x	x	Benzaldehyde	x	x
Acetic acid, glacial	x	x	Benzene	x	x
Acetic anhydride	x	x	Benzenesulfonic acid 10%	180	82
Acetone	x	x	Benzoic acid	200	93
Acetyl chloride	x	x	Benzyl alcohol	x	x
Acrylic acid	x	x	Benzyl chloride	x	x
Acrylonitrile	x	x	Borax	200	93
Adipic acid	200	93	Boric acid	210	99
Allyl alcholol 96%	200	90	Bromine gas, dry	x	x
Allyl chloride	x	x	Bromine gas, moist	x	x
Alum	200	93	Bromine, liquid	x	x
Aluminum acetate	100	38	Butadiene	150	66
Aluminum chloride, aqueous	200	93	Butyl acetate	x	x
Aluminum chloride, dry	180	82	Butyl alcohol	140	60
Aluminum fluoride	200	93	n-Butylamine	x	x
Aluminum hydroxide	200	93	Butyric acid	140	60
Aluminum nitrate	200	93	Calcium bisulfide	180	82
Aluminum oxychloride	200	93	Calcium bisulfite	210	99
Aluminum sulfate	200	93	Calcium carbonate	210	99
Ammonia gas, dry	200	93	Calcium chlorate	180	82
Ammonium bifluoride	140	60	Calcium chloride	180	82
Ammonium carbonate	200	93	Calcium hydroxide 10%	170	77
Ammonium chloride 10%	180	82	Calcium hydroxide, sat.	210	99
Ammonium chloride 50%	180	82	Calcium hypochlorite	200	93
Ammonium chloride, sat.	200	93	Calcium nitrate	180	82
Ammonium fluoride 10%	200	93	Calcium oxide	180	82
Ammonium fluoride 25%	200	93	Calcium sulfate	180	82
Ammonium hydroxide 25%	x	x	Caprylic acid	180	82
Ammonium hydroxide, sat.	x	x	Carbon bisulfide	x	x
Ammonium nitrate	200	93	Carbon dioxide, dry	210	99
Ammonium persulfate	200	93	Carbon dioxide, wet	160	71
Ammonium phosphate	200	93	Carbon disulfide	x	x
Ammonium sulfate 10–40%	200	93	Carbon monoxide	210	99
Ammonium sulfide	200	93	Carbon tetrachloride	x	x
Ammonium sulfite	160	71	Carbonic acid	180	82
Amyl acetate	x	x	Cellosolve	180	82
Amyl alcohol	130	54	Chloroacetic acid, 50% water	100	38
Amyl chloride	x	x	Chloroacetic acid	x	x
Aniline	x	x	Chlorine gas, dry	140	60
Antimony trichloride	200	93	Chlorine gas, wet	x	x
Aqua regia 3:1	80	27	Chlorine, liquid	x	x
Barium carbonate	200	93	Chlorobenzene	x	x
Barium chloride	180	82	Chloroform	x	x

(Continued)

TABLE 2.77 Continued

Chemical	Maximum temp. °F	°C	Chemical	Maximum temp. °F	°C
Chlorosulfonic acid	x	x	Manganese chloride	180	82
Chromic acid 10%	210	99	Methyl chloride	x	x
Chromic acid 50%	210	99	Methyl ethyl ketone	x	x
Chromyl chloride	180	82	Methyl isobutyl ketone	x	x
Citric acid 15%	180	82	Muriatic acid	170	77
Citric acid, concd	180	82	Nitric acid 5%	180	82
Copper acetate	80	27	Nitric acid 20%	160	71
Copper carbonate	180	82	Nitric acid 70%	180	82
Copper chloride	210	99	Nitric acid, anhydrous	x	x
Copper cyanide	180	82	Nitrous acid, concd	80	27
Copper sulfate	210	99	Oleum	x	x
Cresol	x	x	Perchloric acid 10%	180	82
Cupric chloride 5%	180	82	Perchloric acid 70%	180	82
Cupric chloride 50%	180	82	Phenol	140	60
Cyclohexane	x	x	Phosphoric acid 50–80%	180	82
Cyclohexanol	x	x	Picric acid	x	x
Dichloroacetic acid, 20%	100	38	Potassium bromide 30%	180	82
Dichloroethane	x	x	Salicylic acid	x	x
(ethylene dichloride)			Silver bromide 10%	170	77
Ethylene glycol	210	99	Sodium carbonate	210	99
Ferric chloride	210	99	Sodium chloride	210	99
Ferric chloride 50% in water	180	82	Sodium hydroxide 10%	190	88
Ferric nitrate 10–50%	180	82	Sodium hydroxide 50%	180	82
Ferrous chloride	210	99	Sodium hydroxide, concd	190	88
Ferrous nitrate	180	82	Sodium hypochlorite 20%	190	88
Fluorine gas, dry	x	x	Sodium hypochlorite, concd	180	82
Fluorine gas, moist	80	27	Sodium sulfide to 50%	180	82
Hydrobromic acid, dil	130	54	Stannic chloride	180	82
Hydrobromic acid 20%	180	82	Stannous chloride	180	82
Hydrobromic acid 50%	190	88	Sulfuric acid 10%	180	82
Hydrochloric acid 20%	180	82	Sulfuric acid 50%	180	82
Hydrochloric acid 38%	170	77	Sulfuric acid 70%	200	93
Hydrocyanic acid 10%	80	27	Sulfuric acid 90%	x	x
Hydrofluoric acid 30%	x	x	Sulfuric acid 98%	x	x
Hydrofluoric acid 70%	90	32	Sulfuric acid 100%	x	x
Hydrofluoric acid 100%	x	x	Sulfuric acid, fuming	x	x
Hypochlorous acid	180	82	Sulfurous acid	180	82
Ketones, general	x	x	Thionyl chloride	x	x
Lactic acid 25%	180	82	Toluene	x	x
Lactic acid, concd	100	38	Trichloroacetic acid, 20%	140	60
Magnesium chloride	230	110	White liquor	180	82
Malic acid	180	82	Zinc chloride	180	82

[a] The chemicals listed are in the pure state or in a saturated solution unless otherwise indicated. Compatibility is shown to the maximum allowable temperature for which data are available. Incompatibility is shown by an x.

Source: Schweitzer, PA. Corrosion Resistance Tables, 4th ed. Vols. 1–3. New York: Marcel Dekker, 1995.

TABLE 2.78 Physical and Mechanical Properties of CPE

Property	
Specific gravity	1.4
Water absorption (24 hr at 73°F/23°C) (%)	0.01
Dielectric strength, short-term (V/mil)	
Tensile strength at break (psi)	6,000
Tensile modulus ($\times 10^3$ psi)	160
Elongation at break (%)	
Compressive strength (psi)	9,000
Flexural strength (psi)	5,000
Compressive modulus ($\times 10^3$ psi)	
Flexural modulus ($\times 10^3$ psi) at 73°F/23°C	
200°F/93°C	
250°F/121°C	
Izod impact (ft-lb/in. of notch)	0.4
Hardness, Rockwell	
Coefficient of thermal expansion (10^{-6} in./in./°F)	44
Thermal conductivity (10^{-4} cal-cm/sec-cm^2°C	
or Btu/hr/ft^2/°F/in.)	0.91
Deflection temperature at 264 psi (°F)	184
at 66 psi (°F)	
Max. operating temperature (°F/°C)	225/107
Limiting oxygen index (%)	
Flame spread	
Underwriters Lab. rating (Sub. 94)	

DD. Polyacrylonitrile (PAN)

PAN is a member of the polyolefin family. In general, it is somewhat similar to the other polyolefins such as polyethylene and polypropylene in terms of appearance, general chemical characteristics, and electrical properties. PAN has the following structural formula:

$$\begin{array}{c} \text{H} \quad \text{H} \\ | \quad\quad | \\ -\text{C}-\text{C}- \\ | \quad\quad | \\ \text{H} \quad \text{C}{=}\text{N} \end{array}$$

1. Physical and Mechanical Properties

The differences between PAN and the other olefins are more notably in physical and thermal stability properties. Basically, the polyolefins are all waxlike in appearance and extremely inert chemically, and they exhibit decreases in physical strength at somewhat lower temperatures than the high-performance engineering

TABLE 2.79 Compatibility of CPE with Selected Corrodents[a]

Chemical	Maximum temp.		Chemical	Maximum temp.	
	°F	°C		°F	°C
Acetaldehyde	130	54	Borax	140	60
Acetic acid 10%	250	121	Boric acid	250	121
Acetic acid 50%	250	121	Bromine	140	60
Acetic acid 80%	250	121	Bromine, liquid	x	
Acetic acid, glacial	250	121	Butadiene	250	121
Acetic anhydride	150	66	Butyl acetate	130	54
Acetone	90	32	Butyl alcohol	220	104
Acetyl chloride	x		Butyric acid	230	110
Adipic acid	250	121	Calcium bisulfate	250	121
Allyl alcohol	250	121	Calcium cabonate	250	121
Allyl chloride	90	32	Calcium chlorate	150	66
Alum	250	121	Calcium chloride	250	121
Aluminum chloride, aqueous	250	121	Calcium hydroxide	250	121
Aluminum fluoride	250	121	Calcium hypochlorite	180	82
Aluminum hydroxide	250	121	Calcium nitrate	250	121
Aluminum oxychloride	220	104	Calcium oxide	250	121
Aluminum sulfate	250	121	Calcium sulfate	250	121
Ammonia gas, dry	220	104	Carbon bisulfide	x	
Ammonium bifluoride	220	104	Carbon dioxide, dry	250	121
Ammonium carbonate	250	121	Carbon dioxide, wet	250	121
Ammonium chloride, sat.	250	121	Carbon disulfide	x	
Ammonium fluoride 10%	250	121	Carbon monoxide	250	121
Ammonium fluoride 25%	250	121	Carbon tetrachloride	200	93
Ammonium hydroxide 25%	250	121	Carbonic acid	250	121
Ammonium hydroxide, sat.	250	121	Cellosolve	220	104
Ammonium nitrate	250	121	Chloroacetic acid	220	104
Ammonium persulfate	180	82	Chlorine gas, dry	100	38
Ammonium phosphate	250	121	Chlorine gas, wet	80	27
Ammonium sulfate 10–40%	250	121	Chlorine, liquid	x	
Ammonium sulfide	250	121	Chlorobenzene	150	66
Amyl acetate	180	82	Chloroform	80	27
Amyl alcohol	220	104	Chlorosulfonic acid	x	
Amyl chloride	220	104	Chromic acid 10%	250	121
Aniline	150	66	Chromic acid 50%	250	121
Antimony trichloride	220	104	Citric acid 15%	250	121
Aqua regia 3:1	80	27	Citric acid 25%	250	121
Barium carbonate	250	121	Copper chloride	250	121
Barium chloride	250	121	Copper cyanide	250	121
Barium hydroxide	250	121	Copper sulfate	250	121
Barium sulfate	250	121	Cresol	140	60
Barium sulfide	220	104	Cyclohexane	220	104
Benzaldehyde	80	27	Cyclohexanol	220	104
Benzenesulfonic acid 10%	220	104	Dichloroacetic acid		
Benzoic acid	250	121	Dichloroethane	x	
Benzyl alcohol			Ethylene glycol	220	104
Benzyl chloride	80	27	Ferric chloride	250	121

TABLE 2.79 Continued

Chemical	Maximum temp. °F	°C	Chemical	Maximum temp. °F	°C
Ferric nitrate	250	121	Phosphoric acid 50–80%	250	121
Ferous chloride	250	121	Picric acid	150	66
Hydrobromic acid 20%	250	121	Potassium bromide 30%	250	121
Hydrochloric acid 20%	250	121	Sodium carbonate	250	121
Hydrochloric acid 38%	300	149	Sodium chloride	250	121
Hydrochloric acid 70%			Sodium hydroxide 10%	180	82
Hydrochloric acid 100%			Sodium hydroxide, concd	180	82
Hypochlorous acid	150	66	Sodium hypochlorite 20%	180	82
Ketones, general	x		Sodium hypochlorite, concd	200	93
Lactic acid 25%	250	121	Sodium sulfide to 50%	230	110
Magnesium chloride	250	121	Stannic chloride	250	121
Malic acid	250	121	Stannous chloride	250	121
Methyl chloride	230	110	Sulfuric acid 10%	230	110
Methyl ethyl ketone	90	32	Sulfuric acid 50%	230	110
Methyl isobutyl ketone	80	27	Sulfuric acid 70%	230	110
Muriatic acid			Sulfuric acid 90%	130	54
Nitric acid 5%	180	82	Sulfuric acid 98%	x	
Nitric acid 20%	80	27	Sulfuric acid 100%	x	
Nitric acid 70%	80	27	Sulfuric acid, fuming		
Nitric acid, anhydrous	x		Sulfurous acid	200	93
Oleum			Thionyl chloride	x	
Perchloric acid 10%	140	60	Toluene	200	93
Perchloric acid 70%	x		White liquor	200	93
Phenol			Zinc chloride	250	121
Phosgene gas, dry					

[a] The chemicals listed are in the pure state or in a saturated solution unless otherwise indicated. Compatibility is shown to the maximum allowable temperature for which data are available. Incompatibility is shown by an x. A blank space indicates that data are unavailable. *Source:* Schweitzer, PA. Corrosion Resistance Tables, 4th ed. Vols. 1–3. New York: Marcel Dekker, 1995.

thermoplasts. Refer to Table 2.80 for the physical and mechanical properties of PAN.

Polyacrylonitriles cannot be joined by solvent cementing but can be joined by thermal fusion and adhesive bonding. Prior to adhesive bonding, the surface must be etched with a sodium dichromate–sulfuric acid solution at elevated temperatures. Epoxy and nitrile–phenolic adhesives are used.

2. Corrosion Resistance Properties

PAN has corrosion resistance properties similar to those of polypropylene and polyethylene.

TABLE 2.80   Physical and Mechanical Properties of High-Impact PAN

Property	
Specific gravity	1.11
Water absorption (24 hr at 73°F/23°C) (%)	
Dielectric strength, short-term (V/mil)	220–240
Tensile strength at break (psi)	
Tensile modulus ($\times 10^3$ psi)	450–500
Elongation at break (%)	3–4
Compressive strength (psi)	11,500
Flexural strength (psi)	13,700
Compressive modulus ($\times 10^3$ psi)	
Flexural modulus ($\times 10^3$ psi) at 73°F/23°C	390
200°F/93°C	
250°F/121°C	
Izod impact (ft-lb/in. of notch)	9.0
Hardness, Rockwell	M45
Coefficient of thermal expansion (10^{-6} in./in./°F)	66
Thermal conductivity (10^{-4} cal-cm/sec-cm^2°C or Btu/hr/ft^2/°F/in.)	6.1
Deflection temperature at 264 psi (°F)	151
at 66 psi (°F)	160
Max. operating temperature (°F/°C)	
Limiting oxygen index (%)	
Flame spread	
Underwriters Lab. rating (Sub. 94)	

3. Typical Applications

Polyacrylonitrile is used to produce molded appliance parts, automotive parts, garden hose, vending machine tubing, chemical apparatus, typewriter cases, bags, luggage shells, and trim.

EE.   Polyurethane (PUR)

Polyurethanes are produced from either polyethers or polyesters. Those produced from polyether are more resistant to hydrolysis and have higher resilience, good energy-absorption characteristics, good hysteresis characteristics, and good all-around chemical resistance. The polyester-based urethanes are generally stiffer and will have higher compression and tensile moduli, higher tear strength and cut resistance, higher operating temperature, lower compression set, optimum abrasion resistance, and good fuel and oil resistance. Refer to Figure. 2.2 for the structural formula.

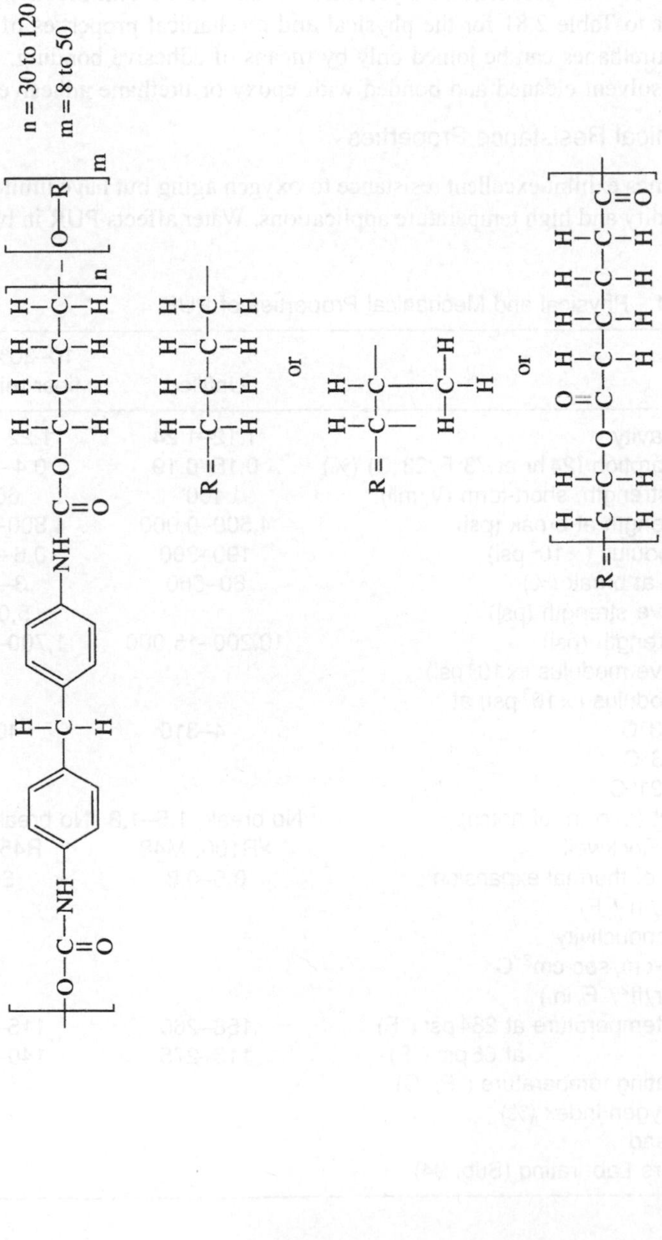

FIGURE 2 Chemical structure of PUR.

1. Physical and Mechanical Properties

Polyurethanes can be formulated to produce a range of materials from elastomers as soft as a Shore A of 5 to tough solids with a Shore D of 90. PURs can have extremely high abrasion resistance and tear strength, excellent shock absorption, and good electrical properties. Polyurethanes can also be reinforced with glass fiber. Refer to Table 2.81 for the physical and mechanical properties of PUR.

Polyurethanes can be joined only by means of adhesive bonding. Surfaces should be solvent cleaned and bonded with epoxy or urethane adhesives.

2. Chemical Resistance Properties

Polyurethanes exhibit excellent resistance to oxygen aging but have limited life in high-humidity and high temperature applications. Water affects PUR in two ways,

TABLE 2.81 Physical and Mechanical Properties of PUR

Property	Unfilled	10–20% glass fiber reinforced
Specific gravity	1.12–1.24	1.22–1.36
Water absorption (24 hr at 73°F/23°C) (%)	0.15–0.19	0.4–0.55
Dielectric strength, short-term (V/mil)	400	600
Tensile strength at break (psi)	4,500–9,000	4,800–7,500
Tensile modulus ($\times 10^3$ psi)	190–300	0.6–1.40
Elongation at break (%)	60–560	3–70
Compressive strength (psi)		5,000
Flexural strength (psi)	10,200–15,000	1,700–6,200
Compressive modulus ($\times 10^3$ psi)		
Flexural modulus ($\times 10^3$ psi) at		
73°F/23°C	4–310	40–90
200°F/93°C		
250°F/121°C		
Izod impact (ft-lb/in. of notch)	No break, 1.5–1.8	No break, 10–14
Hardness, Rockwell	>R100, M48	R45–55
Coefficient of thermal expansion (10^{-6} in./in./°F)	0.5–0.8	34
Thermal conductivity (10^{-4} cal-cm/sec-cm^2°C or Btu/hr/ft^2/°F/in.)		
Deflection temperature at 264 psi (°F)	158–260	115–130
at 66 psi (°F)	115–275	140–145
Max. operating temperature (°F/°C)		
Limiting oxygen index (%)		
Flame spread		
Underwriters Lab. rating (Sub. 94)		

temporary plasticization and permanent degradation. Moisture plasticization results in a slight reduction in hardness and tensile strength. When the absorbed water is removed, the original properties are restored. Hydrolytic degradation causes a permanent reduction in physical and electrical properties.

Since polyurethane is a polar material, it is resistant to nonpolar organic fluids such as oils, fuels, and greases but will be readily attacked and even dissolved by polar organic fluids such as dimethylformamide, and dimethyl sulfoxide.

Table 2.82 provides the compatibility of PUR with selected corrodents. Reference 1 provides a more detailed listing.

3. Typical Applications

Polyurethanes are used where the properties of good abrasion resistance and low coefficient of friction are required. They are also used for high-impact car panels and other parts. Polyurethanes are also used for toys, gears, bushings, and pulleys, golf balls, ski goggle frames, shoe components, and ski boots. The medical field makes use of polyurethane for diagnostic devices, tubing, and catheters.

FF. Polybutylene Terephthalate (PBT)

PBT is also known as a thermoplastic polyester. These thermoplastic polyesters are highly crystalline with a melting point of approximately 430°F/221°C. It has a structural formula as follows:

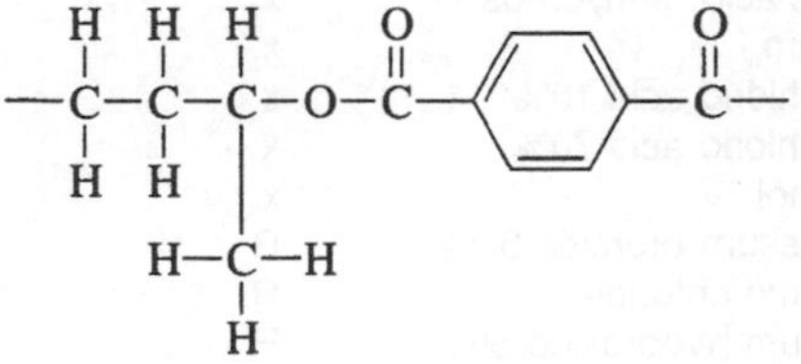

1. Physical and Mechanical Properties

PBT is fairly translucent in thin molded sections and opaque in thick sections but can be extruded into a thin transparent film. It is available in both unreinforced and reinforced formulations. Unreinforced resins generally

1. Are hard, strong, and extremely tough
2. Have good chemical resistance, very low moisture absorption, and resistance to cold flow
3. Have high abrasion resistance and a low coefficient of friction
4. Have good stress crack and fatigue resistance

TABLE 2.82 Compatibility of PUR with Selected Corrodents[a]

Chemical		Chemical	
Acetaldehyde	x	Carbonic acid	R
Acetamine	x	Chloroacetic acid	x
Acetic acid 10%	x	Chlorine gas, dry	x
Acetic acid 50%	x	Chlorine gas, wet	x
Acetic acid 80%	x	Chlorobenzene	x
Acetic acid, glacial	x	Chloroform	x
Acetic anhydride	x	Chlorosulfonic acid	x
Acetone	x	Chromic acid 10%	x
Acetyl chloride	x	Chromic acid 50%	x
Ammonium carbonate	R	Copper chloride	R
Ammonium chloride 10%	R	Copper cyanide	R
Ammonium chloride 50%	R	Copper sulfate	R
Ammonium chloride, sat.	R	Cyclohexane	R
Ammonium hydroxide 25%	R	Cresol	x
Ammonium hydroxide, sat.	R	Ethylene glycol	R
Ammonium persulfate	x	Ferric chloride	R
Amyl acetate	x	Ferric chloride 50%	R
Amyl alcohol	x	Ferric nitrate 10–50%	R
Aniline	x	Hydrochloric acid 20%	R
Aqua regia 3 : 1	x	Hydrdochloric acid 38%	x
Barium chloride	R	Magnesium chloride	R
Barium hydroxide	R	Methyl chloride	x
Barium sulfide	R	Methyl ethyl ketone	x
Benzaldehyde	x	Methyl isobutyl ketone	x
Benzene	x	Nitric acid 5%	x
Benzenesulfonic acid	x	Nitric acid 20%	x
Benzoic acid	x	Nitric acid 70%	x
Benzyl alcohol	x	Nitric acid, anhydrous	x
Benzyl chloride	x	Oleum	x
Borax	R	Perchloric acid 10%	x
Boric acid	R	Perchloric acid 70%	x
Bromine, liquid	x	Phenol	x
Butadiene	x	Potassium bromide 30%	R
Butyl acetate	x	Sodium chloride	R
Carbon bisulfide	R	Sodium hydroxide 50%	R
Calcium chloride	R	Sodium hypochlorite, concd	x
Calcium hydroxide 10%	R	Sulfuric acid 10%	x
Calcium hydroxide, sat.	R	Sulfuric acid 50%	x
Calcium hypochlorite	x	Sulfuric acid 70%	x
Calcium nitrate	R	Sulfuric acid 90%	x
Carbon dioxide, dry	R	Sulfuric acid 98%	x
Carbon dioxide, wet	R	Sulfuric acid 100%	x
Carbon monoxide	R	Toluene	x
Carbon tetrachloride	x		

[a] The chemicals listed are in the pure state or in a saturated solution unless otherwise indicated. Compatibility at 90°F/32°C is shown by an R. Incompatibility is shown by an x.

5. Have good electrical properties
6. Have good surface appearance.

Electrical properties are good up to the rated temperature limits.

The glass-reinforced resins are unique in that they are the first thermoplastics that can compare with, or are better than, thermoset polymers in electrical, mechanical, dimensional, and creep properties at elevated temperatures (approximately 300°F/149°C) while having superior impact properties. Refer to Table 2.83 for the physical and mechanical properties of PBT.

PBT parts are normally joined by adhesives. Surface treatments recommended include abrasion and solvent cleaning with toluene. If maximum strength is required, then chemical etch or gas plasma treatments should be used. Adhesives to be used are isocyanate-cured polyesters, epoxies, and urethanes.

2. Corrosion Resistance Properties

PBT exhibits good chemical resistance in general to dilute mineral acids, aliphatic hydrocarbons, aromatic hydrocarbons, ketones, and esters with limited resistance to hot water and washing soda. It is not resistant to chlorinated hydrocarbons and alkalies.

PBT has good weatherability and is resistent to UV degradation. Refer to Table 2.84 for the compatibility of PBT with selected corrodents.

3. Typical Applications

The unreinforced resins are used in housings that require excellent impact resistance and in moving parts such as gears, bearings, pulleys, and writing instruments. Flame retardant formulations find application as television, radio, electronics, business machine, and pump components.

Reinforced resins find application in the automotive, electrical and electronic, and general industrial areas.

GG. Acetals

Acetals are a group of high-performance engineering polymers that resembles the polyamide somewhat in appearance but not in properties. The general repeating chemical structure is

$$\left[\begin{array}{c} H \\ | \\ -C-O- \\ | \\ H \end{array} \right]_n$$

TABLE 2.83   Physical and Mechanical Properties of PBT

Property	Unfilled	30% glass fiber reinforced	30% glass fiber reinforced flame-retardant grade
Specific gravity	1.30–1.38	1.48–1.58	1.63
Water absorption (24 hr at 73°F/23°C) (%)	0.08–0.09	0.06–0.08	0.06–0.07
Dielectric strength, short-term (V/mil)	550	460–560	490
Tensile strength at break (psi)	8,200–8,700	1,400–19,000	17,400–20,000
Tensile modulus ($\times 10^3$ psi)	280–435	1,300–1,450	1,490–1,700
Elongation at break (%)	50–300	2–4	2.0–3.0
Compressive strength (psi)	8,600–14,500	18,000–23,500	18,000
Flexural strength (psi)	12,000–16,700	22,000–29,000	30,000
Compressive modulus ($\times 10^3$ psi)	375	700	
Flexural modulus ($\times 10^3$ psi) at 73°F/23°C 200°F/93°C 250°F/121°C	330–400	850–1,200	1,300–1,500
Izod impact (ft-lb/in. of notch)	0.7–1.0	0.9–2.0	1.3–1.6
Hardness, Rockwell	M68–78	M90	M88–90
Coefficient of thermal expansion (10^{-6} in./in./°F)	60–90	15–25	
Thermal conductivity (10^{-4}cal-cm/sec-cm^2°C or Btu/hr/ft^2/°F/in.)	4.2–6.9	7.0	
Deflection temperature at 264 psi (°F) at 66 psi (°F)	122–185 240–375	385–437 421–500	400–450 425–490
Max. operating temperature (°F/°C)			
Limiting oxygen index (%)			
Flame spread			
Underwriters Lab. rating (Sub. 94)			

TABLE 2.84 Compatibility of PBT with Selected Corrodents[a]

Chemical		Chemical	
Acetic acid 5%	R	Hydrogen peroxide 0–5%	R
Acetic acid 10%	R	Ink	R
Acetone	R	Linseed Oil	R
Ammonia 10%	x	Methanol	R
Benzene	R	Methyl ethyl ketone	R
Butyl acetate	R	Methylene chloride	x
Calcium chloride	R	Motor oil	R
Calcium disulfide	R	Nitric acid 2%	R
Carbon tetrachloride	R	Paraffin oil	R
Chlorobenzene	x	Phosphoric acid 10%	R
Chloroform	x	Potassium hydroxide 50%	x
Citric acid 10%	R	Potassium dichromate 10%	R
Diesel oil	R	Potassium permanganate 10%	R
Dioxane	R	Silicone oil	R
Edible oil	R	Soap solution	R
Ethanol	R	Sodium bisulfite	R
Ether	R	Sodium carbonate	R
Ethyl acetate	R	Sodium chloride 10%	R
Ethylene chloride	x	Sodium hydroxide 50%	x
Ethylene glycol	R	Sodium nitrate	R
Formic acid	R	Sodium thiosulfate	R
Fruit juice	R	Sulfuric acid 98%	x
Fuel oil	R	Sulfuric acid 2%	R
Gasoline	R	Toluene	R
Glycerine	R	Vaseline	R
Heptane/hexane	R	Water, cold	R
Hydrochloric acid 38%	x	Water, hot	R
Hydrochloric acid 2%	R	Wax, molten	R
Hydrogen peroxide 30%	R	Xylene	R

[a] R = Material resistant at 73°F/20°C; x = material not resistant.

1. Physical and Mechanical Properties

The acetals are strong and rigid, but not brittle, and have good moisture and heat resistance. There are two basic types of acetals. DuPont produces the homopolymers, while Celanese produces the copolymers.

The homopolymers are harder and have higher tensile and flexural strengths with lower elongation than the copolymers.

The copolymers are more stable in long-term high-temperature service and are more resistant to hot water.

The most outstanding properties of the acetals are high tensile strength and stiffness, good recovery from deformation under load, toughness under repeated impact, and a high degree of resilience. Because of their excellent long-term load-carrying properties and dimensional stability, they can be used for precision parts.

The polymer surface is hard, smooth, and glossy with low static and dynamic coefficients of friction.

Refer to Table 2.85 for the physical and mechanical properties of acetals.

TABLE 2.85 Physical and Mechanical Properties of Acetals

Property	Homopolymer	Copolymer	Impact-modified homopolymer
Specific gravity	1.42	1.40	1.32–1.39
Water absorption (24 hr at 73°F/23°C) (%)	0.25–1	0.2–0.7	0.3–0.44
Dielectric strength, short-term (V/mil)	400–500	500	400–460
Tensile strength at break (psi)	9,700–10,000		8,000–8,400
Tensile modulus ($\times 10^3$ psi)	400–520	374–484	190–360
Elongation at break (%)	10–75	15–75	60–200
Compressive strength (psi)	15,000–18,000	18,000	7,600–11,900
Flexural strength (psi)	13,000–16,000	13,000	5,800–10,000
Compressive modulus ($\times 10^3$ psi)	670	450	
Flexural modulus ($\times 10^3$ psi) at			
73°F/23°C	380–490	370–450	150–350
200°F/93°C	120–138		50–100
250°F/121°C	75–90		30–60
Izod impact (ft-lb/in. of notch)	1.1–2.3	0.8–1.5	2.0–17
Hardness, Rockwell	M92–94	M75–90	M68–79
Coefficient of thermal expansion (10^{-6} in./in./°F)	50–112	185–250	140–185
Thermal conductivity (10^{-4} cal-cm/sec-cm^2°C or Btu/hr/ft^2/°F/in.)	5.5	5.5	
Deflection temperature			
at 264 psi (°F)	263–277	185–250	145–185
at 66 psi (°F)	324–342	311–330	243–336
Max. operating temperature (°F/°C)			
Limiting oxygen index (%)			
Flame spread			
Underwriters Lab. rating (Sub. 94)			

TABLE 2.86 Compatibility of Acetals with Selected Corrodents[a]

Chemical	Maximum temp. °F	°C	Chemical	Maximum temp. °F	°C
Acetaldehyde	70	23	Ethyl chloride, wet	70	23
Acetic acid 5%	70	23	Ethyl ether	70	23
Acetic acid 10%	x		Ethylene chloride	70	23
Acetone	70	23	Ethylenediamine	70	23
Acid mine water	100	38	Fatty acid	95	35
Ammonia, wet	x		Ferric chloride	x	
Ammonium hydroxide, concd	x		Ferric nitrate	x	
Aniline	70	23	Ferric sulfate	x	
Aqua regia	x		Ferrous chloride	x	
Benzene	70	23	Ferrous sulfate 10%	x	
Benzyl chloride	70	23	Formic acid 3%	x	
n-Butylamine	x		Heptane	70	23
Calcium bisulfite	x		Hexane	70	23
Calcium chloride	x		Hydrochloric acid 1–20%	x	
Calcium chlorate 5%	70	23	Hydrofluoric acid 4%	x	
Calcium hydroxide 10%	70	23	Hydrogen peroxide	x	
Calcium hydroxide 20%	x		Hydrogen sulfide, moist	x	
Calcium hypochlorite 2%	x		Kerosene	140	60
Calcium nitrate	x		Lactic acid	x	
Calcium sulfate	x		Linseed oil	70	23
Calcium disulfide	70	23	Magnesium chloride 10%	70	23
Carbon tetrachloride	x		Magneisum hydroxide 10%	70	23
Carbonic acid	70	23	Magnesium sulfate 10%	70	23
Cellosolve	70	23	Magnesium sulfate	70	23
Chloric acid	x		Mercury	70	23
Chlorine, dry	x		Methyl acetate	x	
Chlorine, moist	70	23	Methyl alcohol	70	23
Chloroacetic acid	x		Methylene chloride	70	23
Chlorobenzene, dry	70	23	Milk	70	23
Chloroform	x		Tetrahydrofuran	70	23
Chlorosulfonic acid	x		Thionyl chloride	x	
Chromic acid	x		Turpentine	140	60
Copper nitrate	x		Mineral oil	70	23
Cupric chloride	x		Motor oil	150	71
Cyclohexane	70	23	Monoethanolamine	x	
Cyclohexanol	70	23	Nitric acid 1%	x	
Cyclohexanne	70	23	Nitric acid	x	
Detergents	70	23	Oxalic acid	x	
Dichloroethane	70	23	Paraffin	70	23
Diethyl ether	70	23	Perchloroethylene	70	23
Dimethyl formamide	140	60	Perchloric acid 10%	x	
Dimethyl sulfoxide	70	23	Phosphoric acid	x	
Dioxane	140	60	Pyridine	70	23
Ethanolamine	70	23	Silver nitrate	70	23
Ethers	70	23	Sodium bicarbonate 50%	70	23
Ethyl acetate 10%	200	93	Sodium carbonate 20%	180	82

(*Continued*)

TABLE 2.86   Continued

Chemical	Maximum temp. °F	°C	Chemical	Maximum temp. °F	°C
Sodium carbonate	70	23	Sodium nitrate 50%	70	23
Sodium chlorate 10%	70	23	Sodium thiosulfate	70	23
Sodium chloride 10%	180	82	Sulfuric acid 10%	x	
Sodium chloride	70	23	Sulfuric acid	x	
Sodium cyanide 10%	70	23	Sulfurous acid 10%	x	
Sodium hydroxide 10%	70	23	Trichloroethylene	x	
Sodium hypochlorite	x		Wax	200	93

[a] The chemicals listed are in the pure state or in a saturated solution unless otherwise indicated. Compatibility is shown to the maximum allowable temperature for which data are available. Incompatibility is shown by an x.

Homopolymers are usually adhesive bonded although they may be thermally fused. They are not usually solvent cemented. Prior to bonding the surface should be chemically treated using a sulfuric–chromic acid treatment followed by a solvent wipe. Epoxies, nitrile, and nitrile–phenolics can be used as adhesives.

Copolymer acetals are commonly joined by thermal fusion and solvent cementing. Adhesive bonding is also used. Prior to adhesive bonding, the surfaces must be treated with a sulfuric–chromic acid solution. Bonding is accomplished using epoxies, isocyanate-cured polyesters, or cyanoacrylates.

2. Corrosion Resistance Properties

Neither the homopolymer nor the copolymer is resistant to strong mineral acids, but the copolymers are resistant to strong bases. The acetals will be degraded by ultraviolet light and gamma rays. They also exhibit poor resistance to oxidizing agents but in general are resistant to the common solvents. Refer to Table 2.86 for the compatibility of acetals with selected corrodents.

Homopolymer materials are usually implied when no identification of acetal type is made.

3. Typical Applications

Acetals are used in the automotive industry for fuel-system components and moving parts such as gears, structural components in the suspension system, and bushings. Applications also include appliance housings and plumbing fixtures.

REFERENCES

1. Schweitzer PA. *Corrosion Resistance Tables*, 4th ed., Vols. 1–3. New York: Marcel Dekker, 1995.
2. Schweitzer PA. *Corrosion Resistant Piping Systems*. New York: Marcel Dekker, 1994.
3. Schweitzer PA. *Corrosion Engineering Handbook*. New York: Marcel Dekker, 1996.

3

Thermosetting Polymers

3.1 INTRODUCTION

Thermoset polymers assume a permanent shape or set once cured. Once set they cannot be reshaped. They are formed by a large amount of cross-linking of linear prepolymers (a small amount of cross-linking will produce elastomers) or by direct formation of networks by the reaction of two monomers. The latter is the more prominent of the two methods. It is a stepwise or condensation method, which has been defined as "the reaction of two monomers to produce a third plus a by-product, usually water or alcohol." Since in some cases a by-product is not produced, this definition is no longer exactly correct. The reaction is now referred to as a "stepwise" polymerization. When the reaction results in a by-product, it is called a "condensation reaction." Table 3.1 lists the principal thermoset polymers.

Although fewer in number than the thermoplastic polymers, they comprise approximately 14% of the total polymer market. Compared to the thermoplasts, they are more brittle, stronger, harder, and generally more temperature resistant. Table 3.2 gives the operating temperature range of thermoset polymers. In addition they offer the advantages of better dimensional stability, creep resistance, chemical resistance, and good electrical properties. Their disadvantages lie in the facts that most are more difficult to process and more expensive.

Phenolics represent about 43% of the thermoset market, making them the most widely used. They are relatively inexpensive and are readily molded with good stiffness. Most contain wood or glass-flour fillers and on occasion glass fibers.

TABLE 3.1 Principal Thermoset
Polymers

Alkyds
Amino resins
Diallyl phthalates (allyls)
Epoxies
Furans
Melamines
Phenol formaldehyde
Phenolics
Polybutadienes
Polyurethanes
Silicones
Unsaturated polyesters
Ureas
Vinyl esters

TABLE 3.2 Operating Temperature Range of Thermoset Polymers[a]

| Polymer | Allowable temp. (°F/°C) | |
	Minimum	Maximum
Epoxy	−423/−252	300/150
Polyesters:		
Bisphenol A–fumarate		250–300/120–150
Halogenated		250–300/120–150
Hydrogenated bisphenol A–bisphenol A		250–300/120–150
Isophthalic		150/70
Terephthalic		250–300/120–150
Vinyl ester		200–280/98–140
Furan		300–400/150–200
Phenolic		230/110
Silicone	−423/−252	500/260

[a] Temperatures might have to be modified depending on the corrodent.

Since in general, unfilled thermosetting polymers tend to be harder, more brittle, and not as tough as thermoplastic polymers, it is common practice to add fillers to thermoset molding materials. A wide variety of fillers are used to produce desired properties in the finished product.

For molded products, usually compression or transfer molding, mineral or cellulose fibers are often used as low-cost, general-purpose fillers, and glass fiber fillers are often used for optimum strength or dimensional stability.

A. Corrosion of Thermosets

Unreinforced, unfilled thermoset polymers can corrode by several mechanisms. The type of corrosion can be divided into two main categories: physical and chemical.

Physical corrosion is the interaction of a thermoset polymer with its environment so that its properties are altered but no chemical reactions take place. The diffusion of a liquid into the polymer is a typical example. In many cases physical corrosion is reversible. Once the liquid is removed, the original properties are restored.

When a polymer absorbs a liquid or gas which results in plasticization or swelling of the thermoset network, physical corrosion has taken place. For a cross-linked thermoset, swelling caused by solvent absorption will be at a maximum when the solvent and polymer solubility parameters are exactly matched.

Chemical corrosion takes place when the bonds in the thermoset are broken by means of a chemical reaction with the polymer's environment. There may be more than one form of chemical corrosion taking place at the same time. Chemical corrosion is usually not reversible.

As a result of chemical corrosion, the polymer itself may be affected in one or more ways. For example, the polymer may be embrittled, softened, charred, crazed, delaminated, discolored, dissolved, blistered, or swollen. All thermosets will be attacked in essentially the same manner. However, certain chemically resistant types suffer negligible attack or exhibit significantly lower rates of attack under a wide range of severely corrosive conditions. This is the result of the unique molecular structure of the resins, including built-in protection of ester groups.

Cure of the resin plays an important part in the chemical resistance of the thermoset. Improper curing will result in a loss of corrosion resistant properties. Construction of the laminate and the type of reinforcing used also affect the corrosion resistance of the laminate. The degree and nature of the bond between the resin and the reinforcement also plays an important role.

The various modes of attack affect the strength of the laminate in different ways, depending upon the environment and other service conditions and the mechanisms or combination of mechanisms that are at work.

Some environments may weaken primary and/or secondary polymer linkages with resulting depolymerization. Other environments may cause swelling or microcracking, while still others may hydrolyze ester groupings or linkages. In certain environments, repolymerization can occur, with a resultant change in structure. Other results may be chain scission and decreases in molecular weight or simple solvent action. Attack or absorption at the interface between the reinforcing material and the resin will result in weakening.

In general, chemical attack on thermoset polymers is a "go/no-go" situation. With an improper environment, attack on the reinforced polymer will occur in a relatively short time. Experience has indicated that if an installation has been soundly engineered and has operated successfully for 12 months, in all probability it will continue to operate satisfactorily for a substantial period of time.

Thermoset polymers are not capable of handling concentrated sulfuric acid (93%) and concentrated nitric acid. Pyrolysis or charring of the resin quickly occurs, so that within a few hours the laminate is destroyed. Tests show that polyesters and vinyl esters can handle up to 70% sulfuric acid for long periods of time.

The attack of aqueous solution on reinforced polymers occurs through hydrolysis, with the water degrading bonds in the backbone of the resin molecules. The ester linkage is the most susceptible.

The attack by solvents is of a different nature. The solvent penetrates the resin matrix of the polymer through spaces between the polymer chains. Penetration between the polymer chains causes the laminate surface to swell, soften, and crack. In the first stages of solvent attack the following occur:

1. Softening: decrease in hardness by a substantial amount.
2. Swelling: the laminate swells considerably.
3. Weight: pronounced weight gain. Anything over 2% is cause for concern.

There are many degrees of attack:

1. If there is less than a 2% weight gain and considerable retention in hardness over a 12 month period, plus little or no swelling, the laminate will do very well.
2. If hardness is retained at a high level, weight gain has stabilized, and swelling has stabilized at the 12 month level, then the resin may do fairly well in limited service.
3. Bisphenol resin laminate in contact with toluene experiences a 19% weight gain in one month, a swelling of 19%, and hardness drops from 43 to 0, typical of a material undergoing total failure. As the solvent attack continues,
   a. Softening: hardness drops to zero;
   b. Swelling: the absorption of the molecule continues, producing mechanical stresses that cause fractures in the laminate. This may occur in the liquid or vapor zone.

Organic compounds with carbon–carbon unsaturated double bonds, such as carbon disulfide, are powerful swelling solvents and show greater swelling action than their saturated counterparts. Smaller solvent molecules can penetrate a polymer matrix more effectively. The degree of similarity between solvent and resin is important. Slightly polar resins, such as the polyesters and the vinyl esters, are attacked by mildly polar solvents.

Generally, saturated, long-chain organic molecules, such as the straight-chain hydrocarbons, are handled well by polyesters. This is why polyester gasoline tanks are so successful.

The polymer's ability to resist attack is improved by

1. The cross-link density of the resin.
2. The ability of the resin to pack into a tight structure.
3. The heat distortion temperature of the resin, which is strongly related to its solvent resistance. The higher the heat distortion temperature, the better the solvent resistance. The latter is not true for all resin systems.

In general, the more brittle resins possess poor solvent resistance while the more resilient resins can withstand a greater degree of solvent absorption.

Orthophthalic, isophthalic, bisphenol, and chlorinated or brominated polyesters exhibit poor resistance to such solvents as acetone, carbon disulfide, toluene, trichloroethylene, trichloroethane, and methyl ethyl ketone. The vinyl esters show improved solvent resistance. Heat-cured epoxies exhibit better solvent resistance. However, the furan resins offer the best all around solvent resistance. They excel in this area. Furan resins are capable of handling solvents in combination with acids and bases.

Stress corrosion is another factor to be considered. The failure rate of glass-reinforced composites can be significant. This is particularly true of composites exposed to the combination of acid and stress. Weakening of the glass fibers upon exposure to acid is believed to be caused by ion exchange between the acid and glass.

Under stress an initial fiber fracture occurs, which is specifically a tensile type of failure. If the resin matrix surrounding the failed fiber fractures, the acid is allowed to attack the next available fiber, which subsequently fractures. The process continues until total failure occurs.

B. Joining of Thermosets

Thermoset polymers cannot be solvent cemented or fusion welded. However, they can be adhesive bonded using epoxies, thermosetting acrylics, and urethanes. Prior to bonding, the surfaces should be abraded and solvent cleaned.

C. Ultraviolet Light Stability

Thermoset resins will be degraded by ultraviolet light. In view of this, it is necessary that UV protection be provided. Polymers installed outdoors will suffer severe degradation in a year or two from ultraviolet exposure. The glass filament will fray. After the initial fraying, the phenomenon will stop far short of having any particular effect on the strength of the material. This can be prevented by adding an ultraviolet inhibitor to the resin. Another approach is to apply a 3-mil thick coat of black epoxy paint, which will also effectively screen out the ultraviolet rays.

With the exception of UV light stability, thermoset polymers resist weathering, being resistant to normal atmosphere pollutants. Refer to Table 3.3 for the atmospheric resistance of fiberglass-reinforced thermoset polymers.

3.2 REINFORCING MATERIALS

Many different reinforcing fibers are used in laminates and reinforced polymers. Which fiber is selected depends upon the cost, the properties required, and the nature of the resin system. Fiberglass is the most often used reinforcement material. Table 3.4 lists the common reinforcements and the properties they impart to the reinforced polymer.

TABLE 3.3 Atmospheric Resistance of Fiberglass-Reinforced Thermoset Polymers

Polymer	UV[a] degradation	Moisture[b] absorption	Weathering	Ozone	SO_2	NO_2	H_2S
Epoxy	R	0.03	R	R	R	R	R
Polyesters							
Bisphenol A– fumarate	RS	0.20	R	R	R	R	R
Halogenated Hydrogenated	RS	0.20	R	R	R	R	R
Bisphenol A– bisphenol A	RS	0.20	R	R	R[c]	R	
Isophthalic	RS	0.20	R	R	R	R	R
Terephthalic	RS	0.20	R	R	R	R	R
Vinyl ester	R	0.20	R	R	R	R	R
Furan	R	2.65	R	R	R	R	R
Phenolic							
Silicone	R	0.02–0.06	R	R	R	R	

[a] R, resistant; RS, resistant only if stabilized with UV protector; x, not resistant.
[b] Water absorption rate 24 h at 73°F/23°C (%).
[c] SO_3 will cause severe attack.

TABLE 3.4 Properties Imparted to Polymer by Reinforcing Materials

Reinforcing fiber	Mechanical strength	Electrical properties	Impact resistance	Corrosion resistance	Machining and punching	Heat resistance	Moisture resistance	Abrasion resistance	Low cost	Stiffness
Glass strands	X		X	X		X	X		X	X
Glass fabric	X	X	X	X		X	X			X
Glass mat			X	X		X	X		X	X
Asbestos		X	X			X				
Paper		X			X				X	
Cotton/linen	X	X	X		X				X	
Nylon		X	X	X						
Short inorganic fibers	X		X							
Organic fibers	X				X					
Ribbons		X								
Polyethylene	X		X		X		X			
Metals	X		X							X
Aramid	X		X				X			X
Boron	X	X	X			X				X
Carbon/graphite	X					X				X
Ceramic	X		X			X				X

A. Glass Fibers

Glass fibers are formed continuously from a melt in a special fiber-forming furnace. There are six formulations which are produced. The most common is E glass. This glass resists moisture and results in products with excellent electrical properties. C glass is designed to be used where optimum chemical resistance is required. D glass has very good electrical properties, particularly the dielectric constant, and is used in electronic applications. S glass is used for its high strength and stiffness while R glass is a lower cost fiber than S glass. Table 3.5 shows the physical and electrical properties of these fibers.

1. E Glass

As discussed previously, E glass is an electrical grade borosilicate glass. It possesses excellent water resistance, strength, low elongation, and is reasonable in cost. Practically all glass mat, continuous filaments, and woven rovings come from this source.

2. C Glass

This is a calcium aluminosilicate glass widely used for surfacing mats, glass flakes, or flake glass linings and for acid resistant cloths. C glass has poor water resistance and carries a premium cost. C veil has been used as a surfacing mat for many years when the standard specification for corrosion resistant equipment was

TABLE 3.5 Properties of Glass Fibers[a]

	Glass Type					
	A	C	D	E	R	S & S2
Specific gravity	2.5	2.49	2.16	2.52 2.61	2.55	2.49
Tensile strength (psi$\times 10^6$)	3.5	0.4	0.35	0.5	0.6	0.665
Tensile elastic modulus ($\times 10^6$)	9.8	10	7.5	10.5	12.4	12.5
Elongation at 70°F (%)				3–4		5.2
Coefficient of thermal expansion (in/in/°F $\times 10^{-7}$)	90	40	17	28–33	74	13–17
Thermal conductivity (Btu/in/hr/ft^2/°F at 72°F)				7.2	6.9	
Dielectric constant @ 72°F (10^6 hertz)	6.9	6.24	3.56	6.1 6.7	6.2	5.34

[a] A = soda lime glass; C = chemical glass; D = yarn; E = roving, fabrics, yarn; R = yarn; S = roving, yarn.

10-mil C veil. The use of 20-mil C veil and brittle resins such as the bisphenols should be avoided as they are easily subject to impact and handling damage. Until the development of synthetic veils, 10-mil C glass was nearly universally used. It is available in thicknesses of 10, 15, 20, and 30 mils.

3. S Glass

Because of its exceptional strength, S glass is widely used in the aerospace industry. It has excellent resistance to acids and water but is several times as expensive as E glass. Because of its cost, it is not used in the corrosion industry. It is comparable in strength to aramid fiber.

4. Glass Filaments

Glass filaments are available in a variety of diameters, as shown in Table 3.6. Filaments in the range of diameters G to T are normally used to reinforce polymers. These filaments are formed into strands with 204, 400, 800, 1000,

TABLE 3.6 Glass Fiber Diameters

Filament	Diameter range ($\times 10^5$ in.)	Nominal diameter ($\times 10^5$ in.)
B	10.0–14.9	12.3
C	15.0–19.9	17.5
D	20.0–24.9	22.5
DE	23.0–27.9	25.0
E	25.0–29.9	27.5
F	30.0–34.9	32.5
G	35.0–39.9	36.0
G	35.0–39.0	37.5
H	40.0–44.9	42.5
J	45.0–49.9	47.5
K	50.0–54.9	52.5
L	55.0–59.9	57.5
M	60.0–64.9	62.5
N	65.0–69.9	67.5
P	70.0–74.9	72.5
Q	75.0–79.9	77.5
R	80.0–84.9	82.5
S	85.0–89.9	87.5
T	90.0–94.9	92.5
U	95.0–99.9	97.5
V	100.0–104.9	102.5
W	105.0–109.9	107.5

2000, 3000, and 4000 filaments to a strand. The filaments are bonded into a strand using a sizing agent applied to the filament. The sizing agent also gives them environmental and abrasive protection. Coupling agents, such as silanes, chrome complexes, and polymers, are added to the finished products to improve adhesion of the resin matrix to the glass fibers. These strands are then used to manufacture the various types of glass reinforcements. Glass fabrics, glass mats, and chopped strands are the most common reinforcements used in reinforced polymers.

5. Chopped Strands

Strands of glass (usually E glass) are mechanically cut into lengths from 0.25 to 2 inches and are used to reinforce molding compounds. The longer lengths are used with thermoset resins while the short fibers are used with thermoplastic polymers.

6. Glass Mats

Glass strands are cut and dropped onto a moving belt where a polymer binder is applied to hold the mat together. Mat weights vary from 0.75 to 3 oz/ft^2 in widths up to 10 feet.

   Continuous strand mats are produced by depositing the uncut strands continuously in a swirling pattern onto a moving belt where a binder is applied. Continuous strand mats exhibit better physical properties than cut strand mats, but the material is less homogenous. Continuous strand mats vary in weight from 0.75 to 4.5 oz/ft^2 and are available in widths up to 6 feet.

   Woven roving is a mat fabric made by weaving multiple strands collected into a roving into a coarse fabric. The physical properties of woven roving are intermediate between mats and fabrics. Various thicknesses are available in widths up to 10 feet. These constructions are used in low-pressure laminations and in pultrusions.

7. Glass Fabrics

Reinforced polymers make use of many different glass fabrics. E glass is used in most, and filament laminates of D, G, H, and K are common. Strands of glass filaments are plied into yarns and woven into fabrics on looms. The machine direction of the loom is called the warp, while the cross section is the welt (also called woof or fill). The number of yarns can be varied in both warp and welt to control the weight, thickness, appearance, and strength of the fabric. When each welt yarn is laced alternately over and under the warp, the type of fabric (called the weave) is known as plain. Other weaves such as basket, satin, eight-harness satin, lino and mock lino, can be produced by having yarns cross two or more adjacent yarns, or by staggering the crossing along the warp. Refer to Table 3.7.

   Table 3.8 lists those glass fabrics used for mechanical laminates for marine applications.

TABLE 3.7 Mechanical-Grade Glass Fabrics

Fabric style no.	Count (yarns/in. yarns/5 cm)	Warp yarn (yarn count TEX)	Welt yarn (yarn count TEX)	Weave	Mass (oz/yd² g/m²)	Thickness (in. mm.)	Breaking strength (lb/in. N/5 cm)
1581	57 × 54	EC6 150-1/2	ECG 150-1/2	8-H satin	8.90	0.009	350 × 340
	112 × 106	EC9 33 1 × 2	EC9 33 1 × 2	8 H satin	3.02	0.228	3065 × 2977
1582	60 × 56	EC6 150-1/3	EC6 150-1/3	8 H satin	13.9	0.0140	490 × 450
	118 × 110	EC9 33 1 × 3	EC9 33 1 × 3	8 H satin	47.1	0.365	4291 × 3940
1583	54 × 48	EC6 150 2/2	EC6 150 2/2	8 H satin	16.10	0.0160	650 × 590
	106 × 94	EC9 33 2 × 2	EC9 33 2 × 2	8 H satin	54.5	0.406	5692 × 5106
1584	44 × 35	EC6 150 4/2	EC6 150 4/2	8 H satin	26.00	0.0260	950 × 800
	87 × 69	EC9 33 4 × 2	EC9 33 4 × 2	8 H satin	880.0	0.670	8318 × 7005
3706	12 × 6	EC6 37 1/0	EC6 37 1/2	Leno	3.70	0.0086	140 × 120
	24 × 12	EC9 134 1 × 0	EC9 134 1 × 2	Leno	125.0	0.218	125 × 1050
7781	54 × 54	ECDE 75 1/0	ECDE 75 1/0	8 H satin	8.95	0.0090	350 × 340
	112 × 106	EC6 66 1 × 0	EC6 66 1 × 0	8 H satin	304.0	0.228	3065 × 2977
7826	32 × 32	EC6 75 1/0	EC6 75 1/0	Plain	5.40	0.0066	225 × 200
	67 × 63	EC9 66 1 × 0	EC9 66 1 × 0	Plain	183.0	0.168	1970 × 1731
181	57 × 54	ECD 225 1/3	ECD 225 1/3	Satin	8.9	0.009	350 × 340
	22 × 21	ECS 22 1 × 3	ECS 22 1 × 3	Satin	302.0	0.229	3065 × 2971

TABLE 3.8 Marine-Grade Glass Fabrics

Fabric style no.	Count (yarns/in.) (yarns/5 cm)	Warp yarn (yarn count Tex)	Welt yarn (yarn count Tex)	Weave	Mass (oz/yd^2) (g/m^2)	Thickness (in.) (mm)	Breaking Strength (lb/in.) (N/S cm)
1800	16 × 14	ECK 18 1/0	ECK 18 1/0	Plain	9.60	0.0130	450 × 350
	31 × 28	EC 13 215 1 × 0	EC13 275 1 × 0	Plain	326.0	0.330	3940 × 3065
2532	16 × 14	ECH 25 1/0	ECH 25 1/0	Plain	7.25	0.010	300 × 280
	31 × 28	EC10 200 1 × 0	EC10 200 1 × 0	Plain	246.0	0.254	2627 × 2432
7500	16 × 14	ECG 75 2/2	ECG 75 2/2	Plain	9.66	0.014	480 × 410
	31 × 28	EC9 66 2 × 2	EC9 66 2 × 2	Plain	327.0	0.356	3940 × 3590
7533	18 × 18	ECG 75 1/2	ECG 75 1/2	Plain	5.80	0.008	250 × 220
	35 × 35	EC9 66 1 × 2	EC9 66 1 × 2	Plain	197.0	0.243	2189 × 1926
7544	24 × 14	ECG 75 2/2	ECG 75 2/2	Basket	18.00	0.022	750 × 750
	58 × 28	EC9 66 2 × 2	EC9 66 2 × 2	Basket	610.0	0.559	6597 × 6597
7587	40 × 21	ECG 75 2/2	ECG 75 2/2	Mock leno	20.5	0.030	750 × 450
	79 × 41	EC9 66 2 × 2	EC9 66 2 × 2	Mock leno	695	0.761	6597 × 3940

B. Polyester

Polyester is used primarily for surfacing mat for the resin-rich inner surface of filament-wound or custom contact-molded structures. It may also be used in conjunction with C glass surfacing mats.

The tendency of C glass to bridge and form voids is reduced or eliminated by overwinding a C glass surfacing mat with a polyester mat under tension. Nexus (registered trademark of Burlington Industries) surfacing veils are also used on interior and exterior surfaces of pultruded products. The Nexus surfacing veil possesses a relatively high degree of elongation that makes it very compatible with the higher elongation resins and reduces the risk of checking, crazing, and cracking in temperature-cycling operations.

The Nexus surfacing veil exhibits excellent resistance to alcohols, bleaching agents, water, hydrocarbons, and aqueous solutions of most weak acids at boiling. Being a polyester derivative, it is not resistant to strong acids, such as 93% sulfuric acid.

Examination shows that the Nexus surfacing veil has an improved surface finish over C glass, better impact strength, and increased resistance to flexural fatigue before cracking. This is particularly pronounced when used with higher elongation resins, such as vinyl esters.

C. Carbon Fiber

Carbon fiber is available in mat form, typically 0.2, 0.5, 1, and 2 oz/yd^2. It is also available in a 0.5 oz/yd^2 mat as a blend with 33% glass fiber. Other blends are also available such as 25% carbon fiber–75% glass fiber, and 50% aluminized glass–50% carbon fiber.

Carbon fibers are also available in continuous roving and as chopped fibers in sizes 1/8 to 2 in.

The carbon fiber mat, either alone or supplemented with a ground carbon or graphite filler, provides in-depth grounding systems and static control in hazardous areas where static sparks may result in fires or explosions.

D. Aramid Fibers

Aramid is a generic name for aromatic polyamide (Nylon). It is sold under the tradename Kevlar, manufactured by DuPont. The fiber is used in the same manner as glass fibers to reinforce polymers. Because of its great tensile strength and consistency coupled with low density, it essentially revolutionized pressure vessel technology. Aramid composites, though still widely used for pressure vessels, have been largely replaced by the very high strength graphite fibers.

Aramid fibers are woven into fabrics for reinforcing polymers. The fabrics range in size from thin 2-mil (1 oz/yd^2) to thick 30-mil (6 oz/yd^2) cloths. Kevlar

TABLE 3.9   Properties of Aramid Fibers

Property	Type				
	29	49	149	68	129
Tensile strength ($\times 10^3$ psi)	525	525	500	525	610
Tensile modulus ($\times 10^6$ psi)	12	18	25	16	16
Yarns					
No. size	5	7	3	3	3
Denier range	200–30,000	55–2,840	380–1,420	1,420–2,840	840–1,420
Rovings					
No. size	2	6	1	1	
Denier range	9,000–15,000	4,320–22,720	7,100	7,100	

is available in surfacing mats and cloth form. Surfacing mats can be had in weights of 0.4 and 1 oz/yd^2.

The properties of aramid fibers are given in Table 3.9. Careful design is required when using aramid fiber composites in structural applications that involve bending since they have relatively poor shear and compression properties.

Aramid fiber finds varied applications in aircraft and aerospace components because of its low density. This light weight is also advantageous in the fabrication of laminated canoes and kayaks. Other sport applications include downhill skis, tennis racquets, and golf shafts. Boat hulls are stronger and lighter and ride better because of the vibration-dampening qualities of the fiber.

Aramid reinforcement is also available as a paper and sold under the tradenames of Nomex by DuPont and TP Technora from Teijin America. Aramid paper is used in circuit boards to improve crack resistance and to impart a smooth surface. The paper is limited by such drawbacks as a high coefficient of thermal expansion, poor resin adhesion, and difficult drilling and machining characteristics. The properties of aramid fibers used in paper are shown in Table 3.10.

E. Polyethylene Fibers

Allied Signal, Inc. produces a high-strength, high-modulus polyethylene fiber under the tradename Spectra. The polyethylene is an ultrahigh molecular weight polymer, up to 5 million, compared to conventional polyethylene which has a molecular weight of about 200,000. Two Spectra fibers are produced, the 900 and

TABLE 3.10 Properties of Aramid Paper Fiber

Density, g/cm^3	1.39
Tensile strength ($\times 10^3$ psi)	600
Tensile modulus ($\times 10^6$ psi)	13
Elongation (%)	4.5
Coefficient of thermal expansion/°C	6
Water absorption (%)	2
Specific heat (cal/g°C)	0.26

TABLE 3.11 Polyethylene Fiber Properties

Property	Spectra 900	Grade 1000
Tenacity ($\times 10^3$ psi)	373	435
Modulus ($\times 10^6$ psi)	17.4	24.8
Elongation (%)	3.5	2.7
Dielectric constant	2.0–2.3	

1000 grades. The 1000 grade has higher strength and modulus. Since the density of Spectra is the lowest of all of the fibers, it finds many applications in aerospace laminates such as wing tips and helicopter seats. Spectra laminates must not be exposed to temperatures above 250°F/121°C. Table 3.11 shows the fiber properties of polyethylene. Spectra has a very low dielectric constant and a loss tangent of 0.0002. These properties are useful in radomes. Spectra laminates are virtually transparent to radar.

Table 3.12 compares the corrosion resistance of Spectra and aramid fibers.

F. Paper

Kraft paper is widely used as a reinforcement. When saturated with phenolics, it is made into a common printed wiring board. When combined with melamine, it becomes a decorative high-pressure laminate used in furniture, countertops, and wall panels. Paper reinforcement is inexpensive and easy to machine, drill, and punch. It imparts good electrical properties but is sensitive to moisture and cannot withstand high temperatures.

G. Cotton and Linen

Fabrics of cotton or linen, when impregnated with phenolic resin, are used in several grades of laminates. They have good impact strength and abrasion

TABLE 3.12   Corrosion Resistance of Spectra and Aramid Fibers

	% Retention of Fiber Tensile Strength			
	Spectra		Aramid	
Chemical	6 months	2 years	6 months	2 years
Seawater	100	100	100	98
Hydraulic fluid	100	100	100	87
Kerosene	100	100	100	97
Gasoline	100	100	93	x
Toluene	100	96	72	x
Acetic acid, glacial	100	100	82	x
Hydrochloric acid 1 M	100	100	40	x
Sodium hydroxide 5 M	100	100	42	x
Ammonium hydroxide 29%	100	100	70	x
Perchlorethylene	100	100	75	x
Detergent solution 10%	100	100	91	x
Clorox	91	73	0	0

x = Samples not tested due to physical deterioration.

resistance and machine well. Electrical properties are poor. Laminates with these fabrics have better water resistance than paper-based laminates.

These laminates are used for gears or pulleys.

H. Manufacturing Processes

Reinforced thermoset polymer configurations can be manufactured in several ways. Processes include

> Molding
> Hand lay-up
> Pultrusion
> Filament winding

1. Molding

Bulk molding compounds (BMC) are materials produced by combining a resin and chopped fibers in a mixer and mixing until the fibers are well wetted and the material has the consistency of modeling clay. Fibers are usually E glass. Polyesters make up the bulk of the BMC, although vinyl esters and some epoxies are used. The strength of BMC is lower than other types because the fibers are degraded in the mixer and the fibers orient somewhat in the mold but usually not in the optimum alignment.

Sheet molding compounds (SMC) are produced by compounding a polyester resin with fillers, pigments, catalysts, mold release agents, and special thickeners that react with the polymer to increase the viscosity. The resin mixture is spread on a moving Nylon film and passes under cutters that chip the roving into fibers $\frac{1}{2}$ to 2 inches long. A second film is placed on top, sandwiching the compound inside. This passes through rollers that help the resin to wet the fibers, and then the material is rolled up to mature for 24 to 72 hours. During this period it thickens and reaches a nontacky final consistency of leather. Before molding, the Nylon films are removed. Molding conditions are 400°F/204°C and 2000 psi for 1 to 2 minutes. The mechanical properties are superior to those of BMC since the fibers have not been damaged during the preparation of the compound.

2. Hand Layup

This is the simplest method for producing large reinforced polymer parts. It is frequently employed by boat and swimming pool manufacturers. Hand layup can be used on either male or female molds.

The mold is covered with a release agent such as floor wax or a film-forming polymer like PVA. In many applications, a gel coat is applied to the mold surface. This coating is a thick, pigmented resin that contributes a smooth, colorful outer surface to the molding. A reinforcing web (glass cloth, mat or woven roving) is then placed on the mold, and a low viscosity resin is painted or sprayed onto the reinforcement. The air is carefully removed from the reinforcement using a paint roller or rubber squeegee. Additional layers of reinforcement and resin are added until the desired thickness is reached. Polyethylene, cellophane, or polyester film is used as the final layer to make a smooth surface. When the resin has hardened, the part is removed.

The moldings produced by this process are not of the highest quality and may not have the properties expected. Laminates produced by this method contain more resin than laminates produced by other methods. More voids may be present, and little control can be exercised over the uniformity of the wall thickness.

3. Pultrusion

Pultrusion is a continuous process used to manufacture such items as fishing poles, ladders, I-beams, electrical pole line hardware, and tool handles. The reinforcing fibers used determine the physical properties of the pultruded product. Fiberglass is usually used. With fiberglass, the product has highly directional properties. The tensile and flexural strengths in the fiber direction are very high, exceeding 100,000 psi while the transverse properties are much lower. If glass mat, glass cloth, or chopped fibers are used, the transverse properties can be improved. The physical properties of a 60% by weight fiberglass polyester pultruded part are as follows:

Flexural strength	100,000 psi
Tensile strength	120,000 psi
Compressive strength	40,000 psi
Impact strength	40 ft-lb/in. of notch
Water absorption	0.1%

4. Filament Winding

Filament winding consists of wrapping resin-impregnated continuous fiber around a mandrel forming a surface of revolution. When sufficient layers have been wound, the part is cured, and the mandrel is removed. The continuous filaments are usually tapes or rovings of fiber glass, but other reinforcements are also used. Epoxy resins predominate, although others are also used. Typical properties of glass-epoxy filament-wound parts are shown in Table 3.13.

Filament-wound parts contain a high percentage of reinforcing fiber, and the properties of the filament-wound parts approach the properties of the fiber. The resin is used to bind the fibers together and to transmit the stress from fiber to fiber.

Filament winding is used to produce chemical storage tanks, pipelines, cherry-picker booms, sporting equipment, shotgun barrels, and circuit-breaker parts.

I. Mechanical Properties

The mechanical properties of thermosets are determined by the quantity and type of reinforcing material used. The corrosion resistance is a combination of the resistance of the polymer and of the reinforcing material.

TABLE 3.13 Properties of Filament-Wound Glass-Reinforced Epoxy

Specific gravity	2.0
Thermal conductivity (Btu/hr ft^2/°F/in.)	2.2
Thermal expansion (per °F × 10^{-6})	7.0
Specific heat (Btu/lb/°F)	0.227
Maximum use temperature (°F)	≥ 400
Compressive strength (psi)	70,000
Flexural strength (psi)	100.000
Bearing strength (psi)	35,000
Modulus of elasticity (tension × 10^{-6})	6.0

3.3 POLYESTERS

Polyesters were first prepared in 1929 by William H. Carothers. The manufacturing techniques which he developed are still in use today. It was not until the late 1930s that Ellis discovered that polyester's rate of cross-linking increased around 30 times in the presence of unsaturated monomers. Large scale commercialization of unsaturated polyesters resulted from the discovery by the U.S. Rubber Company in 1942 that adding glass fibers to polyesters greatly improved their physical properties. Widespread use of polyesters had been hampered because of their inherent brittleness. The first successful commercial use of glass reinforced polyesters was the production of radomes for military aircraft during World War II. The first fiberglass-reinforced polyester boat hulls were made in 1946 and is still a major application for unsaturated polyesters.

Of all the reinforced thermoset polymers used for corrosion resistant service, unsaturated polyesters are the most widely used.

Unsaturated polyesters are produced via a condensation reaction between a dibasic organic acid or anhydride and a difunctional alcohol, as shown in Figure 3.1. At least one of these components contributes sites of unsaturation to the oligomer chain. The oligomer or prepolymer is then dissolved in an unsaturated monomer such as styrene. To initiate cross-linking, a free radical source such as an organic peroxide is added to the liquid resin. The cross-linking reaction is a free radical copolymerization between the resin oligomer and the unsaturated monomer. The types and ratio of components used to manufacture the oligomers, the manufacturing procedure, and the molecular weight of the oligomer will determine the properties of the cured polyester.

Maleic anhydride or fumaric acid provides the source of unsaturation for almost all of the polyester resins currently produced, with maleic anhydride being the most common. Propylene glycol is the most common diol used since it yields polyesters with the best overall properties. Figure 3.2 lists the more common components used in the manufacture of unsaturated polyesters.

The primary unsaturated polyesters are

General purpose (orthophthalic)
Isophthalic
Bisphenol A–fumarate
Hydrogenated bisphenol-A–bisphenol A

$$HOn-R-OH + nHO-\overset{\overset{O}{\|}}{C}-R-\overset{\overset{O}{\|}}{C}-OH \longrightarrow \left[R-O-\overset{\overset{O}{\|}}{C}-R-\overset{}{\underset{\underset{O}{\|}}{C}}-O \right]_n + nH_2O$$

FIGURE 3.1 Condensation reaction.

Glycols

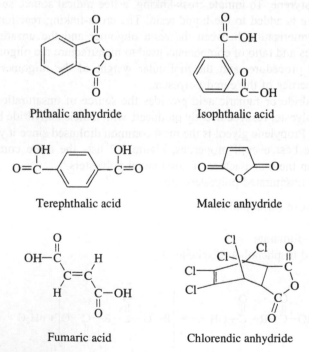

HO−CH₂−CH−CH₃ with OH on CH

Propylene glycol

CH₂−CH₂−O−CH₂−CH₂ with OH groups

Diethylene glycol

HO−CH₂−C−−−−C−OH (with CH₃, CH₃, CH₃ and C, H groups)

2,2,4 Trimethyl-1,3 pentanediol

HO—(ring)—C(CH₃)₂—(ring)—OH

Bisphenol A

HO−CH₂−C(CH₃)₂−CH₂−OH

Neopentyl glycol

Acids

Phthalic anhydride

Isophthalic acid

Terephthalic acid

Maleic anhydride

Fumaric acid

Chlorendic anhydride

FIGURE 3.2 Components of unsaturated polyesters.

Halogenated polyester
Terephthalate (PET)

The strength of a laminate is determined primarily by the type and amount of reinforcing used. The specific polyester resin will determine operating temperatures and corrosion resistance. Table 3.14 lists the physical and mechanical properties of polyester laminates with different reinforcing materials.

A. General Purpose Polyesters

These resins have the following formula:

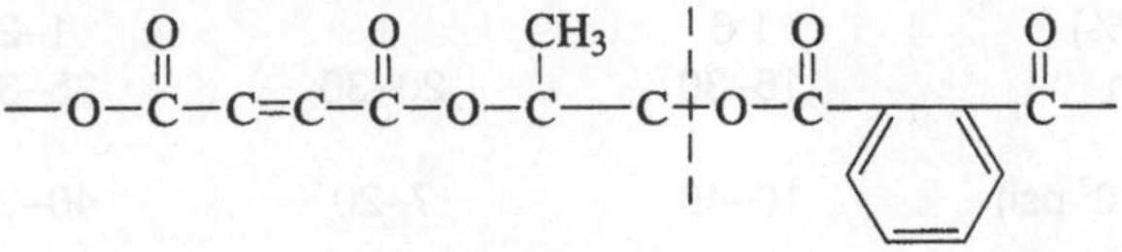

$$-O-\overset{\overset{O}{\|}}{C}-C=C-\overset{\overset{O}{\|}}{C}-O-\overset{\underset{\underset{|}{CH_3}}{|}}{C}-C+O-\overset{\overset{O}{\|}}{C}\underset{\bigcirc}{}\overset{\overset{O}{\|}}{C}-$$

General purpose polyesters are based on phthalic anhydride as the saturated monomer and are the lowest cost class of resin. These general purpose resins are normally not recommended for use in corrosive service. They are adequate for use with nonoxidizing mineral acids and corrodents that are relatively mild. Tests have indicated that general purpose resins will provide satisfactory service with the following materials up to 125°F/52°C.

Acids

Acetic acid 10%	Oleic acid
Citric acid	Benzoic acid
Fatty acids	Boric acid
Lactic acid 1%	

Salts

Aluminum sulfate	Ferrous chloride
Ammonium chloride	Magnesium chloride
Ammonium sulfate 10%	Magnesium sulfate
Calcium chloride (saturated)	Nickel chloride
Calcium sulfate	Nickel nitrate
Copper sulfate	Nickel sulfate
Ferric chloride	Potassium chloride
Ferric nitrate	Potassium sulfate
Ferric sulfate	Sodium chloride 10%

TABLE 3.14 Physical and Mechanical Properties of Polyester Laminates

Property	Preformed chopped roving	Premixed chopped glass	Woven cloth
Specific gravity	1.38–2.3	1.66–2.3	1.5–2.1
Water absorption (24 hr at 73°F/23°C)(%)	1.1–1.0	0.06–0.29	0.05–0.5
Dielectric strength, short-term (V/mil)	350–500	345–420	350–500
Tensile strength at break ($\times 10^3$ psi)	15–30	3–10	30–60
Tensile modulus ($\times 10^3$ psi)	800–2000	1,000–2,500	1,500–4,500
Elongation at break (%)	1.6	1	1–2
Compressive strength ($\times 10^3$ psi)	15–30	20–30	25–30
Flexural strength ($\times 10^3$ psi)	10–40	7–20	40–80
Compressive modulus ($\times 10^3$ psi)			
Flexural modulus ($\times 10^3$ psi) at			
73°F/23°C	1,020–3,000	1,000–2,000	1,000–3,000
200°F/93°C			600
250°F/121°C			450
Izod impact (ft-lb/in. of notch)	2–20	1.5–16	5–30
Hardness, Barcol	50–60	50–60	60–80
Coefficient of thermal expansion (10^{-6} in./in./°F)	20–50	20–33	15–30
Thermal conductivity (10^{-4} cal-cm/sec-cm^2/°C or Btu/hr/ft^2/°F/in.)			
Deflection temperature at 264 psi (°F) at 66 psi (°F)	> 400	> 400	> 400
Max. operating temperature (°F/°C)			
Limiting oxygen index (%)			
Flame spread			
Underwriters Lab. rating (Sub. 94)			

Solvents

Amyl alcohol	Kerosene
Glycerine	Naphtha

General purpose resins are not satisfactory in contact with the following:

Oxidizing acids
Alkaline solutions such as calcium hydroxide, sodium hydroxide, and
   sodium carbonate
Bleach solutions such as 5% sodium hypochlorite
Solvents such as carbon disulfide, carbon tetrachloride, and gasoline.

This resin predominates in boat building. It exhibits excellent resistance to all types of water, including seawater. For the most part, these resins find application in building panels, boats, radomes, television satellite dishes, tote boxes, fishing poles, building materials, car and truck bodies, and any application where exposure is confined to ambient conditions.

B. Isophthalic Polyesters

Isophthalic polyesters have the following chemical formula:

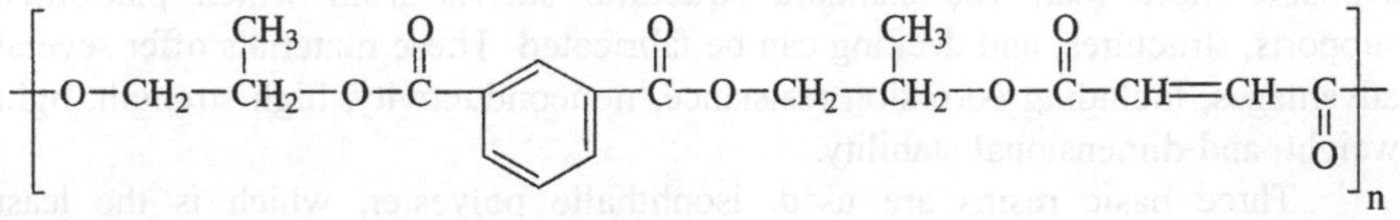

$$\left[-O-CH_2-\underset{CH_3}{\overset{|}{CH}}-CH_2-O-\overset{O}{\overset{\|}{C}}-\underset{}{\bigcirc}-\overset{O}{\overset{\|}{C}}-O-CH_2-\underset{CH_3}{\overset{|}{CH}}-CH_2-O-\overset{O}{\overset{\|}{C}}-CH=CH-\overset{O}{\overset{\|}{C}}- \right]_n$$

These polyester resins use isophthalic acid in place of phthalic anhydride as the saturated monomer. This increases the cost to produce but improves physical properties and chemical resistance.

The standard corrosion grade isophthalic polyesters are made with a 1 : 1 molar ratio of isophthalic acid to maleic anhydride or fumaric acid with propylene glycol. In general, the properties of isophthalic polyesters are superior to the lower cost general purpose polyesters, not only in chemical resistance but physical properties as well. They can be formulated to be fire retardant or to protect against the effects of ultraviolet light.

1. Corrosion Resistance

Isophthalic polyesters have a relatively wide range of corrosion resistance. They are satisfactory for use up to 125°F/52°C in such acids as 10% acetic, benzoic, boric, citric, oleic, 25% phosphoric, tartaric, 10–25% sulfuric, and fatty acids. Most inorganic salts are also compatible with isophthalic polyesters. Solvents such as amyl alcohols, ethylene glycol, formaldehyde, gasoline, kerosene, and naphtha are also compatible.

The isophthalic polyester resins are not resistant to acetone, amyl acetate, benzene, carbon disulfide, solutions of alkaline salts of potassium and sodium, hot distilled water, or higher concentrations of oxidizing acids.

Refer to Table 3.15 for the compatibility of isophthalic polyesters with selected corrodents. Reference 1 provides a more detailed listing.

2. Typical Applications

Isophthalic polyesters provide corrosion resistance in a wide variety of end uses. Recent excavations of 25 year old FRP underground gasoline storage tanks offer evidence to the resins' resistance to internal chemical attack or external attack by water in the surrounding soil. Isopolyester pipes and containment vaults also provide protection against failure from corrosion by acid media. Reference 2 provides information on FRP piping systems.

Isopolyesters used in above-ground tanks and tank linings for chemical storage provide similar protection. In food contact applications, these resins withstand acids and corrosive salts encountered in foods and food handling.

The range of successful commercial applications include support structures, ladders, grating, platforms, and guard rails. At the present time, there are available more than 100 standard structural shapes from which platforms, supports, structures, and decking can be fabricated. These materials offer several advantages, including corrosion resistance, nonconductivity, high strength, light weight, and dimensional stability.

Three basic resins are used: isophthalic polyester, which is the least expensive and is used in most general applications for which it is not necessary that the material have fire-retardant properties; isophthalic polyester resin with a fire-retardant additive that has the same general properties and area of application as the basic isophthalic resin; and vinyl ester resin, which is the most expensive but has the greatest corrosion resistance, superior strength and physical properties, and is also fire retardant. These materials are resistant to most corrodents found in the atmosphere and to a wide range of materials that may spill on the surface.

The glass-reinforced polyester and vinyl ester resin structures are nonconductive. As such, they are neither thermally nor electrically conductive. Consequently, they are ideally suited to be used as cable trays, utility truck booms, ladders, support structures for electrical devices and systems, and power junction boxes.

On a pound-for-pound basis, these fiber-reinforced polymer (FRP) materials are stronger than steel. Shapes have been used to form the superstructures of multistoried buildings and floating bridges. Joining is accomplished by the use of fiberglass studs and nuts. The threads are standardized to be compatible with metal studs. Table 3.16 shows the typical mechanical and physical properties of

TABLE 3.15 Compatibility of Isophthalic Polyester with Selected Corrodents[a]

Chemical	Maximum temp.		Chemical	Maximum temp.	
	°F	°C		°F	°C
Acetaldehyde	x	x	Benzaldehyde	x	x
Acetic acid 10%	180	82	Benzene	x	x
Acetic acid 50%	110	43	Benzenesulfonic acid 10%	180	82
Acetic acid 80%	x	x	Benzoic acid	180	82
Acetic acid, glacial	x	x	Benzyl alcohol	x	x
Acetic anhydride	x	x	Benzyl chloride	x	x
Acetone	x	x	Borax	140	60
Acetyl chloride	x	x	Boric acid	180	82
Acrylic acid	x	x	Bromine gas, dry	x	x
Acrylonitrile	x	x	Bromine gas, moist	x	x
Adipic acid	220	104	Bromine, liquid	x	x
Allyl alcohol	x	x	Butyl acetate	x	x
Allyl chloride	x	x	Butyl alcohol	80	27
Alum	250	121	n-Butylamine	x	x
Aluminum chloride, aqueous	180	82	Butyric acid 25%	129	49
Aluminum chloride, dry	170	77	Calcium bisulfide	160	71
Aluminum fluoride 10%	140	60	Calcium bisulfite	150	66
Aluminum hydroxide	160	71	Calcium carbonate	160	71
Aluminum nitrate	160	71	Calcium chlorate	160	71
Aluminum sulfate	180	82	Calcium chloride	180	82
Ammonia gas	90	32	Calcium hydroxide 10%	160	71
Ammonium carbonate	x	x	Calcium hydroxide, sat.	160	71
Ammonium chloride 10%	160	71	Calcium hypochlorite 10%	120	49
Ammonium chloride 50%	160	71	Calcium nitrate	140	60
Ammonium chloride, sat.	180	82	Calcium oxide	160	71
Ammonium fluoride 10%	90	32	Calcium sulfate	160	71
Ammonium fluoride 25%	90	32	Caprylic acid	160	71
Ammonium hydroxide 25%	x	x	Carbon bisulfide	x	x
Ammonium hydroxide, sat.	x	x	Carbon dioxide, dry	160	71
Ammonium nitrate	160	71	Carbon dioxide, wet	160	71
Ammonium persulfate	160	71	Carbon disulfide	x	x
Ammonium phosphate	160	71	Carbon monoxide	160	71
Ammonium sulfate 10%	180	82	Carbon tetrachloride	x	x
Ammonium sulfide	x	x	Carbonic acid	160	71
Ammonium sulfite	x	x	Cellosolve	x	x
Amyl acetate	x	x	Chloroacetic acid, 50% water	x	x
Amyl alcohol	160	71	Chloroacetic acid to 25%	150	66
Amyl chloride	x	x	Chlorine gas, dry	160	71
Aniline	x	x	Chlorine gas, wet	160	71
Antimony trichloride	160	71	Chlorine, liquid	x	x
Aqua regia 3 : 1	x	x	Chlorobenzene	x	x
Barium carbonate	190	88	Chloroform	x	x
Barium chloride	140	60	Chlorosulfonic acid	x	x
Barium hydroxide	x	x	Chromic acid 10%	x	x
Barium sulfate	160	71	Chromic acid 50%	x	x
Barium sulfide	90	32	Chromyl chloride	140	60

(Continued)

TABLE 3.15 Continued

Chemical	Maximum temp. °F	°C	Chemical	Maximum temp. °F	°C
Citric acid 15%	160	71	Methyl isobutyl ketone	x	x
Citric acid, concd	200	93	Muriatic acid	160	71
Copper acetate	160	71	Nitric acid 5%	120	49
Copper chloride	180	82	Nitric acid 20%	x	x
Copper cyanide	160	71	Nitric acid 70%	x	x
Copper sulfate	200	93	Nitric acid, anhydrous	x	x
Cresol	x	x	Nitrous acid, concd	120	49
Cupric chloride 5%	170	77	Oleum	x	x
Cupric chloride 50%	170	77	Perchloric acid 10%	x	x
Cyclohexane	80	27	Perchloric acid 70%	x	x
Dichloroacetic acid	x	x	Phenol	x	x
Dichloroethane			Phosphoric acid 50–80%	180	82
(ethylene dichloride)	x	x	Picric acid	x	x
Ethylene glycol	120	49	Potassium bromide 30%	160	71
Ferric chloride	180	82	Salicylic acid	100	38
Ferric chloride 50%			Sodium carbonate 20%	90	32
in water	160	71	Sodium chloride	200	93
Ferric nitrate 10–50%	180	82	Sodium hydroxide 10%	x	x
Ferrous chloride	180	82	Sodium hydroxide 50%	x	x
Ferrous nitrate	160	71	Sodium hydroxide, concd	x	x
Fluorine gas, dry	x	x	Sodium hypochlorite 20%	x	x
Fluorine gas, moist	x	x	Sodium hypochlorite, concd	x	x
Hydrobromic acid, dilute	120	49	Sodium sulfide to 50%	x	x
Hydrobromic acid 20%	140	60	Stannic chloride	180	82
Hydrobromic acid 50%	140	60	Stannous chloride	180	82
Hydrochloric acid 20%	160	71	Sulfuric acid 10%	160	71
Hydrochloric acid 38%	160	71	Sulfuric acid 50%	150	66
Hydrocyanic acid 10%	90	32	Sulfuric acid 70%	x	x
Hydrofluoric acid 30%	x	x	Sulfuric acid 90%	x	x
Hydrofluoric acid 70%	x	x	Sulfuric acid 98%	x	x
Hydrofluoric acid 100%	x	x	Sulfuric acid 100%	x	x
Hypochlorous acid	90	32	Sulfuric acid, fuming	x	x
Ketones, general	x	x	Sulfurous acid	x	x
Lactic acid 25%	160	71	Thionyl chloride	x	x
Lactic acid, concd	160	71	Toluene	110	43
Magnesium chloride	180	82	Trichloroacetic acid 50%	170	77
Malic acid	90	32	White liquor	x	x
Methyl ethyl ketone	x	x	Zinc chloride	180	82

[a] The chemicals listed are in the pure state or in a saturated solution unless otherwise indicated. Compatibility is shown to the maximum allowable temperature for which data are available. Incompatibility is shown by an x. A blank space indicates that data are unavailable.

TABLE 3.16 Typical Mechanical Properties of Fiberglass-Reinforced Standard Shapes

	Resin system	
Mechanical property	Polyester	Vinyl ester
Longitudinal direction		
Ultimate tensile strength (psi)	35,000	37,500
Ultimate compressive strength (psi)	40,000	45,000
Ultimate flexural strength (psi)	37,500	37,500
Tensile modulus (psi $\times 10^6$)	3.0	3.0
Compressive modulus (psi $\times 10^6$)	3.0	3.0
Flexural modulus (psi $\times 10^6$)	2.0	2.0
Ultimate shear strength (psi)	6,000	7,000
Ultimate bearing stress (psi)	35,000	37,000
Izod impact (ASTM D-256)	30	30
(ft-lb/in. of notch sample thickness 1/8 in. except 1/4 in. for rod)		
Traverse direction		
Ultimate tensile strength (psi)	8,000	10,000
Ultimate compressive strength (psi)	17,500	20,000
Ultimate flexural strength (psi)	11,000	14,000
Tensile modulus (psi $\times 10^6$)	1.0	1.0
Compressive modulus (psi $\times 10^6$)	1.2	1.2
Flexural modulus (psi $\times 10^6$)	1.0	1.0
Ultimate shear strength (psi)	5,500	5,500
Ultimate bearing stress (psi)	30,000	30,000
Izod impact (ASTM D-256)	50	50
(ft-lb/in. of notch Barcol hardness)		
Full section in bending		
Modulus of elasticity (psi $\times 10^6$)	2.8	2.8
Tensile strength (psi)	25,000	25,000
Compressive strength (psi)	25,000	25,000
Electrical		
Electric strength, short-term in oil, 1/8 in. (ASTM D-149) (V pm)	200	225
Electric strength, short-term in oil (kV/in.)[b]	50	50
Dielectric constant, 60 Hz (ASTM D-150)[a]	5.2	5.2
Dissipation factor 60 Hz (ASTM D-150)[a]	0.03	0.03
Arc resistance (ASTM D-495)(sec[c])	120	120
Thermal		
Thermal coefficient of expansion (ASTM D-696)(in./in./°F)[c]	5×10^{-6}	5×10^{-6}
Thermal conductivity (Btu/ft^2 hr/°F/in.$^{-1}$)	4	4

(Continued)

TABLE 3.16 Continued

Mechanical property	Resin system	
	Polyester	Vinyl ester
Specific heat (Btu/lb/°F/in.$^{-1}$)	0.28	0.28
Other		
Density (ASTM D-702)(lb/in.)^c	0.065 (S-525)	0.065 (S-525)
Solid shapes		
Hollow shapes	0.065 (S-525)	0.065 (S-525)
Specific gravity (ASTM D-792)	1.80 (S-525)	1.80 (S-525)
Solid shapes		
Hollow shapes	1.83 (S-525)	1.83 (S-525)
Water absorption, 24 hr immersion, ASTM D-570 (max.% by wt)	0.50	0.50
Flame-retardant properties, isophthalic polyester with flame-retardant chemical additive flame resistance (FTMS 406-2023), ign/burn (sec)	75/75	75/75
Intermittent flame test (HLT-15) rating	100	100
Flammability test (ASTM D-635)	< 5	< 5
Average time of burning (sec)		
Average extent of burning (min)	15	15
Surface burning characteristics (ASTM E-84), max.	25	25
U.L. approved (yellow card) U.L. 94 flammability classification	V-0	V-0
temperature index (°C)	130	130

^{a} Specimen tested perpendicular to laminate face.
^{b} One-inch-long specimen tested parallel to laminate face using 2-in.-diameter electrodes.
^{c} Reported value measured in longitudinal direction.

fiberglass-reinforced standard shapes. Table 3.17 shows the typical properties of FRP studs.

FRP structural shapes are very light in weight, which permits them to be easily transported, handled, and set into place. They weigh 80% less than corresponding steel shapes and 30% less than corresponding aluminum shapes.

When properly designed, these materials will not sag or creep under prolonged deformation or loading. Their coefficient of thermal expansion is slightly less than that of steel but considerably less than that of aluminum or wood.

Considering the foregoing properties, these materials are ideal for structural supports and decking to resist atmospheric corrosion. These shapes, handrails, deckings, and such, are available in color, with the color completely penetrating

TABLE 3.17 Typical Mechanical Properties of Fiberglass Studs[a]

Property	Size (UNC)				
	1/8-16	1/2-13	5/8-11	3/4-10	1-8
Thread shear strength using fiberglass nut in tension (lb)	1,250	2,200	3,100	4,500	6,500
Traverse shear on double-threaded rod, double shear, ASTM B-565 (load/lb)	3,000	5,000	7,500	12,000	22,000
Traverse shear on threaded rod, single shear (load/lb)	1,600	2,600	3,800	6,200	15,000
Compressive strength, longitudinal, ASTM D-695 (psi)	54,000	54,000	54,000	54,000	65,000
Flexural strength, ASTM D-790 (psi)	55,000	55,000	55,000	55,000	60,000
Torque strength using fiberglass nut lubricated with SAE 16W30 motor oil (ft/lb)	8	18	35	50	110
Dielectric strength, ASTM D-149 (kV/in.)	35	35	35	35	35
Water absorption, 24-hr immersion threaded, D-570 (%)	1	1	1	1	1
Coefficient of thermal expansion, longitudinal (in./in./°F × 10^{-6})	5	5	5	5	5
Max. recommended operating temperature based on 50% retention of ultimate thread shear strength (°F/°C)	All 210/100				
Stud weight (lb/ft)	0.07	0.12	0.18	0.28	0.50

[a]Test results are for bolts with single nuts only. Properties do not apply when fiberglass is used with a metal nut.

the material. This eliminates the need for periodic painting for protection or aesthetic reasons. This makes the maintenance needs of these materials greatly reduced over that of their metallic counterparts. Initial purchase price may be slightly higher than the purchase price for steel, but the final installed cost, including maintenance costs, makes the total installation highly competitive with less expensive materials.

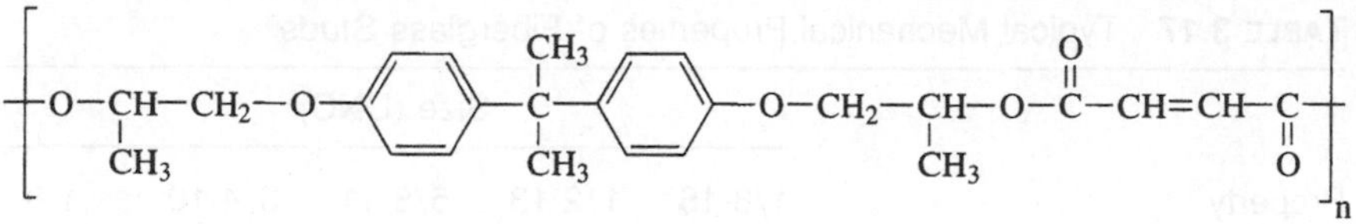

FIGURE 3.3   Chemical structure of bisphenol A–fumarate resins.

Typical uses to which FRP materials have been put include

Ladders	Handrails
Support structures	Walkways
Operating platforms	Roof supports and decking
Pipe supports	Stairs
Cable trays	Gratings
Enclosures	

The water resistant properties of the isophthalates also provide advantages in appearance and performance in cast countertops, as well as pool, tub, and spa applications. Other applications include their use in cooling towers and use as rail cars to transport automobiles.

C.   Bisphenol-A Fumarate Polyesters

This is a premium corrosion-resistant grade resin. It costs approximately twice as much as a general purpose resin and approximately one-third more than an isophthalic resin. The structural formula is shown in Figure 3.3.

Standard bisphenol-A polyester resins are derived from the propylene glycol or oxide diether of bisphenol-A and fumaric acid. The aromatic structure contributed by the bisphenol-A provides several benefits. Thermal stability is improved, and the heat distortion point of the resin is raised, mainly from the more rigid nature of the aromatic structure. The number of interior chain ester groups is reduced, so the resistance to hydrolysis and saponification is increased. Bisphenol-A fumarate polyesters have the best hydrolysis resistance of any commercial unsaturated polyester.

1.   Corrosion Resistance

The bisphenol polyesters are superior in their corrosion resistant properties to the isophthalic polyesters. They show good performance with moderate alkaline solutions and excellent resistance to the various categories of bleaching agents. The bisphenol polyesters will break down in highly concentrated acids or alkalies. These resins can be used in the handling of the following materials:

Acids (to 200°F/93°C)

Acetic	Fatty acids	Stearic
Benzoic	Hydrochloric 10%	Sulfonic 30%
Boric	Lactic	Tannic
Butyric	Maleic	Tartaric
Chloroacetic 15%	Oleic	Trichloroacetic 50%
Chromic 5%	Oxalic	Rayon spin bath
Citric	Phosphoric 80%	

Salt (solution to 200°F/93°C)

All aluminum salts	Copper salts
Most ammonium salts	Iron salts
Calcium salts	Zinc salts
	Most plating solutions

Alkalies

Ammonium hydroxide 5% to 160°F/71°C
Calcium hydroxide 25% to 160°F/71°C
Calcium hypochlorite 20% to 200°F/93°C
Chlorine dioxide 15% to 200°F/93°C
Potassium hydroxide 25% to 160°F/71°C
Sodium hydroxide 25% to 160°F/71°C
Sodium chlorite to 200°F/93°C
Sodium hydrosulfite to 200°F/93°C

Solvents Isophthalics are resistant to all solvents plus

Sour crude oil	Alcohols at ambient temperatures
Glycerine	Linseed oil

Gases to 200°F/93°C

Carbon dioxide	Sulfur dioxide, dry
Carbon monoxide	Sulfur dioxide, wet
Chlorine, dry	Sulfur trioxide
Chlorine, wet	Rayon waste gases 150°F/66°C

Solvents such as benzene, carbon disulfide, ether, methyl ethyl ketone, toluene, xylene, trichloroethylene, and trichloroethane will attack the resin. Sulfuric acid above 70% sodium hydroxide and 30% chromic acid will also attack the resin. Refer to Table 3.18 for the compatibility of bisphenol-A fumarate polyester resin with selected corrodents. Table 3.19 lists the compatibility of

TABLE 3.18 Compatibility of Bisphenol A—Fumarate Polyester with Selected Corrodents[a]

Chemical	Maximum temp.		Chemical	Maximum temp.	
	°F	°C		°F	°C
Acetaldehyde	x	x	Barium sulfide	140	60
Acetic acid 10%	220	104	Benzaldehyde	x	x
Acetic acid 50%	160	171	Benzene	x	x
Acetic acid 80%	160	171	Benzenesulfonic acid 10%	200	93
Acetic acid, glacial	x	x	Benzoic acid	180	82
Acetic anhydride	110	43	Benzyl alcohol	x	x
Acetone	x	x	Benzyl chloride	x	x
Acetyl chloride	x	x	Borax	220	104
Acrylic acid	100	38	Boric acid	220	104
Acrylonitrile	x	x	Bromine gas, dry	90	32
Adipic acid	220	104	Bromine gas, moist	100	38
Allyl alcohol	x	x	Bromine, liquid	x	x
Allyl chloride	x	x	Butyl acetate	80	27
Alum	220	104	Butyl alcohol	80	27
Aluminum chloride, aqueous	200	93	n-Butylamine	x	x
Aluminum fluoride 10%	90	32	Butyric acid	220	93
Aluminum hydroxide	160	71	Calcium bisulfite	180	82
Aluminum nitrate	200	93	Calcium carbonate	210	99
Aluminum sulfate	200	93	Calcium chlorate	200	93
Ammonia gas	200	93	Calcium chloride	220	104
Ammonium carbonate	90	32	Calcium hydroxide 10%	180	82
Ammonium chloride 10%	200	93	Calcium hydroxide, sat.	160	71
Ammonium chloride 50%	220	104	Calcium hypochlorite 10%	80	27
Ammonium chloride, sat.	220	104	Calcium nitrate	220	93
Ammonium fluoride 10%	180	82	Calcium sulfate	220	93
Ammonium fluoride 25%	120	49	Caprylic acid	160	71
Ammonium hydroxide 25%	100	38	Carbon bisulfide	x	x
Ammonium hydroxide, 20%	140	60	Carbon dioxide, dry	350	177
Ammonium nitrate	220	104	Carbon dioxide, wet	210	99
Ammonium persulfate	180	82	Carbon disulfide	x	x
Ammonium phosphate	80	27	Carbon monoxide	350	177
Ammonium sulfate 10–40%	220	104	Carbon tetrachloride	110	43
Ammonium sulfide	110	43	Carbonic acid	90	32
Ammonium sulfite	80	27	Cellosolve	140	60
Amyl acetate	80	27	Chloroacetic acid,		
Amyl alcohol	200	93	50% water	140	60
Amyl chloride	x	x	Chloroacetic acid		
Aniline	x	x	to 25%	80	27
Antimony trichloride	220	104	Chlorine gas, dry	200	93
Aqua regia 3:1	x	x	Chlorine gas, wet	200	93
Barium carbonate	200	93	Chlorine, liquid	x	x
Barium chloride	220	104	Chlorobenzene	x	x
Barium hydroxide	150	66	Chloroform	x	x
Barium sulfate	220	104	Chlorosulfonic acid	x	x

TABLE 3.18 Continued

Chemical	Maximum temp.		Chemical	Maximum temp.	
	°F	°C		°F	°C
Chromic acid 10%	x	x	Methyl isobutyl ketone	x	x
Chromic acid 50%	x	x	Muriatic acid	130	54
Chromyl chloride	150	66	Nitric acid 5%	160	71
Citric acid 15%	220	104	Nitric acid 20%	100	38
Citric acid, concd	220	104	Nitric acid 70%	x	x
Copper acetate	180	82	Nitric acid, anhydrous	x	x
Copper chloride	220	104	Oleum	x	x
Copper cyanide	220	104	Phenol	x	x
Copper sulfate	220	104	Phosphoric acid 50–80%	220	104
Cresol	x	x	Picric acid	110	43
Cyclohexane	x	x	Potassium bromide 30%	200	93
Dichloroacetic acid	100	38	Salicylic acid	150	66
Dichloroethane	x	x	Sodium carbonate	160	71
(ethylene dichloride)	220	104	Sodium chloride	220	104
Ethylene glycol	220	104	Sodium hydroxide 10%	130	54
Ferric chloride	220	104	Sodium hydroxide 50%	220	104
Ferric chloride 50% in water	220	104	Sodium hydroxide, concd	200	93
Ferric nitrate 10–50%	220	104	Sodium hypochlorite 20%	x	x
Ferrous chloride	220	104	Sodium sulfide to 50%	210	99
Ferrous nitrate	220	104	Stannic chloride	200	93
Fluorine gas, moist			Stannous chloride	220	104
Hydrobromic acid, dilute	220	104	Sulfuric acid 10%	220	104
Hydrobromic acid 20%	220	104	Sulfuric acid 50%	220	104
Hydrobromic acid 50%	160	71	Sulfuric acid 70%	160	71
Hydrochloric acid 20%	190	88	Sulfuric acid 90%	x	x
Hydrochloric acid 38%	x	x	Sulfuric acid 98%	x	x
Hydrocyanic acid 10%	200	93	Sulfuric acid 100%	x	x
Hydrofluoric acid 30%	90	32	Sulfuric acid, fuming	x	x
Hypochlorous acid 20%	90	32	Sulfurous acid	110	43
Iodine solution 10%	200	104	Thionyl chloride	x	x
Lactic acid 25%	210	99	Toluene	x	x
Lactic acid, concd	220	104	Trichloroacetic acid 50%	180	82
Magnesium chloride	220	104	White liquor	180	82
Malic acid	160	71	Zinc chloride	250	121
Methyl ethyl ketone	x	x			

[a] The chemicals listed are in the pure state or in a saturated solution unless otherwise indicated. Compatibility is shown to the maximum allowable temperature for which data are available. Incompatibility is shown by an x. A blank space indicates that data are unavailable.
Source: Schweitzer PA. Corrosion Resistance Tables, 4th ed., Vols. 1–3. New York: Marcel Dekker, 1995.

TABLE 3.19 Compatibility of Hydrogenated Bisphenol A–Bisphenol A Polyester with Selected Corrodents[a]

Chemical	Maximum temp.		Chemical	Maximum temp.	
	°F	°C		°F	°C
Acetic acid 10%	200	93	Citric acid 15%	200	93
Acetic acid 50%	160	71	Citric acid, concd	210	99
Acetic anhydride	x	x	Copper acetate	210	99
Acetone	x	x	Copper chloride	210	99
Acetyl chloride	x	x	Copper cyanide	210	99
Acrylonitrile	x	x	Copper sulfate	210	99
Aluminum acetate			Cresol	x	x
Aluminum chloride, aqueous	200	93	Cyclohexane	210	99
Aluminum fluoride	x	x	Dichloroethane		
Aluminum sulfate	200	93	(ethylene dichloride)	x	x
Ammonium chloride, sat.	200	93	Ferric chloride	210	99
Ammonium nitrate	200	93	Ferric chloride 50%		
Ammonium persulfate	200	93	in water	200	93
Ammonium sulfide	100	38	Ferric nitrate 10–50%	200	93
Amyl acetate	x	x	Ferrous chloride	210	99
Amyl alcohol	200	93	Ferrous nitrate	210	99
Amyl chloride	90	32	Hydrobromic acid 20%	90	32
Aniline	x	x	Hydrobromic acid 50%	90	32
Antimony trichloride	80	27	Hydrochloric acid 20%	180	82
Aqua regia 3 : 1	x	x	Hydrochloric acid 38%	190	88
Barium carbonate	180	82	Hydrocyanic acid 10%	x	x
Barium chloride	200	93	Hydrofluoric acid 30%	x	x
Benzaldehyde	x	x	Hydrofluoric acid 70%	x	x
Benzene	x	x	Hydrofluoric acid 100%	x	x
Benzoic acid	210	99	Hypochlorous acid 50%	210	99
Benzyl alcohol	x	x	Lactic acid 25%	210	99
Benzyl chloride	x	x	Lactic acid, concd	210	99
Boric acid	210	99	Magnesium chloride	210	99
Bromine, liquid	x	x	Methyl ethyl ketone	x	x
Butyl acetate	x	x	Methyl isobutyl ketone	x	x
n-Butylamine	x	x	Muriatic acid	190	88
Butyric acid	x	x	Nitric acid 5%	90	32
Calcium bisulfide	120	49	Oleum	x	x
Calcium chlorate	210	99	Perchloric acid 10%	x	x
Calcium chloride	210	99	Perchloric acid 70%	x	x
Calcium hypochlorite 10%	180	82	Phenol	x	x
Carbon bisulfide	x	x	Phosphoric acid 50–80%	210	99
Carbon disulfide	x	x	Sodium carbonate 10%	100	38
Carbon tetrachloride	x	x	Sodium chloride	210	99
Chloroacetic acid,			Sodium hydroxide 10%	100	38
50% water	90	32	Sodium hydroxide 50%	x	x
Chlorine gas, dry	210	99	Sodium hydroxide, concd	x	x
Chlorine gas, wet	210	99	Sodium hypochlorite 10%	160	71
Chloroform	x	x	Sulfuric acid 10%	210	99
Chromic acid 50%	x	x	Sulfuric acid 50%	210	99

TABLE 3.19 Continued

Chemical	Maximum temp.		Chemical	Maximum temp.	
	°F	°C		°F	°C
Sulfuric acid 70%	90	32	Sulfurous acid 25%	210	99
Sulfuric acid 90%	x	x	Toluene	90	32
Sulfuric acid 98%	x	x	Trichloroacetic acid	90	32
Sulfuric acid 100%	x	x	Zinc chloride	200	93
Sulfuric acid, fuming	x	x			

a The chemicals listed are in the pure state or in a saturated solution unless otherwise indicated. Compatibility is shown to the maximum allowable temperature for which data are available. Incompatibility is shown by an x. A blank space indicates that data are unavailable. *Source:* Schweitzer PA. Corrosion Resistance Tables, 4th ed., Vols. 1–3. New York: Marcel Dekker, 1995.

hydrogenated bisphenol-A–bisphenol-A polyesters with selected corrodents. Reference 1 provides a more detailed listing.

2. Typical Applications

Bisphenol-A resins find applications similar to the isophthalic resins but in areas requiring greater resistance to corrosion.

D. Halogenated Polyesters

Halogenated resins consist of chlorinated or brominated polymers. The chlorinated polyester resins cured at room temperature and reinforced with fiber glass possess unique physical, mechanical, and corrosion resistant properties. These are also known as chlorendic polyesters.

These resins have a very high heat distortion point, and the laminates show very high retention of physical strength at elevated temperature. Refer to Table 3.20 for the performance of chlorinated polyester laminates at elevated temperature. This permits them to survive high-temperature upsets in flue gas desulfurization scrubbers, some of which may reach a temperature of 400°F/204°C. These laminates are routinely used as chimney liners at temperatures of 240–280°F/116–138°C.

These resins have the highest heat resistance of any chemically resistant polyester. They are also inherently fire retardant. A noncombustible rating of 20 can be achieved, making this the safest possible polyester for stacks, hoods, fans, ducts, or wherever a fire hazard might exist. This fire retardancy is achieved by the addition of antimony trioxide.

TABLE 3.20   Elevated Temperature Performance of Chlorinated
Polyester Laminates

Temperature (°F/°C)	Retention tensile strength (%)	Retention flexural modulus (%)
75/24		100
125/52		96
175/79	100	88
225/107	98	77
275/135	92	65
325/163	84	46
375/191	73	25
425/218	60	

1. Corrosion Resistance

Excellent resistance is exhibited in contact with oxidizing acids and solutions, such as 35% nitric acid at elevated temperatures and 70% nitric acid at room temperature, 40% chromic acid, chlorine water, wet chlorine, and 15% hypochlorites. They also resist neutral and acid salts, nonoxidizing acids, organic acids, mercaptans, ketones, aldehydes, alcohols, glycols, organic esters, and fats and oils. Table 3.21 is a general application guide.

These polyesters are not resistant to highly alkaline solutions of sodium hydroxide, concentrated sulfuric acid, alkaline solutions with pH greater than 10, aliphatic, primary and aromatic amines, amides, and other alkaline organics, phenol, and acid halides. Table 3.22 provides the compatibility of halogenated polyesters with selected corrodents.

2. Typical Applications

Halogenated polyesters are widely used in the pulp and paper industry in bleach atmospheres where they outperform stainless steel and high nickel alloys. Applications are also found for ductwork, fans, and other areas where potential fire hazards may be present. They are also used for high-temperature applications such as chimney liners, chemical storage tanks, and chemical piping, among other applications.

E.   Terephthalate Polyesters (PET)

Terephthalate polyesters are based on terephthalic acid, the para isomer of phthalic acid. The properties of cured terephthalic-based polyesters are similar to those of isophthalic polyesters, with the terephthalics having higher heat distortion temperatures and being somewhat softer at equal saturation levels.

TABLE 3.21 General Application Guide for Chlorinated Polyester Laminates

Environment	Comments
Acid halides	Not recommended
Acids, mineral nonoxidizing	Resistant to 250°F/121°C
Acids, organic	Resistant to 250°F/121°C; glacial acetic acid to 120°F/49°C
Alcohols	Resistant to 180°F/82°C
Aldehydes	Resistant to 180°F/82°C
Alkaline solutions pH >10	Not recommended for continuous exposure
Amines, aliphatic, primary aromatic	Can cause severe attack
Amides, other alkaline organics	Can cause severe attack
Esters, organic	Resistant to 180°F/82°C
Fats and oils	Resistant to 200°F/95°C
Glycols	Resistant to 180°F/82°C
Ketones	Resistant to 180°F/82°C
Mercaptans	Resistant to 180°F/82°C
Phenol	Not recommended
Salts, acid	Resistant to 250°F/121°C
Salts, neutral	Resistant to 250°F/121°C
Water, demineralized, distilled, deionized, steam and condensate	Resistant to 212°F/100°C; Lowest absorption of any polyester

1. Corrosion Resistance

Corrosion resistance of the PETs is fairly similar to that of the isophthalics. Testing has indicated that the benzene resistance of comparably formulated resins is lower for PET versus isophthalic polyesters. This trend is also followed where retention of flexural modulus is elevated for various terephthalic resins versus the standard corrosion grade isophthalic resin. The PET's loss of properties in gasoline is greater than the isophthalics at the same level of unsaturation, but as the unsaturation increases the gasoline resistance reverses, with the PET performing better. The trend was seen only at unsaturated acid levels of greater than 50 mol%. This was achieved with a reversal of performance in 10% sodium hydroxide; where the PET with lower unsaturation was better than the isophthalic level. This follows a general trend for thermosets, that as crosslink density increases, solvent resistance increases.

Refer to Table 3.23 for the compatibility of terephthalate polyester with selected corrodents. Reference 1 provides a more detailed listing.

TABLE 3.22 Compatibility of Halogenated Polyester with Selected Corrodents[a]

Chemical	Maximum temp.		Chemical	Maximum temp.	
	°F	°C		°F	°C
Acetaldehyde	x	x	Barium sulfide	x	x
Acetic acid 10%	140	60	Benzaldehyde	x	x
Acetic acid 50%	90	32	Benzene	90	32
Acetic acid, glacial	110	43	Benzenesulfonic		
Acetic anhydride	100	38	acid 10%	120	49
Acetone	x	x	Benzoic acid	250	121
Acetyl chloride	x	x	Benzyl alcohol	x	x
Acrylic acid	x	x	Benzyl chloride	x	x
Acrylonitrile	x	x	Borax	190	88
Adipic acid	220	104	Boric acid	180	82
Allyl alcohol	x	x	Bromine gas, dry	100	38
Allyl chloride	x	x	Bromine gas, moist	100	38
Alum 10%	200	93	Bromine, liquid	x	x
Aluminum chloride,			Butyl acetate	80	27
aqueous	120	49	Butyl alcohol	100	38
Aluminum fluoride 10%	90	32	n-Butylamine	x	x
Aluminum hydroxide	170	77	Butyric acid 20%	200	93
Aluminum nitrate	160	71	Calcium bisulfide	x	x
Aluminum oxychloride			Calcium bisulfite	150	66
Aluminum sulfate	250	121	Calcium carbonate	210	99
Ammonia gas	150	66	Calcium chlorate	250	121
Ammonium carbonate	140	60	Calcium chloride	250	121
Ammonium chloride 10%	200	93	Calcium hydroxide, sat.	x	x
Ammonium chloride 50%	200	93	Calcium hypochlorite 20%	80	27
Ammonium chloride, sat.	200	93	Calcium nitrate	220	104
Ammonium fluoride 10%	140	60	Calcium oxide	150	66
Ammonium fluoride 25%	140	60	Calcium sulfate	250	121
Ammonium hydroxide 25%	90	32	Caprylic acid	140	60
Ammonium hydroxide, sat.	90	32	Carbon bisulfide	x	x
Ammonium nitrate	200	93	Carbon dioxide, dry	250	121
Ammonium persulfate	140	60	Carbon dioxide, wet	250	121
Ammonium phosphate	150	66	Carbon disulfide	x	x
Ammonium sulfate 10–40%	200	93	Carbon monoxide	170	77
Ammonium sulfide	120	49	Carbon tetrachloride	120	49
Ammonium sulfite	100	38	Carbonic acid	160	71
Amyl acetate	190	85	Cellosolve	80	27
Amyl alcohol	200	93	Chloroacetic acid,		
Amyl chloride	x	x	50% water	100	38
Aniline	120	49	Chloroacetic acid 25%	90	32
Antimony trichloride 50%	200	93	Chlorine gas, dry	200	93
Aqua regia 3:1	x	x	Chlorine gas, wet	220	104
Barium carbonate	250	121	Chlorine, liquid	x	x
Barium chloride	250	121	Chlorobenzene	x	x
Barium hydroxide	x	x	Chloroform	x	x
Barium sulfate	180	82	Chlorosulfonic acid	x	x

TABLE 3.22 Continued

Chemical	Maximum temp. °F	Maximum temp. °C	Chemical	Maximum temp. °F	Maximum temp. °C
Chromic acid 10%	180	82	Muriatic acid	190	88
Chromic acid 50%	140	60	Nitric acid 5%	210	99
Chromyl chloride	210	99	Nitric acid 20%	80	27
Citric acid 15%	250	121	Nitric acid 70%	80	27
Citric acid, concd	250	121	Nitrous acid, concd	90	32
Copper acetate	210	99	Oleum	x	x
Copper chloride	250	121	Perchloric acid 10%	90	32
Copper cyanide	250	121	Perchloric acid 70%	90	32
Copper sulfate	250	121	Phenol 5%	90	32
Cresol	x	x	Phosphoric acid 50–80%	250	121
Cyclohexane	140	60	Picric acid	100	38
Dibutyl phthalate	100	38	Potassium bromide 30%	230	110
Dichloroacetic acid	100	38	Salicylic acid	130	54
Dichloroethane			Sodium carbonate 10%	190	88
(ethylene dichloride)	x	x	Sodium chloride	250	121
Ethylene glycol	250	121	Sodium hydroxide 10%	110	43
Ferric chloride	250	121	Sodium hydroxide 50%	x	x
Ferric chloride 50%			Sodium hydroxide, concd	x	x
in water	250	121	Sodium hypochlorite 20%	x	x
Ferric nitrate 10–50%	250	121	Sodium hypochlorite, concd	x	x
Ferrous chloride	250	121	Sodium sulfide to 50%	x	x
Ferrous nitrate	160	71	Stannic chloride	80	27
Hydrobromic acid, dil	200	93	Stannous chloride	250	121
Hydrobromic acid 20%	160	71	Sulfuric acid 10%	260	127
Hydrobromic acid 50%	200	93	Sulfuric acid 50%	200	93
Hydrochloric acid 20%	230	110	Sulfuric acid 70%	190	88
Hydrochloric acid 38%	180	82	Sulfuric acid 90%	x	x
Hydrocyanic acid 10%	150	66	Sulfuric acid 98%	x	x
Hydrofluoric acid 30%	120	49	Sulfuric acid 100%	x	x
Hypochlorous acid 10%	100	38	Sulfuric acid, fuming	x	x
Lactic acid 25%	200	93	Sulfurous acid 10%	80	27
Lactic acid, concd	200	93	Thionyl chloride	x	x
Magnesium chloride	250	121	Toluene	110	43
Malic acid 10%	90	32	Trichloroacetic acid 50%	200	93
Methyl chloride	80	27	White liquor	x	x
Methyl ethyl ketone	x	x	Zinc chloridex	200	93
Methyl isobutyl ketone	80	27			

[a] The chemicals listed are in the pure state or in a saturated solution unless otherwise indicated. Compatibility is shown to the maximum allowable temperature for which data are available. Incompatibility is shown by an x. A blank space indicates that data are unavailable. *Source:* Schweitzer PA. Corrosion Resistance Tables, 4th ed., Vols. 1–3. New York: Marcel Dekker, 1995.

TABLE 3.23 Compatibility of Polyester Terephthalate (PET) with Selected Corrodents[a]

Chemical	Maximum temp.		Chemical	Maximum temp.	
	°F	°C		°F	°C
Acetic acid 10%	300	149	Cyclohexane	80	27
Acetic acid 50%	300	149	Dichloroethane	x	x
Acetic anhydride	x	x	Ferric chloride	250	121
Acetone	x	x	Ferric nitrate 10–50%	170	77
Acetyl chloride	x	x	Ferrous chloride	250	121
Acrylonitrile	80	26	Hydrobromic acid 20%	250	121
Aluminum chloride			Hydrobromic acid 50%	250	121
aqueous	170	77	Hydrochloric acid 20%	250	121
Aluminum sulfate	300	149	Hydrochloric acid 38%	90	32
Ammonium chloride, sat.	170	77	Hydrocyanic acid 10%	80	27
Ammonium nitrate	140	70	Hydrofluoric acid 30%	x	x
Ammonium persulfate	180	82	Hydrofluoric acid 70%	x	x
Amyl acetate	80	26	Hydrofluoric acid 100%	x	x
Amyl alcohol	250	121	Lactic acid 25%	250	121
Aniline	x	x	Lactic acid, concd	250	121
Antimony trichloride	250	121	Magnesium chloride	250	121
Aqua regia 3 : 1	80	26	Methyl ethyl ketone	250	121
Barium carbonate	250	121	Methyl isobutyl ketone	x	x
Barium chloride	250	121	Muriatic acid	90	32
Benzaldehyde	x	x	Nitric acid 5%	150	66
Benzene	x	x	Perchloric acid 10%	x	x
Benzoic acid	250	121	Perchloric acid 70%	x	x
Benzyl alcohol	80	27	Phenol	x	x
Benzyl chloride	250	121	Phosphoric acid 50–80%	250	121
Boric acid	200	93	Sodium carbonate 10%	250	121
Bromine liquid	80	27	Sodium chloride	250	121
Butyl acetate	250	121	Sodium hydroxide 10%	150	66
Butyric acid	250	121	Sodium hydroxide 50%	x	x
Calcium chloride	250	121	Sodium hydroxide, concd	x	x
Calcium hypochlorite	250	121	Sodium hypochlorite 20%	80	27
Carbon tetrachloride	250	121	Sulfuric acid 10%	160	71
Chloroacetic acid 50%	x	x	Sulfuric acid 50%	140	60
Chlorine gas, dry	80	27	Sulfuric acid 70%	x	x
Chlorine gas, wet	80	27	Sulfuric acid 90%	x	x
Chloroform	250	121	Sulfuric acid 98%	x	x
Chromic acid 50%	250	121	Sulfuric acid 100%	x	x
Citric acid 15%	250	121	Sulfuric acid, fuming	x	x
Citric acid, concd	150	66	Toluene	250	121
Copper chloride	170	77	Trichloroacetic acid	250	121
Copper sulfate	170	77	Zinc chloride	250	121
Cresol	x	x			

[a] The chemicals listed are in the pure state or in a saturated solution unless otherwise indicated. Compatibility is shown to the maximum allowable temperature for which data are available. Incompatibility is shown by an x. A blank space indicates that data are unavailable.
Source: Schweitzer PA. Corrosion Resistance Tables, 4th ed., Vols. 1–3. New York: Marcel Dekker, 1995.

2. Typical Applications

The terephthalates are used in applications similar to those of the isophthalates but where corrosive conditions and temperatures are more moderate.

3.4 EPOXY ESTERS

Epoxy-based thermosets are the most widely used and versatile thermosets. They dominated the reinforced piping field until the introduction of the vinyl esters and are still widely used. They find usage in many fields of application including adhesives, coatings, sealants, casting, encapsulants, tooling compounds, composites, and molding compounds. Their versatility is due to the wide latitude in properties that can be achieved by formulation. A large variety of epoxy resins, modifiers, and curing agents are available, which permits the epoxy formulator to tailor the epoxy system to meet the needs of each application.

A. Resin Types

The epoxide or oxirane functionality is a three membered carbon–oxygen–carbon ring. The simplest 1,2-epoxide is ethylene oxide:

$$\overset{\displaystyle O}{\overset{\diagup\diagdown}{CH_2-CH_2-}}$$

A common term used in naming epoxy resins is the term glycidyl. The terminology for the glycidyl group:

$$\overset{\displaystyle O}{\overset{\diagup\diagdown}{CH_2-CHCH_2-}}$$

comes from the trivially named glycidyl

$$\overset{\displaystyle O}{\overset{\diagup\diagdown}{CH_2-CHCH_2OH-}}$$

and glycidic acid:

$$\overset{\displaystyle O}{\overset{\diagup\diagdown}{CH_2-CHCOOH}}$$

Figure 3.4 illustrates the structures of the most common epoxy resins in use today. Two or more epoxide groups per molecule are required in order to form a cross-linked network.

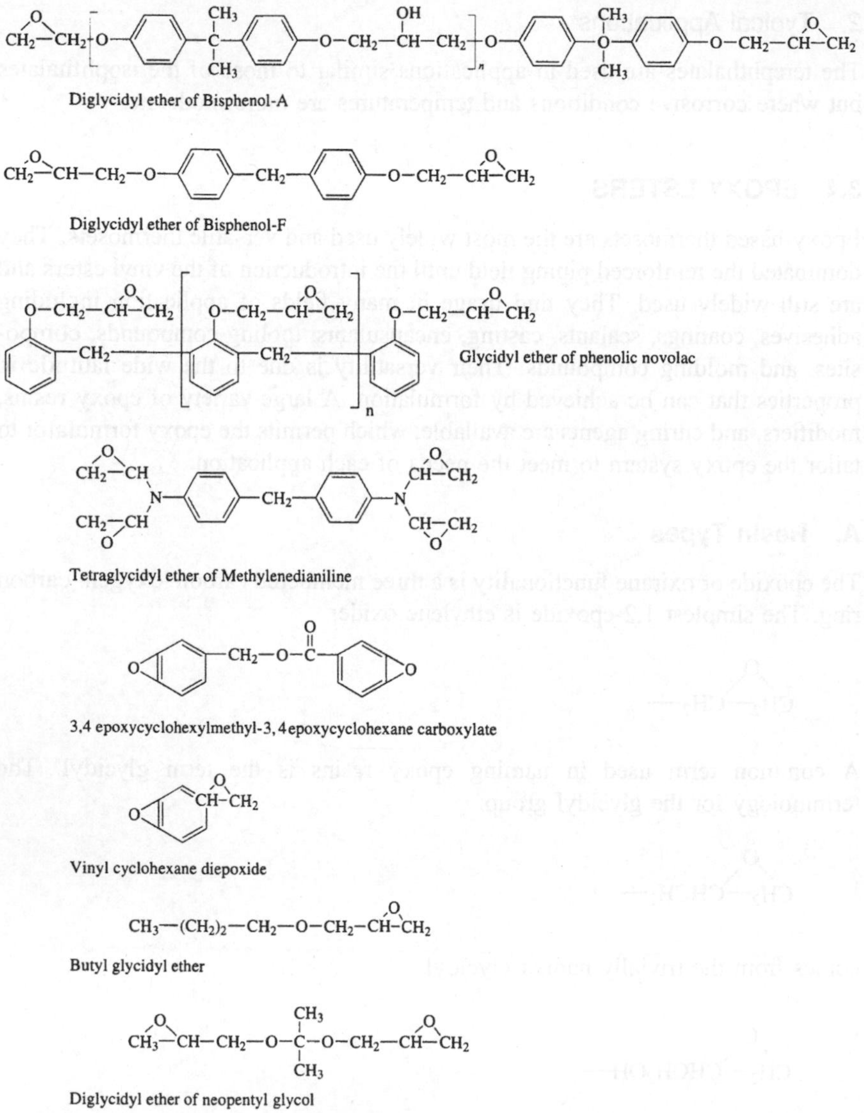

Diglycidyl ether of Bisphenol-A

Diglycidyl ether of Bisphenol-F

Glycidyl ether of phenolic novolac

Tetraglycidyl ether of Methylenedianiline

3,4 epoxycyclohexylmethyl-3,4 epoxycyclohexane carboxylate

Vinyl cyclohexane diepoxide

Butyl glycidyl ether

Diglycidyl ether of neopentyl glycol

FIGURE 3.4 Common commercial epoxy resins.

The most widely used thermoset epoxy is the diglycidyl ether of bisphenol-A (DGEBA). It is available in both liquid and solid forms. The chemical structure is shown in Figure 3.4. Table 3.24 lists the physical and mechanical properties of glass fiber reinforced and mineral-filled DGEBA, while Table 3.25 does likewise for DGEBA molding compounds.

TABLE 3.24 Physical and Mechanical Properties of DGEBA Reinforced Epoxy Resins

Property	Glass fiber reinforced	Bisphenol mineral filled
Specific gravity	1.6–2.0	1.6–2.1
Water absorption (24 hr at 73°F/23°C (%)	0.04–0.20	0.03–0.20
Dielectric strength, short-term (V/mil)	250–400	250–420
Tensile strength at break (psi)	5,000–20,000	4,000–10,800
Tensile modulus ($\times 10^3$ psi)	3,000	350
Elongation at break (%)	4	
Compressive strength (psi)	18,000–40,000	18,000–40,000
Flexural strength (psi)	8,000–30,000	6,000–18,000
Compressive modulus ($\times 10^3$ psi)		650
Flexural modulus ($\times 10^3$ psi) at		
73°F/23°C,	2,000–4,500	1,400–2,000
200°F/93°C		
250°F/121°C		
Izod impact (ft-lb/in. of notch)	0.3–10.0	0.03–0.5
Hardness, Rockwell	M100–112	M100–112
Coefficient of thermal expansion (10^{-6} in./in./°F)	11–50	20–60
Thermal conductivity (10^{-4} cal-cm/sec-cm^2 °C or Btu/hr/ft^2/°F/in.)	4.0–10.0	4–35
Deflection temperature at 264 psi (°F) at 66 psi (°F)	225–500	225–500
Max. operating temperature (°F/°C)		
Limiting oxygen index (%)		
Flame spread		
Underwriters Lab. rating (Sub. 94)		

By controlling operating conditions and varying the ratio of epichlorohydrin to bisphenol-A, products of different molecular weight can be produced. For liquid resins the n in the structural formula is generally less than 1; for solid resins n is 2 or greater. Solids with very high melting points have n values as high as 20.

The novolacs are another class of epoxy resins. They are produced by reacting a novolac resin, usually formed by the reaction of o-cresol or phenol and formaldehyde with epichlorohydrin. Figure 3.5 shows the general structure. These materials find application as transfer molding powders, electrical laminates, and parts where superior thermal properties and high resistance to solvents and chemicals are required. Refer to Tables 3.26 and 3.27 for the physical and mechanical properties of novolac epoxy formulations. Table 3.28 lists the

TABLE 3.25 Physical and Mechanical Properties of DGEBA Molding Compounds

Property	Glass filled reinforced	Mineral filled	Glass fiber reinforced
Specific gravity	1.5–2	1.5–2	0.1
Water absorption (24 hr at 73°F/23°C) (%)	0.04–0.2	0.03–0.2	1.4
Dielectric strength, short-term (V/mil)	250–400	250–420	
Tensile strength at break (psi)	5,000–20,000	4,000–10,000	20,000–35,000
Tensile modulus ($\times 10^3$ psi)	3,000	350	2,000–4,000
Elongation at break (%)	4		0.5–2.0
Compressive strength (psi)	18,000–40,000	18,000–40,000	20,000–30,000
Flexural strength (psi)	8,000–30,000	6,000–18,000	50,000–70,000
Compressive modulus ($\times 10^3$ psi)		600	
Flexural modulus ($\times 10^3$ psi) at 73°F/23°C	2,000–4,500	1,400–2,000	2,000–3,000
200°F/93°C			1,500–2,500
250°F/121°C			
Izod impact (ft-lb/in. of notch)	0.3–10	0.3–0.5	30–40
Hardness, Rockwell	M100–112	M100–112	Shore B55–65
Coefficient of thermal expansion (10^{-6} in./in./°F)	11–50	20–40	12
Thermal conductivity (10^{-4} cal-cm/sec-cm^2°C or Btu/hr/ft^2/°F/in.)	4–10	4–35	1.7–1.9
Deflection temperature at 264 psi (°F)	225–500	225–500	500
at 66 psi (°F)			
Max. operating temperature (°F/°C)			
Limiting oxygen index (%)			
Flame spread			
Underwriters Lab. rating (Sub. 94)			

physical and mechanical properties of epoxy casting resins, while Table 3.29 lists the properties of sheet molding compounds.

1. Curing

Epoxy resins must be cured with cross-linking agents (hardeners) or catalysts to develop desired properties. Cross-linking takes place at the epoxy and hydroxyl

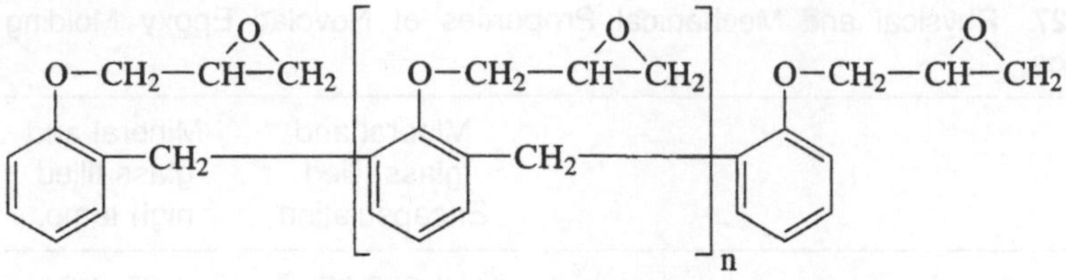

Structure of epoxy novolac

FIGURE 3.5 General structures of novolac epoxies.

TABLE 3.26 Physical and Mechanical Properties of Novolac Epoxy Resins

Property	
Specific gravity	1.6–2.05
Water absorption (24 hr at 73°F/23°C) (%)	0.04–0.29
Dielectric strength, short-term (V/mil)	325–450
Tensile strength at break (psi)	5,000–12,500
Tensile modulus ($\times 10^3$ psi)	2,100
Elongation at break (%)	
Compressive strength (psi)	24,000–48,000
Flexural strength (psi)	10,000–21,800
Compressive modulus ($\times 10^3$ psi)	
Flexural modulus ($\times 10^3$ psi) at 73°F/23°C	1,400–2,400
200°F/93°C	
250°F/121°C	
Izod impact (ft-lb/in. of notch)	0.3–0.5
Hardness, Rockwell	M115
Coefficient of thermal expansion (10^{-6} in./in./°F)	18–43
Thermal conductivity (10^{-4} cal-cm/sec-cm^2°C or Btu/hr/ft^2/°F/in.)	10–31
Deflection temperature at 264 psi (°F)	300–500
at 66 psi (°F)	
Max. operating temperature (°F/°C)	
Limiting oxygen index (%)	
Flame spread	
Underwriters Lab. rating (Sub. 94)	

TABLE 3.27 Physical and Mechanical Properties of Novolac Epoxy Molding Compounds

Property	Mineral and glass filled Encapsulation	Mineral and glass filled high temp.
Specific gravity	1.6–2.05	1.85–1.94
Water absorption (24 hr at 73°F/23°C) (%)	0.04–0.29	0.15–0.17
Dielectric strength, short-term (V/mil)	325–450	440–450
Tensile strength at break (psi)	5,000–12,000	6,000–15,000
Tensile modulus ($\times 10^3$ psi)	2,100	2,300–2,400
Elongation at break (%)		
Compressive strength (psi)	24,000–40,000	30,000–48,000
Flexural strength (psi)	10,000–21,500	10,000–21,500
Compressive modulus ($\times 10^3$ psi)		600
Flexural modulus ($\times 10^3$ psi) at 73°F/23°C		
200°F/93°C		
250°F/121°C		
Izod impact (ft-lb/in. of notch)	0.3–0.5	0.4–0.45
Hardness, Rockwell	M115	
Coefficient of thermal expansion (10^{-6} in./in./°F)	18–43	35
Thermal conductivity (10^{-4} cal-cm/sec-cm^2 °C or Btu/hr/ft^2/°F/in.)	10–31	17–34
Deflection temperature at 264 psi (°F)	300–500	500
at 66 psi (°F)		
Max. operating temperature (°F/°C)		
Limiting oxygen index (%)		
Flame spread		
Underwriters Lab. rating (Sub. 94)		

groups which are the reaction sites. Useful agents are amines, anhydrides, aldehyde condensation products, and Lewis acid catalysts. In order to achieve a balance of application properties and initial handling characteristics, careful selection of the proper curing agent is required. The primary types of curing agents are aromatic amines, aliphatic amines, catalytic curing agents, and acid anhydrides.

Aromatic Amines Aromatic amines usually require an elevated temperature cure. Epoxies cured with aromatic amines usually have a longer working life than epoxies cured with aliphatic amines. These curing agents are relatively difficult to use since they are solids and must be melted into the epoxy. However, the allowable operating temperatures for epoxies cured with aromatic amines are higher than those of epoxies cured with aliphatic amines.

TABLE 3.28 Physical and Mechanical Properties of Epoxy Casting Resins

Property	Silica Filled	Al Filled
Specific gravity	1.6–2.0	1.4–1.8
Water absorption (24 hr at 73°F/23°C) (%)	0.04–0.1	0.1–4.0
Dielectric strength, short-term (V/mil)	300–550	
Tensile strength at break (psi)	7,000–13,000	7,000–12,000
Tensile modulus ($\times 10^3$ psi)		
Elongation at break (%)	1–3	0.5–3
Compressive strength (psi)	15,000–35,000	15,000–33,000
Flexural strength (psi)	8,000–14,000	8,000–24,000
Compressive modulus ($\times 10^3$ psi)		
Flexural modulus ($\times 10^3$ psi) at 73°F/23°C		
200°F/93°C		
250°F/121°C		
Izod impact (ft-lb/in. of notch)	0.3–0.45	0.4–1.6
Hardness, Rockwell	M85–120	M55–85
Coefficient of thermal expansion		
(10^{-6} in./in./°F)	20–40	5.5
Thermal conductivity		
(10^{-4} cal-cm/sec-cm^2 °C or	20–30	15–25
Btu/hr/ft^2/°F/in.)		
Deflection temperature at 264 psi (°F)	160–550	150–600
at 66 psi (°F)		
Max. operating temperature (°F/°C)		
Limiting oxygen index (%)		
Flame spread		
Underwriters Lab. rating (Sub. 94)		

Aliphatic Amines   These are widely used since the curing of the epoxies takes place at room temperature. High exothermic temperatures develop during the curing reaction which limit the mass of material that can be cured. The electrical and physical properties of these aliphatic-cured resins have the greatest tendency toward degradation of electrical and physical properties at elevated temperatures.

Typical aliphatic amines used include diethylene triamine (DETA) and triethylene tetramine (TETA).

Catalytic Curing Agents   Catalytic curing agents require a temperature of 200°F/93°C or higher to react. These epoxy formulations have a longer working life than the aliphatic amine cured epoxies. The exothermic reaction may be critically affected by the mass of the resin mixture.

TABLE 3.29 Physical and Mechanical Properties of Epoxy Sheet Molding Compounds

Property	Glass fiber reinforced	Carbon fiber reinforced
Specific gravity	0.1	0.1
Water absorption (24 hr at 73°F/23°C) (%)	1.4	1.6
Dielectric strength, short-term (V/mil)		
Tensile strength at break (psi)	20,000–30,000	40,000–50,000
Tensile modulus ($\times 10^3$ psi)	2,000–4,000	10,000
Elongation at break (%)	0.5–2.0	0.5–2.0
Compressive strength (psi)	20,000–30,000	30,000–40,000
Flexural strength (psi)	50,000–70,000	75,000–95,000
Compressive modulus ($\times 10^3$ psi)		
Flexural modulus ($\times 10^3$ psi) at 73°F/23°C	2,000–3,000	5,000
200°F/93 °C	1,500–2,500	
250°F/121°C		
Izod impact (ft-lb/in. of notch)	30–40	15–20
Hardness, Shore	B56–65	B55–65
Coefficient of thermal expansion (10^{-6} in./in./°F)		3
Thermal conductivity (10^{-4} cal-cm/sec-cm^2 °C or Btu/hr/ft^2/°F/in.)	1.7–19	1.4–1.5
Deflection temperature at 264 psi (°F)	550	550
at 66 psi (°F)		
Max. operating temperature (°F/°C)		
Limiting oxygen index (%)		
Flame spread		
Underwriters Lab. rating (Sub. 94)		

Typical materials used include piperidine, boron trifluoride ethylamine complex, and benzyl dimethylamine (BDMA).

Acid Anhydrides These curing agents are becoming more widely used since they are easy to work with, have minimum toxicity problems compared with amines, and offer optimum high-temperature properties to the cured resin.

Typical acid anhydrides used include nadic methyl anhydride (NMA), dodecenyl succinic anhydride (DDSA), hexahydrophthalic anhydride (HHPA), and alkendic anhydride.

2. Corrosion Resistance

The epoxy resin family exhibits good resistance to alkalies, nonoxidizing acids, and many solvents. Typically, epoxies are compatible with the following materials at 200°F/93°C unless otherwise noted:

Acids

Acetic acid 10% to 150°F/66°C	Hydrochloric acid 10%
Benzoic acid	Sulfuric acid 20% to 180°F/82°C
Butyric acid	Rayon spin bath
Fatty acids	Oxalic acid

Bases

Sodium hydroxide, 50% to 180°F/82°C
Sodium sulfide 10%
Calcium hydroxide
Trisodium phosphate
Magnesium hydroxide

Salts Metallic salts:

Aluminum	Potassium
Calcium	Sodium
Iron	Most ammonium salts
Magnesium	

Solvents Alcohols:

Methyl	Benzene to 150°F/66°C
Ethyl	Ethylacetate to 150°F/66°C
Isopropyl to 150°F/66°C	Naphtha
	Toluene
	Xylene

Miscellaneous

Distilled water	Jet fuel
Seawater	Gasoline
White liquor	Diesel fuel
Sour crude oil	Black liquor

Epoxies are not satisfactory for use with

Bromine water	Hydrogen peroxide
Chromic acid	Sulfuric acid above 70%

Bleaches   Wet chlorine gas
Fluorine   Wet sulfur dioxide
Methylene chloride

Refer to Table 3.30 for the compatibility of epoxies with selected corrodents. Reference 1 provides a more complete listing.

3. Typical applications

Epoxy resins find many applications in the chemical process industry as piping. Refer to Reference 2 for more details. They are also widely used in the electronics field because of the wide variety of formulations possible. Formulations range from flexible to rigid in the cured state and from thin liquids to thick pastes and molding powders in the uncured state. Conversion from uncured to cured state is made by use of hardeners, or heat, or both. Embedding applications (potting, casting, encapsulating, and impregnating) in molded parts and laminated construction are the predominant uses.

3.5 VINYL ESTERS

The vinyl ester class of resins was developed during the late 1950s and early 1960s. Vinyl esters were first used as dental fillings. They had improved toughness and bonding ability over the acrylic materials that were being used at the time. Over the next several years, changes in the molecular structure of the vinyl esters produced resins that found extensive use in corrosion resistant equipment.

Present day vinyl esters possess several advantages over unsaturated polyesters. They provide improved toughness in the cured polymer while maintaining good thermal stability and physical properties at elevated temperatures. This improved toughness permits the resins to be used in castings, as well as in reinforced products. The structural formulas for typical vinyl ester resins are shown in Figure 3.6. These resins have improved bonding to inorganic fillers and reinforcements as a result of the internal hydroxyl group. Composites produced from vinyl esters have improved bonding to the reinforcements and improved damage resistance. Refer to Table 3.31 for the physical and mechanical properties of the vinyl esters.

Vinyl ester resins are also available in halogenated modifications for ductwork and stack construction where fire retardance and ignition resistance are major concerns. Vinyl esters have a number of basic advantages among which are

1. They cure rapidly and give high early strength and superior creep resistance as a result of their molecular structure.

TABLE 3.30 Compatibility of Epoxy with Selected Corrodents[a]

Chemical	Maximum temp. °F	°C	Chemical	Maximum temp. °F	°C
Acetaldehyde	150	66	Barium sulfide	300	149
Acetamide	90	32	Benzaldehyde	x	x
Acetic acid 10%	190	88	Benzene	160	71
Acetic acid 50%	110	43	Benzenesulfonic		
Acetic acid 80%	110	43	acid 10%	160	71
Acetic anhydride	x	x	Benzoic acid	200	93
Acetone	110	43	Benzyl alcohol	x	x
Acetyl chloride	x	x	Benzyl chloride	60	16
Acrylic acid	x	x	Borax	250	121
Acrylonitrile	90	32	Boric acid 4%	200	93
Adipic acid	250	121	Bromine gas, dry	x	x
Allyl alcohol	x	x	Bromine gas, moist	x	x
Allyl chloride	140	60	Bromine, liquid	x	x
Alum	300	149	Butadiene	100	38
Aluminum chloride,			Butyl acetate	170	77
aqueous 1%	300	149	Butyl alcohol	140	60
Aluminum chloride, dry	90	32	n-Butylamine	x	x
Aluminum fluoride	180	82	Butyric acid	210	99
Aluminum hydroxide	180	82	Calcium bisulfide		
Aluminum nitrate	250	121	Calcium bisulfite	200	93
Aluminum sulfate	300	149	Calcium carbonate	300	149
Ammonia gas, dry	210	99	Calcium chlorate	200	93
Ammonium bifluoride	90	32	Calcium chloride 37.5%	190	88
Ammonium carbonate	140	60	Calcium hydroxide, sat.	180	82
Ammonium chloride, sat.	180	82	Calcium hypochlorite 70%	150	66
Ammonium fluoride 25%	150	66	Calcium nitrate	250	121
Ammonium hydroxide 25%	140	60	Calcium sulfate	250	121
Ammonium hydroxide, sat.	150	66	Caprylic acid	x	x
Ammonium nitrate 25%	250	121	Carbon bisulfide	100	38
Ammonium persulfate	250	121	Carbon dioxide, dry	200	93
Ammonium phosphate	140	60	Carbon disulfide	100	38
Ammonium sulfate			Carbon monoxide	80	27
10–40%	300	149	Carbon tetrachloride	170	77
Ammonium sulfite	100	38	Carbonic acid	200	93
Amyl acetate	80	27	Cellosolve	140	60
Amyl alcohol	140	60	Chloroacetic acid,		
Amyl chloride	80	27	92% water	150	66
Aniline	150	66	Chloroacetic acid	x	x
Antimony trichloride	180	82	Chlorine gas, dry	150	66
Aqua regia 3 : 1	x	x	Chlorine gas, wet	x	x
Barium carbonate	240	116	Chlorobenzene	150	66
Barium chloride	250	121	Chloroform	110	43
Barium hydroxide 10%	200	93	Chlorosulfonic acid	x	x
Barium sulfate	250	121	Chromic acid 10%	110	43

(*Continued*)

TABLE 3.30 Continued

Chemical	Maximum temp. °F	Maximum temp. °C	Chemical	Maximum temp. °F	Maximum temp. °C
Chromic acid 50%	x	x	Methyl ethyl ketone	90	32
Citric acid 15%	190	88	Methyl isobutyl ketone	140	60
Citric acid 32%	190	88	Muriatic acid	140	60
Copper acetate	200	93	Nitric acid 5%	160	71
Copper carbonate	150	66	Nitric acid 20%	100	38
Copper chloride	250	121	Nitric acid 70%	x	x
Copper cyanide	150	66	Nitric acid, anhydrous	x	x
Copper sulfate 17%	210	99	Nitrous acid, concd	x	x
Cresol	100	38	Oleum	x	x
Cupric chloride 5%	80	27	Perchloric acid 10%	90	32
Cupric chloride 50%	80	27	Perchloric acid 70%	80	27
Cyclohexane	90	32	Phenol	x	x
Cyclohexanol	80	27	Phosphoric acid 50–80%	110	43
Dichloroacetic acid	x	x	Picric acid	80	27
Dichloroethane (ethylene dichloride)	x	x	Potassium bromide 30%	200	93
			Salicylic acid	140	60
Ethylene glycol	300	149	Sodium carbonate	300	149
Ferric chloride	300	149	Sodium chloride	210	99
Ferric chloride 50% in water	250	121	Sodium hydroxide 10%	190	88
Ferric nitrate 10–50%	250	121	Sodium hydroxide 50%	200	93
Ferrous chloride	250	121	Sodium hypochlorite 20%	x	x
Ferrous nitrate			Sodium hypochlorite, concd	x	x
Fluorine gas, dry	90	32	Sodium sulfide to 10%	250	121
Hydrobromic acid, dilute	180	82	Stannic chloride	200	93
Hydrobromic acid 20%	180	82	Stannous chloride	160	71
Hydrobromic acid 50%	110	43	Sulfuric acid 10%	140	60
Hydrochloric acid 20%	200	93	Sulfuric acid 50%	110	43
Hydrochloric acid 38%	140	60	Sulfuric acid 70%	110	43
Hydrocyanic acid 10%	160	71	Sulfuric acid 90%	x	x
Hydrofluoric acid 30%	x	x	Sulfuric acid 98%	x	x
Hydrofluoric acid 70%	x	x	Sulfuric acid 100%	x	x
Hydrofluoric acid 100%	x	x	Sulfuric acid, fuming	x	x
Hypochlorous acid	200	93	Sulfurous acid 20%	240	116
Ketones, general	x	x	Thionyl chloride	x	x
Lactic acid 25%	220	104	Toluene	150	66
Lactic acid, concd	200	93	Trichloroacetic acid	x	x
Magnesium chloride	190	88	White liquor	90	32
Methyl chloride	x	x	Zinc chloride	250	121

[a] The chemicals listed are in the pure state or in a saturated solution unless otherwise indicated. Compatibility is shown to the maximum allowable temperature for which data are available. incompatibility is shown by an x. A blank space indicates that the data are unavailable.

Source: Schweitzer, PA. Corrosion Resistance Tables, 4th ed., Vols. 1–3. New York: Marcel Dekker, 1995.

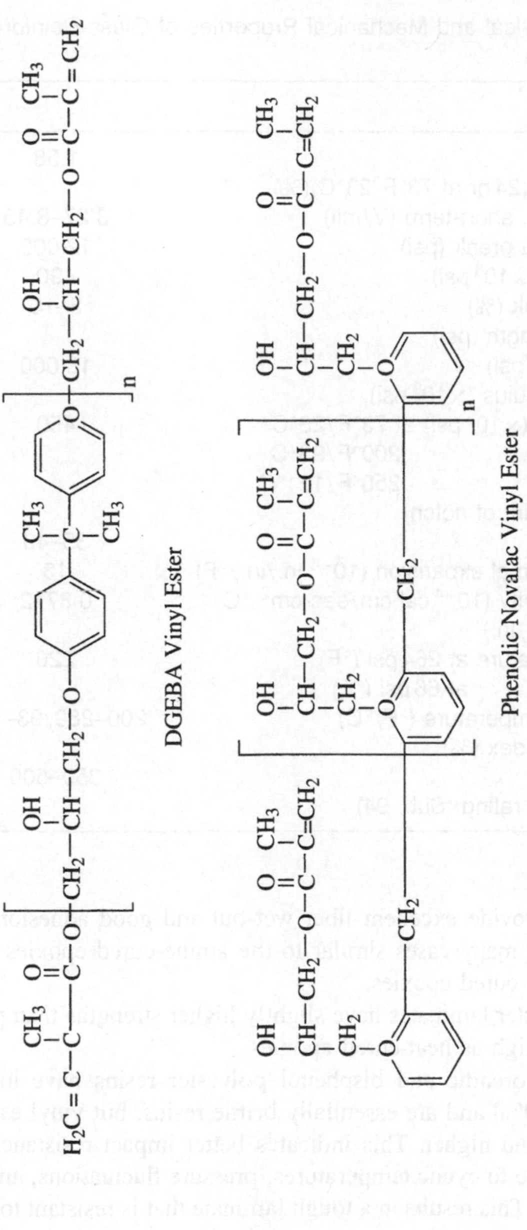

DGEBA Vinyl Ester

Phenolic Novalac Vinyl Ester

FIGURE 3.6 Structural formula for typical vinyl ester resins.

TABLE 3.31 Physical and Mechanical Properties of Glass-Reinforced Vinyl Ester Resins

Property	
Specific gravity	1.58
Water absorption (24 hr at 73°F/23°C)(%)	
Dielectric strength, short-term (V/mil)	3.27–3.45
Tensile strength at break (psi)	12,000
Tensile modulus ($\times 10^3$ psi)	490
Elongation at break (%)	3–10
Compressive strength (psi)	
Flexural strength (psi)	18,000
Compressive modulus ($\times 10^3$ psi)	
Flexural modulus ($\times 10^3$ psi) at 73°F/23°C	450
200°F/93°C	
250°F/121°C	
Izod impact (ft-lb/in. of notch)	
Hardness, Barcol	35–40
Coefficient of thermal expansion (10^{-6} in./in./°F)	13
Thermal conductivity (10^{-4} cal-cm/sec-cm^2 °C or Btu/hr/ft^2/°F/in.)	0.8712
Deflection temperature at 264 psi (°F)	220
at 66 psi (°F)	
Max. operating temperature (°F/°C)	200–280/93–140
Limiting oxygen index (%)	
Flame spread	350–500
Underwriters Lab. rating (Sub. 94)	

2. They provide excellent fiber wet-out and good adhesion to the glass fiber, in many cases similar to the amine-cured epoxies but less than the heat-cured epoxies.
3. Vinyl ester laminates have slightly higher strengths than polyesters but not as high as heat-cured epoxies.
4. The chlorendic and bisphenol polyester resins have low elongation (1.5–2.0%) and are essentially brittle resins, but vinyl ester resins run 4–6% and higher. This indicates better impact resistance and greater tolerance to cyclic temperatures, pressure fluctuations, and mechanical shocks. This results in a tough laminate that is resistant to cracking and crazing.
5. Because of the basic structure of the vinyl ester molecule, it is more resistant to hydrolysis and oxidation or halogenation than the polyesters.

TABLE 3.32 Elevated Temperature Performance of a Vinyl Ester Laminate

| Temperature (°F/°C) | Retention | |
	Physical strength (%)	Flexural modulus (%)
70/23	100	100
150/66	121	98
200/93	105	83
225/107	88	48
250/121	57	22
300/149	37	22

In order to have maximum retention of physical properties at elevated temperatures, the upper temperature limit should be approximately 225°F/107°C; refer to Table 3.32. Exceptions to this are vinyl ester resins having a novolac backbone. These resins can be used at 325–350°F/163–177°C. Flexural modulus must be examined at all elevated temperatures, particularly important in vacuum applications. Note in Table 3.32 that the vinyl esters have lost up to half of their flexural strength at 225°F/107°C.

1. Corrosion Resistance

In general, vinyl esters can be used to handle most hot, highly chlorinated, and acid mixtures at elevated temperatures. They also provide excellent resistance to strong mineral acids and bleaching solutions. Vinyl esters excel in alkaline and bleach environments and are used extensively in the very corrosive conditions found in the pulp and paper industry.

The family of vinyl esters includes a wide variety of formulations. As a result there can be differences in the compatibility of formulations among manufacturers. When one checks compatibility in a table, one must keep in mind that all formulations may not act as shown. An indication that vinyl ester is compatible generally means that at least one formulation is compatible. This is the case in Table 3.33 which shows the compatibility of vinyl ester laminates with selected corrodents. The resin manufacturer must be consulted to verify the resistance.

2. Typical Applications

Vinyl ester laminates find application in industrial equipment and scrubbers such as absorption towers, process vessels, storage tanks, piping, hood scrubbers, ducts and exhaust stacks, all handling highly corrosive materials. Reference 2 provides detailed information concerning vinyl esters as piping materials.

TABLE 3.33 Compatibility of Vinyl Ester with Selected Corrodents[a]

Chemical	Maximum temp.		Chemical	Maximum temp.	
	°F	°C		°F	°C
Acetaldehyde	x	x	Barium sulfate	200	93
Acetamide			Barium sulfide	180	82
Acetic acid 10%	200	93	Benzaldehyde	x	x
Acetic acid 50%	180	82	Benzene	x	x
Acetic acid 80%	150	66	Benzenesulfonic		
Acetic acid, glacial	150	66	acid 10%	200	93
Acetic anhydride	100	38	Benzoic acid	180	82
Acetone	x	x	Benzyl alcohol	100	38
Acetyl chloride	x	x	Benzyl chloride	90	32
Acrylic acid	100	38	Borax	210	99
Acrylonitrile	x	x	Boric acid	200	93
Adipic acid	180	82	Bromine gas, dry	100	38
Allyl alcohol	90	32	Bromine gas, moist	100	38
Allyl chloride	90	32	Bromine, liquid	x	x
Alum	240	116	Butadiene		
Aluminum acetate	210	99	Butyl acetate	80	27
Aluminum chloride,			Butyl alcohol	120	49
aqueous	260	127	n-Butylamine	x	x
Aluminum chloride, dry	140	60	Butyric acid	130	54
Aluminum fluoride	100	38	Calcium bisulfide		
Aluminum hydroxide	200	93	Calcium bisulfite	180	82
Aluminum nitrate	200	93	Calcium carbonate	180	82
Aluminum oxychloride			Calcium chlorate	260	127
Aluminum sulfate	250	121	Calcium chloride	180	82
Ammonia gas	100	38	Calcium hydroxide 10%	180	82
Ammonium bifluoride	150	66	Calcium hydroxide, sat.	180	82
Ammonium carbonate	150	66	Calcium hypochlorite	180	82
Ammonium chloride 10%	200	93	Calcium nitrate	210	99
Ammonium chloride 50%	200	93	Calcium oxide	160	71
Ammonium chloride, sat.	200	93	Calcium sulfate	250	116
Ammonium fluoride 10%	140	60	Caprylic acid	220	104
Ammonium fluoride 25%	140	60	Carbon bisulfide	x	x
Ammonium hydroxide 25%	100	38	Carbon dioxide, dry	200	93
Ammonium hydroxide, sat.	130	54	Carbon dioxide, wet	220	104
Ammonium nitrate	250	121	Carbon disulfide	x	x
Ammonium persulfate	180	82	Carbon monoxide	350	177
Ammonium phosphate	200	93	Carbon tetrachloride	180	82
Ammonium sulfate 10–40%	220	104	Carbonic acid	120	49
Ammonium sulfide	120	49	Cellosolve	140	60
Ammonium sulfite	220	104	Chloroacetic acid,		
Amyl acetate	110	38	50% water	150	66
Amyl alcohol	210	99	Chloroacetic acid	200	93
Amyl chloride	120	49	Chlorine gas, dry	250	121
Aniline	x	x	Chlorine gas, wet	250	121
Antimony trichloride	160	71	Chlorine, liquid	x	x
Aqua regia 3 : 1	x	x	Chlorobenzene	110	43
Barium carbonate	260	127	Chloroform	x	x
Barium chloride	200	93	Chlorosulfonic acid	x	x
Barium hydroxide	150	66	Chromic acid 10%	150	66

TABLE 3.33 Continued

Chemical	Maximum temp. °F	Maximum temp. °C	Chemical	Maximum temp. °F	Maximum temp. °C
Chromic acid 50%	x	x	Manganese chloride	210	99
Chromyl chloride	210	99	Methyl chloride		
Citric acid 15%	210	99	Methyl ethyl ketone	x	x
Citric acid, concd	210	99	Methyl isobutyl ketone	x	x
Copper acetate	210	99	Muriatic acid	180	82
Copper carbonate			Nitric acid 5%	180	82
Copper chloride	220	104	Nitric acid 20%	150	66
Copper cyanide	210	99	Nitric acid 70%	x	x
Copper sulfate	240	116	Nitric acid, anhydrous	x	x
Cresol	x	x	Nitrous acid 10%	150	66
Cupric chloride 5%	260	127	Oleum	x	x
Cupric chloride 50%	220	104	Perchloric acid 10%	150	66
Cyclohexane	150	66	Perchloric acid 70%	x	x
Cyclohexanol	150	66	Phenol	x	x
Dibutyl phthalate	200	93	Phosphoric acid 50–80%	210	99
Dichloroacetic acid	100	38	Picric acid	200	93
Dichloroethane			Potassium bromide 30%	160	71
(ethylene dichloride)	110	43	Salicylic acid	150	66
Ethylene glycol	210	99	Silver bromide 10%		
Ferric chloride	210	99	Sodium carbonate	180	82
Ferric chloride 50% in water	210	99	Sodium chloride	180	82
Ferric nitrate 10–50%	200	93	Sodium hydroxide 10%	170	77
Ferrous chloride	200	93	Sodium hydroxide 50%	220	104
Ferrous nitrate	200	93	Sodium hydroxide, concd		
Fluorine gas, dry	x	x	Sodium hypochlorite 20%	180	82
Fluorine gas, moist	x	x	Sodium hypochlorite, concd	100	38
Hydrobromic acid, dil	180	82	Sodium sulfide to 50%	220	104
Hydrobromic acid 20%	180	82	Stannic chloride	210	99
Hydrobromic acid 50%	200	93	Stannous chloride	200	93
Hydrochloric acid 20%	220	104	Sulfuric acid 10%	200	93
Hydrochloric acid 38%	180	82	Sulfuric acid 50%	210	99
Hydrocyanic acid 10%	160	71	Sulfuric acid 70%	180	82
Hydrofluoric acid 30%	x	x	Sulfuric acid 90%	x	x
Hydrofluoric acid 70%	x	x	Sulfuric acid 98%	x	x
Hydrofluoric acid 100%	x	x	Sulfuric acid 100%	x	x
Hypochlorous acid	150	66	Sulfuric acid, fuming	x	x
Iodine solution 10%	150	66	Sulfurous acid 10%	120	49
Ketones, general	x	x	Thionyl chloride	x	x
Lactic acid 25%	210	99	Toluene	120	49
Lactic acid, concd	200	93	Trichloroacetic acid 50%	210	99
Magnesium chloride	260	127	White liquor	180	82
Malic acid 10%	140	60	Zinc chloride	180	82

[a] The chemicals listed are in the pure state or in a saturated solution unless otherwise indicated. Compatibility is shown to the maximum allowable temperature for which data are available. Incompatibility is shown by an x. A blank space indicates that data are unavailable.
Source: Schweitzer PA. Corrosion Resistance Tables, Vols. 1 and 2. New York: Marcel Dekker, 1991.

3.6 FURANS

Furan polymers are derivatives of furfuryl alcohol and furfural. Using an acid catalyst, polymerization occurs by the condensation route, which generates heat and by-product water. The exotherm must be controlled to prevent the water vapor from blistering and cracking the laminate. Furan resin catalysts should have exotherms above 65°F (18°C) but not over 85°F (30°C).

All furan laminates must be postcured to drive out the reaction "condensate" in order to achieve optimum properties. Curing for a fresh laminate should start with an initial temperature of 150°F (66°C) for 4 hours, which is slowly raised to 180°F (82°C) for 8 hours of curing. Too fast a cure can result in a blistered or cracked laminate. A final Barcol hardness of 40–45 is necessary to develop optimum laminate properties, and Barcols as high as 55 are achieved.

The structural formula for furan resin is

Refer to Table 3.34 for the physical and mechanical properties of a furan glass-reinforced laminate.

1.   Corrosion Resistance

Furan thermosets are noted for their excellent resistance to solvents and they are considered to have the best overall chemical resistance of all of the thermosets. They also have excellent resistance to strong concentrated mineral acids, caustics, and combinations of solvents with acids and bases.

Since there are different formulations of furans, the supplier should be checked as to the compatibility of a particular resin with the corrodents to be encountered.

Typical materials with which the furan resins are compatible are

Solvents

Acetone	Ethanol
Benzene	Ethyl acetate
Carbon disulfide	Methanol
Chlorobenzene	Methyl ethyl ketone

Perchlorethylene   Trichloroethylene
Styrene   Xylene
Toluene

Acids

Acetic acid   Phosphoric acid
Hydrochloric acid   Sulfuric acid 60% to 150°F/66°C
Nitric acid 5%

Bases

Diethylamine   Sodium sulfide
Sodium carbonate   Sodium hydroxide 50%

TABLE 3.34 Physical and Mechanical Properties of a Glass-Reinforced Furan Laminate

Property	
Specific gravity	1.8
Water absorption (24 hr at 73°F/23°C) (%)	2.65
Dielectric strength, short-term (V/mil)	
Tensile strength at break (psi)	3,040
Tensile modulus ($\times 10^3$ psi)	980
Elongation at break (%)	
Compressive strength (psi)	20,300
Flexural strength (psi)	9,260
Compressive modulus ($\times 10^3$ psi)	
Flexural modulus ($\times 10^3$ psi) at 73°F/23°C	
200°F/93°C	
250°F/121°C	
Izod impact (ft-lb/in. of notch)	0.67
Hardness, Rockwell	
Coefficient of thermal expansion (10^{-6} in./in./°F)	13.5
Thermal conductivity (10^{-4} cal-cm/sec-cm^2 °C	1.96
or Btu/hr/ft^2/ °F/in.)	
Deflection temperature at 264 psi (°F)	
at 66 psi (°F)	
Max. operating temperature (°F/°C)	300/149
Limiting oxygen index (%)	
Flame spread	< 25
Underwriters Lab. Rating (Sub. 94)	

Water

Demineralized   Distilled

Others

Pulp mill liquor

Furan resins are not satisfactory for use with oxidizing media, such as chromic or nitric acids, peroxides, hypochlorites, chlorine, phenol, and concentrated sulfuric acid.

Refer to Table 3.35 for the compatibility of furan resins with selected corrodents. Reference 1 provides a more detailed listing.

2. Typical Applications

Furan resins cost approximately 30–50% more than polyester resins and do not have as good impact resistance as the polyesters. They find application as piping, tanks, and special chemical equipment such as scrubbing columns. Refer to Reference 2 for information regarding furan piping systems.

3.7 PHENOLICS

These are the oldest commercial classes of polymers in use today. Although first discovered in 1876, it was not until 1907 after the "heat and pressure" patent was applied for by Leo H. Bakeland that the development of and application of phenolic molding compounds became economical.

Phenolic resin precursors are formed by a condensation reaction, and water is formed as a by-product. The structural formula is

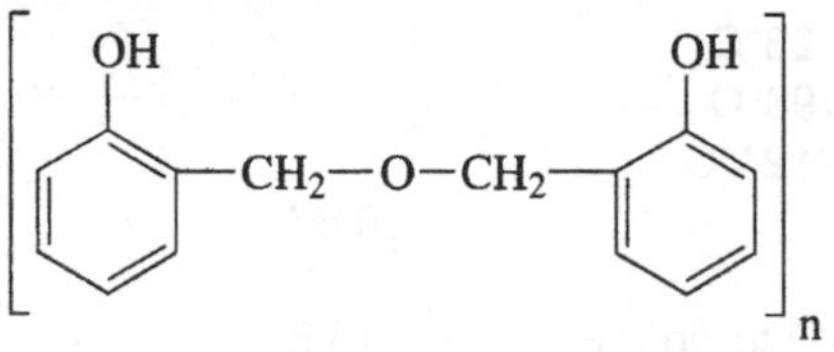

$$\left[\begin{array}{c} OH \\ \underset{}{\bigcirc} -CH_2-O-CH_2- \underset{}{\bigcirc} \overset{OH}{} \end{array} \right]_n$$

Phenolic compounds are available in a large number of variations, depending on the nature of the reactants, the ratios used, and the catalysts, plasticizers, lubricants, fillers, and pigments employed. The phenolics are excellent general-purpose materials with a wide range of mechanical and electrical properties. By selecting the correct reinforcement material, products with many properties may be obtained. Table 3.36 provides physical and mechanical properties on wood flour filled, high-strength glass fiber reinforced, and high-impact cotton-filled phenolics.

TABLE 3.35 Compatibility of Furan Resins with Selected Corrodents[a]

Chemical	Maximum temp.		Chemical	Maximum temp.	
	°F	°C		°F	°C
Acetaldehyde	x	x	Borax	140	60
Acetic acid 10%	212	100	Boric acid	300	149
Acetic acid 50%	160	71	Bromine gas, dry	x	x
Acetic acid 80%	80	27	Bromine gas, moist	x	x
Acetic acid, glacial	80	27	Bromine, liquid		
Acetic anhydride	80	27	3% max.	300	149
Acetone	80	27	Butadiene		
Acetyl chloride	200	93	Butyl acetate	260	127
Acrylic acid	80	27	Butyl alcohol	212	100
Acrylonitrile	80	27	n-Butylamine	x	x
Adipic acid 25%	280	138	Butyric acid	260	127
Allyl alcohol	300	149	Calcium bisulfite	260	127
Allyl chloride	300	149	Calcium chloride	160	71
Alum 5%	140	60	Calcium hydroxide, sat.	260	127
Aluminum chloride, aqueous	300	149	Calcium hypochlorite	x	x
Aluminum chloride, dry	300	149	Calcium nitrate	260	127
Aluminum fluoride	280	138	Calcium oxide		
Aluminum hydroxide	260	127	Calcium sulfate	260	127
Aluminum sulfate	160	71	Caprylic acid	250	121
Ammonium carbonate	240	116	Carbon bisulfide	160	71
Ammonium hydroxide 25%	250	121	Carbon dioxide, dry	90	32
Ammonium hydroxide, sat.	200	93	Carbon dioxide, wet	80	27
Ammonium nitrate	250	121	Carbon disulfide	260	127
Ammonium persulfate	260	127	Carbon tetrachloride	212	100
Ammonium phosphate	260	127	Cellosolve	240	116
Ammonium sulfate 10–40%	260	127	Chloroacetic acid,		
Ammonium sulfide	260	127	50% water	100	38
Ammonium sulfite	240	116	Chloroacetic acid	240	116
Amyl acetate	260	127	Chlorine gas, dry	260	127
Amyl alcohol	278	137	Chlorine gas, wet	260	127
Amyl chloride	x	x	Chlorine, liquid	x	x
Aniline	80	27	Chlorobenzene	260	127
Antimony trichloride	250	121	Chloroform	x	x
Aqua regia 3 : 1	x	x	Chlorosulfonic acid	260	127
Barium carbonate	240	116	Chromic acid 10%	x	x
Barium chloride	260	127	Chromic acid 50%	x	x
Barium hydroxide	260	127	Chromyl chloride	250	121
Barium sulfide	260	127	Citric acid 15%	250	121
Benzaldehyde	80	27	Citric acid, concd	250	121
Benzene	160	71	Copper acetate	260	127
Benzenesulfonic			Copper carbonate		
acid 10%	160	71	Copper chloride	260	127
Benzoic acid	260	127	Copper cyanide	240	116
Benzyl alcohol	80	27	Copper sulfate	300	149
Benzyl chloride	140	60	Cresol	260	127

(*Continued*)

TABLE 3.35 Continued

Chemical	Maximum temp. °F	Maximum temp. °C	Chemical	Maximum temp. °F	Maximum temp. °C
Cupric chloride 5%	300	149	Nitric acid 5%	x	x
Cupric chloride 50%	300	149	Nitric acid 20%	x	x
Cyclohexane	140	60	Nitric acid 70%	x	x
Cyclohexanol			Nitric acid, anhydrous	x	x
Dichloroacetic acid	x	x	Nitrous acid, concd	x	x
Dichloroethane			Oleum	190	88
(ethylene dichloride)	250	121	Perchloric acid 10%	x	x
Ethylene glycol	160	71	Perchloric acid 70%	260	127
Ferric chloride	260	127	Phenol	x	x
Ferric chloride 50%			Phosphoric acid 50%	212	100
in water	160	71	Picric acid		
Ferric nitrate 10–50%	160	71	Potassium bromide 30%	260	127
Ferrous chloride	160	71	Salicylic acid	260	127
Ferrous nitrate			Silver bromide 10%		
Fluorine gas, dry	x	x	Sodium carbonate	212	100
Fluorine gas, moist	x	x	Sodium chloride	260	127
Hydrobromic acid, dilute	212	100	Sodium hydroxide 10%	x	x
Hydrobromic acid 20%	212	100	Sodium hydroxide 50%	x	x
Hydrobromic acid 50%	212	100	Sodium hydroxide, concd	x	x
Hydrochloric acid 20%	212	100	Sodium hypochlorite 15%	x	x
Hydrochloric acid 38%	80	27	Sodium hypochlorite, concd	x	x
Hydrocyanic acid 10%	160	71	Sodium sulfide to 10%	260	127
Hydrofluoric acid 30%	230	110	Stannic chloride	260	127
Hydrofluoric acid 70%	140	60	Stannous chloride	250	121
Hydrofluoric acid 100%	140	60	Sulfuric acid 10%	160	71
Hypochlorous acid	x	x	Sulfuric acid 50%	80	27
Iodine solution 10%	x	x	Sulfuric acid 70%	80	27
Ketones, general	100	38	Sulfuric acid 90%	x	x
Lactic acid 25%	212	100	Sulfuric acid 98%	x	x
Lactic acid, concd	160	71	Sulfuric acid 100%	x	x
Magnesium chloride	260	127	Sulfuric acid, fuming	x	x
Malic acid 10%	260	127	Sulfurous acid	160	71
Manganese chloride	200	93	Thionyl chloride	x	x
Methyl chloride	120	49	Toluene	212	100
Methyl ethyl ketone	80	27	Trichloroacetic acid 30%	80	27
Methyl isobutyl ketone	160	71	White liquor	140	60
Muriatic acid	80	27	Zinc chloride	160	71

[a] The chemicals listed are in the pure state or in a saturated solution unless otherwise indicated. Compatibility is shown to the maximum allowable temperature for which data are available. Incompatibility is shown by an x. A blank space indicates that data are unavailable.
Source: Schweitzer PA. Corrosion Resistance Tables, 4th ed., Vols. 1–3. New York: Marcel Dekker, 1995.

TABLE 3.36 Physical and Mechanical Properties of Filled Phenolic Resins

Property	Wood flour filled	High-strength glass fiber reinforced	High-impact cotton filled
Specific gravity	1.37–1.46	1.69–2.0	1.36–1.42
Water absorption (24 hr at 73°F/23°C) (%)	0.3–1.2	0.3–1.2	
Dielectric strength, short-term (V/mil)	260–400	140–400	200–360
Tensile strength at break (psi)	5,000–9,000	7,000–18,000	6,000–10,000
Tensile modulus ($\times 10^3$ psi)	800–1,700	1,900–3,500	1,100–1,400
Elongation at break (%)	0.4–0.8	0.2	1–2
Compressive strength (psi)	25,000–31,000	16,000–70,000	23,000–31,000
Flexural strength psi	7,000–14,000	12,000–64,000	9,000–13,000
Compressive modulus ($\times 10^3$ psi)		2,704–3,500	
Flexural modulus ($\times 10^3$ psi) at 73°F/23°C 200°F/93°C 250°F/121°C	1,000–1,200	1,150–3,300	800–1,300
Izod impact (ft-lb/in. of notch)	0.2–0.6	0.5–1.8	0.3–1.9
Hardness, Rockwell	M100–115	E54–101	M105–120
Coefficient of thermal expansion (10^{-6} in./in./°F)	30–45	8–34	15–22
Thermal conductivity (10^{-4} cal-cm/sec-cm²°C or Btu/hr/ft²/°F/in.)	4–8	8–14	8–10
Deflection temperature at 264 psi (°F) at 66 psi (°F)	300–370	350–600	300–400
Max. operating temperature (°F/°C)			
Limiting oxygen index (%)			
Flame spread			
Underwriters Lab. rating (Sub. 94)			

An extremely large number of phenolic materials are available as a result of the many resin and filler combinations. Table 3.37 lists the physical and mechanical properties of phenolic casting resins, while Table 3.38 does likewise for sheet molding compounds.

Although the phenolics are not equivalent to diallyl phthalates and epoxies in resistance to humidity and retention of electrical properties in extreme

TABLE 3.37 Physical and Mechanical Properties of Phenolic Casting Resins

Property	Unfilled	Mineral filled
Specific gravity	1.24–1.32	1.66–1.70
Water absorption (24 hr at 73°F/23°C) (%)	0.1–0.38	
Dielectric strength, short-term (V/mil)	250–400	100–250
Tensile strength at break (psi)	5,000–9,000	4,000–9,000
Tensile modulus ($\times 10^3$ psi)	400–700	
Elongation at break (%)	1.5–2	
Compressive strength (psi)	12,000–15,000	20,000–34,000
Flexural strength (psi)	11,000–17,000	9,000–12,000
Compressive modulus ($\times 10^3$ psi)		
Flexural modulus ($\times 10^3$ psi) at 73°F/23°C		
200°F/93°C		
250°F/121°C		
Izod impact (ft-lb/in. of notch)	0.24–0.4	0.38–0.5
Hardness, Rockwell	M93–120	M93–120
Coefficient of thermal expansion (10^{-6} in./in./°F)	68	75
Thermal conductivity (10^{-4} cal-cm/sec-cm^2°C or Btu/hr/ft^2/°F/in.)	3.5	
Deflection temperature at 264 psi (°F)	165–175	150–175
at 66 psi (°F)		
Max. operating temperature (°F/°C)		
Limiting oxygen index (%)		
Flame spread		
Underwriters Lab. rating (Sub. 94)		

environments, they are quite adequate for a large percentage of electrical applications. Formulations are available which have considerable improvement in resistance to humid environments and higher temperatures. Glass-filled, heat-resistant grades maintain stability up to 400°F/204°C and higher, and some formulations are useful up to 500°F/260°C.

1. Corrosion Resistance Properties

Phenolic resins exhibit resistance to most organic solvents, especially aromatics and chlorinated solvents. Organic polar solvents capable of hydrogen bonding, such as alcohols and ketones, can attack phenolics. Although the phenolics have an aromatic character, the phenolic hydroxyls provide sites for hydrogen bonding and attack by caustics. Phenolics are not suitable for use in strong alkaline environments. Strong mineral acids also attack the phenolics, and acids such as

TABLE 3.38 Physical and Mechanical Properties of Phenolic Sheet Molding Compounds

Property	Unfilled	Glass fiber reinforced
Specific gravity	1.24–1.32	1.69–2.0
Water absorption (24 hr at 73°F/23°C) (%)	0.1–0.36	0.03–1.2
Dielectric strength, short-term (V/mil)	250–400	140–400
Tensile strength at break (psi)	5,000–9,000	7,000–18,000
Tensile modulus ($\times 10^3$ psi)	400–700	1,900–3,300
Elongation at break (%)	15–20	0.2
Compressive strength (psi)	12,000–15,000	2,740–3,500
Flexural strength (psi)	11,000–17,000	12,000–60,000
Compressive modulus ($\times 10^3$ psi)		16,000–70,000
Flexural modulus ($\times 10^3$ psi) at 73°F/23°C		1,150–3,300
200°F/93°C		
250°F/121°C		
Izod impact (ft-lb/in. of notch)	0.24–0.4	0.5–18.0
Hardness, Rockwell	M93–120	E54–101
Coefficient of thermal expansion (10^{-6} in./in./°F)	68	8.34
Thermal conductivity (10^{-4} cal-cm/sec-cm^2°C or Btu/hr/ft^2/°F/in.)	3.5	8–14
Deflection temperature at 264 psi (°F) at 66 psi (°F)	165–175	350–600
Max. operating temperature (°F/°C)		
Limiting oxygen index (%)		
Flame spread		
Underwriters Lab. rating (Sub. 94)		

nitric, chromic, and hydrochloric cause severe degradation. Sulfuric and phosphoric acids may be suitable under some conditions. There is some loss of properties when phenolics are in contact with organic acids, such as acetic, formic, and oxalic.

Refer to Table 3.39 for the compatibility of phenolics with selected corrodents. Reference 1 provides a more comprehensive and detailed listing.

2. Typical Applications

In addition to molding compounds, phenolics are used to bond friction materials for automotive brake linings, clutch parts, and transmission bands. They serve as binders for core material in furniture, as the water-resistant adhesive for exterior-

TABLE 3.39 Compatibility of Phenolics with Selected Corrodents[a]

Chemical	Maximum temp.		Chemical	Maximum temp.	
	°F	°C		°F	°C
Acetaldehyde			Barium sulfide		
Acetamide			Benzaldehyde	70	21
Acetic acid 10%	212	100	Benzene	160	71
Acetic acid 50%			Benzenesulfonic	70	21
Acetic acid 80%			acid 10%		
Acetic acid, glacial	70	21	Benzoic acid		
Acetic anhydride	70	21	Benzyl alcohol		
Acetone	x	x	Benzyl chloride	70	21
Acetyl chloride			Borax		
Acrylic acid			Boric acid		
Acrylonitrile			Bromine gas, dry		
Adipic acid			Bromine gas, moist		
Allyl alcohol			Bromine, liquid		
Allyl chloride			Butadiene		
Alum			Butyl acetate	x	x
Aluminum acetate			Butyl alcohol		
Aluminum chloride, aqueous	90	32	n-Butylamine		
Aluminum chloride, dry			Butyl phthalate	160	71
Aluminum fluoride			Butyric acid 25%		
Aluminum hydroxide			Calcium bisulfide		
Aluminum nitrate			Calcium bisulfite		
Aluminum oxychloride			Calcium carbonate		
Aluminum sulfate	300	149	Calcium chlorate		
Ammonia gas	90	32	Calcium chloride	300	149
Ammonia bifluoride			Calcium hydroxide 10%		
Ammonium carbonate	90	32	Calcium hydroxide, sat.		
Ammonium chloride 10%	80	27	Calcium hypochlorite 10%	x	x
Ammonium chloride 50%	80	27	Calcium nitrate		
Ammonium chloride, sat.	80	27	Calcium oxide		
Ammonium fluoride 10%			Calcium sulfate		
Ammonium fluoride 25%			Caprylic acid		
Ammonium hydroxide 25%	x	x	Carbon bisulfide		
Ammonium hydroxide, sat.	x	x	Carbon dioxide, dry	300	149
Ammonium nitrate	160	71	Carbon dioxide, wet	300	149
Ammonium persulfate			Carbon disulfide		
Ammonium phosphate			Carbon monoxide		
Ammonium sulfate 10–40%	300	149	Carbon tetrachloride	200	93
Ammonium sulfide			Carbonic acid	200	93
Ammonium sulfite			Cellosolve		
Amyl acetate			Chloroacetic acid,		
Amyl alcohol			50% water		
Amyl chloride			Chloroacetic acid		
Aniline	x	x	Chlorine gas, dry		
Antimony trichloride			Chlorine gas, wet	x	x
Aqua regia 3 : 1			Chlorine, liquid	x	x
Barium carbonate			Chlorobenzene	260	127
Barium chloride			Chloroform	160	71
Barium hydroxide			Chlorosulfonic acid		
Barium sulfate			Chromic acid 10%		

TABLE 3.39 Continued

Chemical	Maximum temp.		Chemical	Maximum temp.	
	°F	°C		°F	°C
Chromic acid 50%	x	x	Manganese chloride		
Chromyl chloride	x	x	Methyl chloride	160	71
Citric acid 15%	160	71	Methyl ethyl ketone	x	x
Citric acid, concd	160	71	Methyl isobutyl ketone	160	71
Copper acetate			Muriatic acid	300	149
Copper carbonate			Nitric acid 5%	x	x
Copper chloride			Nitric acid 20%	x	x
Copper cyanide			Nitric acid 70%	x	x
Copper sulfate	300	149	Nitric acid, anhydrous	x	x
Cresol			Nitrous acid, concd		
Cupric chloride 5%			Oleum		
Cupric chloride 50%			Perchloric acid 10%		
Cyclohexane			Perchloric acid 70%		
Cyclohexanol			Phenol	x	x
Dibutyl phthalate			Phosphoric acid 50–80%	212	100
Dichloroacetic acid			Picric acid		
Dichloroethane			Potassium bromide 30%		
(ethylene dichloride)			Salicylic acid		
Ethylene glycol	70	21	Silver bromide 10%		
Ferric chloride	300	149	Sodium carbonate		
Ferric chloride			Sodium chloride	300	149
50% in water	300	149	Sodium hydroxide 10%	x	x
Ferric nitrate 10–50%			Sodium hydroxide 50%	x	x
Ferrous chloride 40%			Sodium hydroxide, concd	x	x
Ferrous nitrate			Sodium hypochlorite 15%	x	x
Fluorine gas, dry			Sodium hypochlorite,	x	x
Fluorine gas, moist			concd		
Hydrobromic acid, dilute	200	93	Sodium sulfide to 50%		
Hydrobromic acid 20%	200	93	Stannic chloride		
Hydrobromic acid 50%	200	93	Stannous chloride		
Hydrochloric acid 20%	300	149	Sulfuric acid 10%	250	121
Hydrochloric acid 38%	300	149	Sulfuric acid 50%	250	121
Hydrocyanic acid 10%			Sulfuric acid 70%	200	93
Hydrofluoric acid 30%	x	x	Sulfuric acid 90%	70	21
Hydrofluoric acid 60%	x	x	Sulfuric acid 98%	x	x
Hydrofluoric acid 100%	x	x	Sulfuric acid 100%	x	x
Hypochlorous acid			Sulfuric acid, fuming		
Iodine solution 10%			Sulfurous acid	80	27
Ketones, general			Thionyl chloride	200	93
Lactic acid 25%	160	71	Toluene		
Lactic acid, concd			Trichloroacetic acid 30%		
Magnesium chloride			White liquor		
Malic acid 10%			Zinc chloride	300	149

a The chemicals listed are in the pure state or in a saturated solution unless otherwise indicated. Compatibility is shown to the maximum allowable temperature for which data are available. Incompatibility is shown by an x. A blank space indicates that the data are unavailable.
Source: Schweitzer, PA. Corrosion Resistance Tables, 4th ed., Vols. 1–3. New York: Marcel Dekker, 1995.

grade plywood, binders for wood particle boards, and as the bonding agent for converting organic and inorganic fibers into acoustical and thermal insulation pads, baths, or cushioning for home, industrial, and automotive applications. Decorative or electrical laminates are produced by impregnating paper with phenolic resins.

The excellent water resistance of the phenolics makes them particularly suitable for marine applications. They also find applications as gears, wheels, and pulleys because of their wear and abrasion resistance. Because of their stability in a variety of environments, they are used in printed circuits and terminal blocks.

Their use is limited because of two disadvantages. The laminates are only available in dark colors, usually brown or black, because of the nature of the resin. Secondly, they have somewhat poor resistance to electric arcs, and even though high filler loading can improve the low-power arc resistance, moisture, dirt, or high voltages usually result in complete arcing breakdown. Although they maintain their properties in the presence of water, their water absorption is high, reaching 14% in some paper-based grades of laminates.

Haveg, a division of Ametek, Inc., produces a phenolic piping system using silica filaments and fillers. It is sold under the tradename of Haveg SP. Refer to reference 2 for more details.

3.8 PHENOL-FORMALDEHYDE

As the name implies, these resins are derived from phenol and formaldehyde. The structural formula is shown in Figure 3.7. Phenol-formaldehyde is a cross-linked phenolic resin and in general has approximately the same basic physical and mechanical properties as other phenolic resins. However, it does not have the

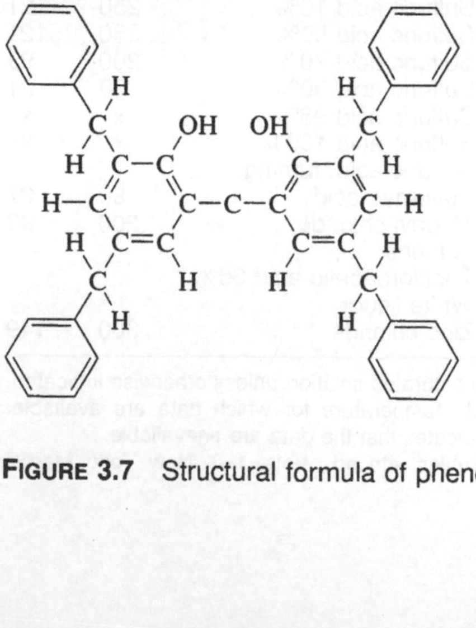

FIGURE 3.7   Structural formula of phenol-formaldehyde.

TABLE 3.40 Physical and Mechanical Properties of Phenol-Formaldehyde Resins

Property	
Specific gravity	1.8
Water absorption (24 hr at 73°F/23°C) (%)	
Dielectric strength, short-term (V/mil)	
Tensile strength at break (psi)	5,540
Tensile modulus ($\times 10^3$ psi)	1,370
Elongation at break (%)	
Compressive strength (psi)	19,300
Flexural strength (psi)	9,840
Compressive modulus ($\times 10^3$ psi)	
Flexural modulus ($\times 10^3$ psi) at 73°F/23°C	
200°F/93°C	
250°F/121°C	
Izod impact (ft-lb/in. of notch)	0.60
Hardness, Rockwell	
Coefficient of thermal expansion (10^{-6} in./in./°F)	8.3
Thermal conductivity (10^{-4} cal-cm/sec-cm^2 °C or Btu/hr/ft^2/°F/in.)	1.96
Deflection temperature at 264 psi (°F)	
at 66 psi (°F)	
Max. operating temperature (°F/°C)	300/149
Limiting oxygen index (%)	
Flame spread	< 50
Underwriters Lab. rating (Sub. 94)	

impact resistance of the polyesters or epoxies. Table 3.40 lists physical and mechanical properties of phenol-formaldehyde resins.

1. Corrosion Resistance Properties

Phenol-formaldehyde laminates are generally used with mineral acids, salts, and chlorinated aromatic hydrocarbons. By using graphite as a filler, the laminate is suitable for use with hydrofluoric acid and certain fluoride salts. Refer to Table 3.41 for the compatibility of phenol-formaldehyde with selected corrodents. Reference 1 provides a more comprehensive and detailed listing.

2. Typical Applications

Applications include gears, wheels, and pulleys, printed circuits, terminal blocks and piping. Reference 2 provides details on phenol-formaldehyde piping systems.

TABLE 3.41 Compatibility of Phenol-Formaldehyde with Selected Corrodents[a]

Chemical	°F	°C	Chemical	°F	°C
Acetaldehyde	x	x	Ethylene glycol	80	27
Acetamide			Ferric chloride 50% in water	300	149
Acetic acid 10%	212	100	Ferric nitrate 10–50%	300	149
Acetic acid 50%	160	71	Ferrous chloride 40%	300	149
Acetic acid 80%	120	49	Hydrobromic acid, dilute	212	100
Acetic acid, glacial	120	49	Hydrobromic acid 20%	212	100
Acetone	x	x	Hydrochloric acid 20%	300	149
Acetyl chloride	x	x	Hydrochloric acid 38%	300	149
Acrylic acid 90%	80	27	Hydrocyanic acid 10%	160	71
Acrylonitrile	x	x	Hydrofluoric acid 30%	x	x
Aluminum acetate			Hydrofluoric acid 70%	x	x
Aluminum chloride, aqueous	300	149	Hydrofluoric acid 100%	x	x
Aluminum chloride, dry	300	149	Hypochlorous acid		
Aluminum sulfate	300	149	Iodine solution 10%	x	x
Ammonium hydroxide 25%	x	x	Lactic acid 25%	160	71
Ammonium hydroxide, sat.	x	x	Lactic acid, concd	160	71
Amyl alcohol	160	71	Methyl chloride	300	149
Aniline	x	x	Methyl ethyl ketone	x	x
Aqua regia 3:1	x	x	Methyl isobutyl ketone	x	x
Benzene	160	71	Muriatic acid	300	149
Benzenesulfonic acid 10%	160	71	Nitric acid 5%	x	x
Benzyl chloride	160	71	Nitric acid 20%	x	x
Boric acid	300	149	Nitric acid 70%	x	x
Bromine, liquid 3% max.	300	149	Nitric acid, anhydrous	x	x
Butyric acid	260	127	Phenol	x	x
Calcium chloride	300	149	Phosphoric acid 50%	212	100
Calcium hydroxide 10%	x	x	Sodium carbonate	x	x
Calcium hydroxide, sat.	x	x	Sodium hydroxide 10%	x	x
Calcium hypochlorite	x	x	Sodium hydroxide 50%	x	x
Carbon bisulfide	160	71	Sodium hydroxide, concd	x	x
Carbon tetrachloride	212	100	Sodium hypochlorite 15%	x	x
Chlorine gas, dry	160	71	Sodium hypochlorite, concd	x	x
Chlorine gas, wet	160	71	Sulfuric acid 10%	300	149
Chlorobenzene			Sulfuric acid 50%	300	149
Chloroform	160	71	Sulfuric acid 70%	250	121
Chlorosulfonic acid	80	27	Sulfuric acid 90%	100	38
Chromic acid 10%	x	x	Sulfurous acid	160	71
Chromic acid 50%	x	x	Thionyl chloride	80	27
Copper sulfate	300	149	Toluene	212	100
Cresol			Trichloroacetic acid 30%	80	27
Cupric chloride 5%	300	149	Zinc chloride	300	149
Cupric chloride 50%	300	149			

[a] The chemicals listed are in the pure state or in a saturated solution unless otherwise indicated. Compatibility is shown to the maximum allowable temperature for which data are available. Incompatibility is shown by an x. A blank space indicates that the data are unavailable.

Source: Schweitzer, PA. Corrosion Resistance Tables, 4th ed., Vols 1–3. New York: Marcel Dekker, 1995.

Phenol-formaldehyde resins have high heat resistance and good char strengths. Because of this, they find application as ablative shields for reentry vehicles, rocket nozzles, nose cones, and rocket motor chambers.

They also produce less smoke and toxic by-products of combustion and are often used as aircraft interior panels.

3.9 SILICONES

Silicon is in the same chemical group as carbon but is a more stable element. The silicones are a family of synthetic polymers that are partly organic and inorganic. They have a backbone structure of alternating silicon and oxygen atoms rather than a backbone of carbon–carbon atoms. The basic structure is

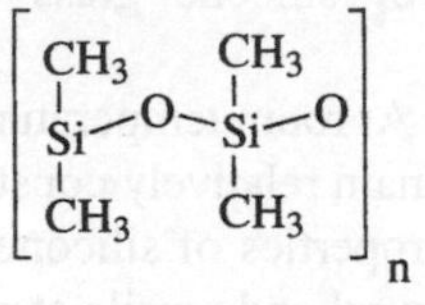

$$\left[\begin{array}{c} CH_3 \quad\ CH_3 \\ | \qquad\ | \\ Si{-}O{-}Si{-}O \\ | \qquad\ | \\ CH_3 \quad\ CH_3 \end{array} \right]_n$$

Typically the silicon atoms will have one or more organic side groups attached to them, generally phenol (C_6H_5-), methyl (CH_3-), or vinyl ($CH_2{=}CH-$) units. These groups impact properties such as solvent resistance, lubricity, and reactivity with organic chemicals and polymers. Silicone polymers may be filled or unfilled depending upon the properties required and the application.

Silicone polymers possess several properties which distinguish them from their organic counterparts. These are

1. Chemical inertness
2. Weather resistance
3. Extreme water repellency
4. Uniform properties over a wide temperature range
5. Excellent electrical properties over a wide range of temperature and frequencies
6. Low surface tension
7. High degree of slip or lubricity
8. Excellent release properties
9. Inertness and compatibility, both physiologically and in electronic applications

Silicone resins and composites produced with silicone resins exhibit outstanding long-term thermal stability at temperatures approaching 572°F/300°C and excellent moisture resistance and electrical properties. These materials are also useful in the cryogenic temperature range. Table 3.42 shows the effect of

TABLE 3.42 Strength of Silicone Glass Fabric Laminate at Cryogenic Temperatures

Transformation (°F/°C)	Tensile strength (psi×10³)	Flexural strength (psi×10³)	Compressive strength (psi×10³)
72/22	30	38	21
−110/−79	47	46	39
−320/−201	70	67	43
−424/−253	76	65	46

cryogenic temperatures on the physical properties of silicone glass fabric laminates.

Silicones also possess useful dielectric properties. At room temperature, the dissipation factor and dielectric constant are low and remain relatively constant to 300°F/149°C. Aging has little effect on the physical properties of silicones, but compared with laminates based on other resins, the flexural and tensile strengths of the silicones are not unusually high. Refer to Table 3.43 for the physical and mechanical properties of mineral- and/or glass-filled silicones.

1. Corrosion Resistance Properties

Silicone laminates can be used in contact with dilute acids and alkalies, alcohols, animal and vegetable oils, and lubricating oils. They are also resistant to aliphatic hydrocarbons, but aromatic solvents such as benzene, toluene, gasoline, and chlorinated solvents will cause excessive swelling. Although they have excellent resistance to water and weathering, they are not resistant to high-pressure, high-temperature steam.

As discussed previously, the silicon atoms may have one or more organic side groups attached. The addition of these side groups has an effect on the corrosion resistance. Therefore, it is necessary to check with the supplier as to the properties of the silicone laminate being supplied. Table 3.44 lists the compatibility of a silicone laminate with methyl groups appended to the silicon atoms.

2. Typical Applications

Silicone laminates find application as radomes, structures in electronics, heaters, rocket components, slot wedges, ablation shields, coil forms, and terminal boards.

TABLE 3.43 Physical and Mechanical Properties of Silicone Laminates

Property	
Specific gravity	1.8–2.03
Water absorption (24 hr at 73°F/23°C) (%)	0.15
Dielectric strength, short-term (V/mil)	200–550
Tensile strength at break (psi)	500–1,500
Tensile modulus (×10³ psi)	
Elongation at break (%)	80–800
Compressive strength (psi)	21,000
Flexural strength (psi)	38,000
Compressive modulus (×10³ psi)	
Flexural modulus (×10³ psi) at 73°F/23°C	
200°F/93°C	
250°F/121°C	
Izod impact (ft-lb/in. of notch)	
Hardness, Shore	A10–80
Coefficient of thermal expansion (10⁻⁶ in./in./°F)	20–50
Thermal conductivity (10⁻⁴ cal-cm/sec-cm²°C or Btu/hr/ft²/°F/in.)	7–18
Deflection temperature at 264 psi (°F)	> 500
at 66 psi (°F)	
Max. operating temperature (°F/°C)	> 550/288
Limiting oxygen index (%)	
Flame spread	
Underwriters Lab. rating (Sub. 94)	

3.10 SILOXIRANE

Siloxirane is the registered trademark for Tankenetics homopolymerized polymer with a cross-linked organic–inorganic (SiO) backbone with oxirane end caps. Siloxirane is a homopolymerized polymer with an ether cross-linkage (carbon–oxygen–carbon) having a very dense, highly cross-linked molecular structure. The absence of the hydroxyl (found in epoxies) and ester (found in vinyl esters) groups eliminates the built-in failure modes of other polymers and provides superior performance.

The end products are extremely resistant to material abrasion and have an operating temperature range of −80 to +500°F/−62 to +260°C. The operating temperature will be tempered by the material being handled.

TABLE 3.44 Compatibility of Methyl Appended Silicone Laminate with Selected Corrodents[a]

Chemical	Maximum temp. °F	Maximum temp. °C	Chemical	Maximum temp. °F	Maximum temp. °C
Acetic acid 10%	90	32	Lactic acid, concd	80	27
Acetic acid 50%	90	32	Magnesium chloride	400	204
Acetic acid 80%	90	32	Methyl alcohol	410	210
Acetic acid, glacial	90	32	Methyl ethyl ketone	x	x
Acetone	100	43	Methyl isobutyl ketone	x	x
Acrylic acid 75%	80	27	Nitric acid 5%	80	23
Acrylonitrile	x	x	Nitric acid 20%	x	x
Alum	220	104	Nitric acid 70%	x	x
Aluminum sulfate	410	210	Nitric acid, anhydrous	x	x
Ammonium chloride 10%	x	x	Oleum	x	x
Ammonium chloride 50%	80	27	Phenol	x	x
Ammonium chloride, sat.	80	27	Phosphoric acid 50–80%	x	x
Ammonium fluoride 25%	80	27	Propyl alcohol	400	204
Ammonium hydroxide 25%	x	x	Sodium carbonate	300	149
Ammonium nitrate	210	99	Sodium chloride 10%	400	204
Amyl acetate	80	27	Sodium hydroxide 10%	90	27
Amyl alcohol	x	x	Sodium hydroxide 50%	90	27
Amyl chloride	x	x	Sodium hydroxide, concd	90	27
Aniline	x	x	Sodium hypochlorite 20%	x	x
Antimony trichloride	80	27	Sodium sulfate	400	204
Aqua regia 3:1	x	x	Stannic chloride	80	27
Benzene	x	x	Sulfuric acid 10%	x	x
Benzyl chloride	x	x	Sulfuric acid 50%	x	x
Boric acid	390	189	Sulfuric acid 70%	x	x
Butyl alcohol	80	27	Sulfuric acid 90%	x	x
Calcium bisulfide	400	204	Sulfuric acid 98%	x	x
Calcium chloride	300	149	Sulfuric acid 100%	x	x
Calcium hydroxide 30%	200	99	Sulfuric acid, fuming	x	x
Calcium hydroxide, sat.	400	204	Sulfurous acid	x	x
Carbon bisulfide	x	x	Tartaric acid	400	204
Carbon disulfide	x	x	Tetrahydrofuran	x	x
Carbon monoxide	400	204	Toluene	x	x
Carbonic acid	400	204	Tributyl phosphate	x	x
Chlorobenzene	x	x	Turpentine	x	x
Chlorosulfonic acid	x	x	Vinegar	400	204
Ethylene glycol	400	204	Water, acid mine	210	99
Ferric chloride	400	204	Water, demineralized	210	99
Hydrobromic acid 50%	x	x	Water, distilled	210	99
Hydrochloric acid 20%	90	32	Water, salt	210	99
Hydrochloric acid 38%	x	x	Water, sea	210	99
Hydrofluoric acid 30%	x	x	Xylene	x	x
Lactic acid, all concn	80	27	Zinc chloride	400	204

[a] The chemicals listed are in the pure state or in a saturated solution unless otherwise indicated. Compatibility is shown to the maximum allowable temperature. Incompatibility is shown by an x.
Source: Schweitzer PA. Corrosion Resistance Tables, 4th ed., Vols. 1–3. New York: Marcel Dekker, 1995.

1. Corrosion Resistance Properties

Siloxirane laminates have a wide range of chemical resistance. Following is a list of some of the chemicals with which siloxirane is compatible:

Acetamide	Hydrofluoric acid 40%
Acetic acid, glacial	Hydrofluoric acid 52%
Acetic anhydride	Iodine
Acetone	Jet fuel
Aluminum chloride	Kerosene
Ammonium chloride	Ketones
Ammonium hydroxide	Latex
Aqua regia	Methanol
Benzene	Methyl ethyl ketone
Benzenesulfonic acid	Methyl isobutyl ketone
Black liquor	Methylene chloride
Bromine hypochlorite	Molten sulfur
Carbon tetrachloride	Monochloroacetic acid
Chloric acid	Nickel plating solutions
Chlorine water	Nitrous oxide
Chloroacetic acid	Phosphoric acid
Chlorobenzene	Phosphoric acid 85%
Chromic acid 10%	Sodium chloride
Chromic acid 50%	Sodium dichromate
Dibutyl phthalate	Sodium hydroxide
Dichlorobenzene	Sodium hypochlorite 17%
Dimethylformamide	Sodium hypochlorite, aged
Ethanol	Sulfate liquor (paper)
Ethyl acetate	Sulfur trioxide
Ferric chloride	Sulfuric acid 1–98%
Formaldehyde	Sulfuric acid, fuming oleum
Furan	Tallow
Furfural alcohol	Thionyl chloride
Gasohol	Toluene
Gasoline	Trichloroethylene
Green liquor	Tricresyl phosphate
Hydraulic oil	Water, deionized
Hydrazine	Water, salt
Hydrochloric acid 1%	White liquor (paper)
Hydrochloric acid 0–37%	

2. Typical Applications

Siloxirane laminates are used as piping, duct work, storage tanks, and tank liners. Vessels produced from Siloxirane have been approved to receive ASME Section X class 11 certification and stamping as code vessels. Class 11 vessels may be a maximum of 144 inches in diameter and the product of pressure (psig) and the diameter in inches may not exceed 7200.

3.11 POLYURETHANES

Polyurethanes are reaction products of isocyanates, polyols, and curing agents. Because of the hazards involved in handling free isocyanate, prepolymers of the isocyanate and the polyol are generally used in casting. Polyurethanes can be formulated to produce a range of materials from elastomers as soft as Shore A of 5 to tough solids with a Shore D of 90. Polyurethane thermosets can be rigid or flexible depending on the formulation. The flexible ones have better toughness and are often used as foamed parts. The physical and mechanical properties of mineral-filled and unfilled polyurethane are shown in Table 3.45.

1. Corrosion Resistance Properties

Polyurethane is resistant to most mineral and vegetable oils. Some 1,4-butane-diol-cured polyurethanes have been FDA approved for use in applications that will come into contact with dry, aqueous, and fatty foods. They are also resistant to greases, fuels, aliphatic, and chlorinated hydrocarbons. This makes these materials particularly suited for service in contact with lubricating oils and automotive fuels.

Aromatic hydrocarbons, polar solvents, esters, and ketones will attack urethane.

2. Typical Applications

Applications include large automotive parts and building components. Foam materials blown with halocarbons have the lowest thermal conductivity of any commercially available insulation. They are used in refrigerators, picnic boxes, and building construction. Flexible foam is also used in furniture, packaging, and shock and vibration mounts.

3.12 MELAMINES

Melamine is a polymer formed by a condensation reaction between formaldehyde and amino compounds containing NH_2 groups. Consequently, they are also referred to as melamine formaldehydes. Their structural formula is shown in Figure 3.8.

TABLE 3.45 Physical and Mechanical Properties of Mineral-Filled and Unfilled Polyurethane

Property	50–65% mineral filled	Unfilled
Specific gravity	1.37–2.1	1.2
Water absorption (24 hr at 73°F/23°C)(%)	0.06–0.52	
Dielectric strength, short-term (V/mil)	500–750	
Tensile strength at break (psi)	1,000–7,000	2,900–10,100
Tensile modulus ($\times 10^3$ psi)		438–870
Elongation at break (%)	5–55	180–300
Compressive strength (psi)		
Flexural strength (psi)		
Compressive modulus ($\times 10^3$ psi)		
Flexural modulus ($\times 10^3$ psi) at 73°F/23°C		
200°F/93°C		
250°F/121°C		
Izod impact (ft-lb/in. of notch)		
Hardness, Shore	A90	D39–64
Coefficient of thermal expansion (10^{-6} in./in./°F)	71–100	
Thermal conductivity (10^{-4} cal-cm/sec-cm^2 °C or Btu/hr/ft^2/°F/in.)	6.8–10	
Deflection temperature at 264 psi (°F)		
at 66 psi (°F)		
Max. operating temperature (°F/°C)		
Limiting oxygen index (%)		
Flame spread		
Underwriters Lab. rating (Sub. 94)		

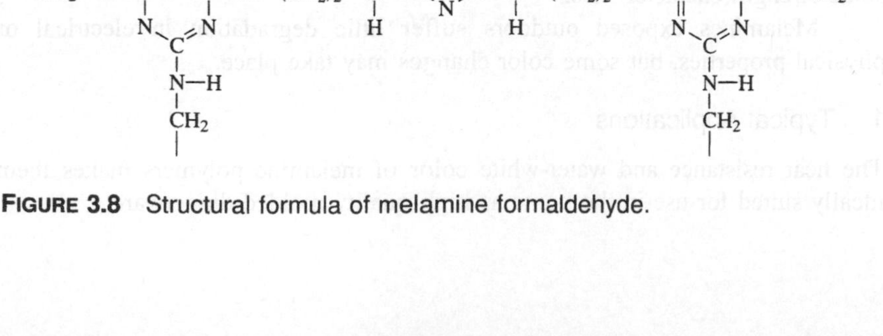

FIGURE 3.8 Structural formula of melamine formaldehyde.

TABLE 3.46 Low Temperature Mechanical Properties of Melamine

Property	Temperature (°F)	Melamine glass	Melamine formaldehyde
Tensile	77	32.5	7.8
strength	10	32.9	6.9
($\times 10^3$ psi)	-40	38.1	6.9
	-65	37.2	6.7
Modulus of	77	2.130	1.270
elasticity	10	2.200	1.640
($\times 10^4$ psi)	-40	1.420	1.730
	-65	1.680	1.880
Elongation	77	2.15	0.62
at break (%)	10	2.17	0.44
	-40	2.36	0.39
	-65	2.75	0.37
Izod impact	77	11.12	0.31
strength	10	12.64	0.28
(ft lb/in.	-40	13.43	0.28
notch	-65	14.68	0.29

Melamine resin can be combined with a variety of reinforcing fibers. However the best properties are obtained when glass cloth is used as the reinforcement material. Low temperatures have relatively little effect on the properties of the melamine. Refer to Table 3.46.

The clarity of melamine resins permits products to be fabricated in virtually any color. Finished melamine products exhibit excellent resistance to moisture, greases, oils, and solvents; are tasteless and odorless; are self-extinguishing; resist scratching and marring; and offer excellent electrical properties.

Table 3.47 provides the physical and mechanical properties of cellulose-filled and glass fiber reinforced melamine.

Melamine formulations filled with cellulose are capable of producing an unlimited range of light-stable colors with a high degree of translucency. The basic material properties are not affected by the addition of color. Prolonged exposure at high temperatures will affect the color, as well as causing a loss of some strength characteristics.

Melamines exposed outdoors suffer little degradation in electrical or physical properties, but some color changes may take place.

1. Typical Applications

The heat resistance and water-white color of melamine polymers makes them ideally suited for use in the home as high-quality molded dinnerware.

TABLE 3.47 Physical and Mechanical Properties of Cellulose-Filled and Glass Fiber Reinforced Melamine Formaldehyde

Property	Cellulose filled	Glass fiber reinforced
Specific gravity	1.47–1.52	1.5–2.0
Water absorption (24 hr at 73°F/23°C)(%)	0.1–0.8	0.09–1.3
Dielectric strength, short-term (V/mil)	275–400	130–370
Tensile strength at break (psi)	5,000–13,000	5,000–10,500
Tensile modulus ($\times 10^3$ psi)		
Elongation at break (%)	0.6–1	0.6
Compressive strength (psi)	33,000–45,000	20,000–35,000
Flexural strength psi	9,000–15,000	14,000–23,000
Compressive modulus ($\times 10^3$ psi)	1,100–1,400	1,600–2,400
Flexural modulus ($\times 10^3$ psi) at 73°F/23°C	1100	
200°F/93°C		
250°F/121°C		
Izod impact (ft-lb/in. of notch)	0.2–0.4	0.5–18
Hardness, Rockwell	M115–125	M115
Coefficient of thermal expansion (10^{-6} in./in./°F)	40–45	15–28
Thermal conductivity (10^{-4} cal-cm/sec-cm²°C or Btu/hr/ft²/°F/in.)	6.5–10	10–11.5
Deflection temperature at 264 psi (°F)	350–390	375–400
at 66 psi (°F)		
Max. operating temperature (°F/°C)		
Limiting oxygen index (%)		
Flame spread		
Underwriters Lab. rating (Sub. 94)		

Melamine-surfaced decorative laminates find wide application in the home. Decorative laminates are high-pressure laminates with a paper base, similar to industrial laminates except for the decorated surface. The laminate is composed of a core of sheets of phenolic-resin impregnated kraft paper, on top of which is placed a special grade of paper with a decorative pattern printed on the surface and impregnated with clear melamine resin. On top of this is placed another sheet of paper called an overlay that is impregnated with melamine resin. The purpose of the overlay is to protect the decorative sheet and provide the unique abrasion and stain resistance of these laminates. The decorative pattern may be a solid color, a wood grain, or any other design. The stack is then placed in a press where the sheets are bonded together at a temperature of 300°F/149°C and a pressure of 1500 psi.

The finished laminate is usually bonded to a substrate such as plywood, chipboard, or composition board. The resistance of these laminates to household preparations and chemicals is excellent. The melamine laminates are compatible with the following chemicals:

Acetone	Mustard
Alcohol	Naphtha
Ammonium hydroxide 10%	Olive oil
Amyl acetate	Soaps
Carbon tetrachloride	Shoe polish
Citric acid	Sodium bisulfite
Coffee	Trisodium phosphate
Detergent	Urine
Fly spray	Water
Gasoline	Wax and crayons
Moth spray	

There are certain chemicals and household preparations that tend to stain the melamine laminate. However, the stain can be removed by buffing with a mild abrasive. Included are such materials as

Beet juice	Mercurochrome solution
Bluing	Phenol (Lysol)
Dyes	Tea
Ink	Vinegar
Iodine solution	

The following chemicals and household preparations may damage melamine laminates:

Berry juices	Mineral acids
Gentian violet	Potassium permanganate
Hydrogen peroxide	Silver nitrate
Hypochlorite bleaches	Silver protein (Argyrol)
Lye solutions	Sodium bisulfate

3.13 ALKYDS

Alkyds are unsaturated resins produced from the reaction of organic alcohols with organic acids. The ability to use any of the many suitable polyfunctional alcohols and acids permits selection of a large variation of repeating units. Formulating can provide resins that demonstrate a wide range of characteristics involving flexibility, heat resistance, chemical resistance, and electrical properties.

Alkyd compounds are chemically similar to polyester compounds but make use of higher viscosity or dry monomers. Alkyd compounds often contain glass fiber filler but may contain clay, calcium carbonate, or alumina. Refer to Table 3.48 for the physical and mechanical properties of glass fiber reinforced alkyd polymer.

The greatest limitations of alkyds are in the extremes of temperature above 350°F/117°C and in high humidity.

The alkyds exhibit poor resistance to solvents and alkalies. Their resistance to dilute acid is fair. However, they have good resistance to weatherability. Although the alkyds are used outdoors, they are not as durable to long-term exposure as the acrylics, and their color and gloss retention are inferior.

1. Typical Applications

Alkyds are used for finishing metal and wood products but not to the degree previously used. Their durability to interior exposure is good, but their durability

TABLE 3.48 Physical and Mechanical Properties of Glass Fiber Reinforced Alkyd Polymer

Property	
Specific gravity	2–2.3
Water absorption (24 hr at 73°F/23°C) (%)	0.03–0.5
Dielectric strength, short-term (V/mil)	259–530
Tensile strength at break (psi)	4,000–9,500
Tensile modulus ($\times 10^3$ psi)	2,000–2,800
Elongation at break (%)	
Compressive strength (psi)	15,000–36,000
Flexural strength (psi)	8,500–26,000
Compressive modulus ($\times 10^3$ psi)	
Flexural modulus ($\times 10^3$ psi) at 73°F/23°C	2,000
200°F/93°C	
250°F/121°C	
Izod impact (ft-lb/in. of notch)	0.5–16
Hardness, Rockwell	E95
Coefficient of thermal expansion (10^{-6} in./in./°F)	15–33
Thermal conductivity (10^{-4} cal-cm/sec-cm^2°C or Btu/hr/ft^2/°F/in.)	15–25
Deflection temperature at 264 psi (°F)	400–500
at 66 psi (°F)	
Max. operating temperature (°F/°C)	
Limiting oxygen index (%)	
Flame spread	
Underwriters Lab. rating (Sub. 94)	

to exterior exposure is only fair. Because of their formulating flexibility, they are used in fillers, sealers, and caulks for wood finishing. They are still used for finishing by the machine tool and other industries. Alkyd-modified acrylic latex paints are excellent architectural finishes.

3.14 UREAS (AMINOS)

Ureas (commonly referred to as aminos) are polymers formed by condensation reactions and do not produce by-products. They are reaction products of formaldehyde with amino compounds containing NH_2 groups. Therefore, they are often referred to as urea formaldehydes. The general chemical structure of urea formaldehyde is shown in Figure 3.9.

Amino polymers are self-extinguishing and have excellent electrical insulation properties. The most commonly used filler for the aminos is alpha cellulose. Table 3.49 shows the physical and mechanical properties of amino compounds.

The addition of alpha cellulose produces an unlimited range of light-stable colors and high degrees of translucency. Basic material properties are unaffected by the addition of color.

When urea moldings are subjected to severe cycling between dry and wet conditions, cracks develop. Certain strength characteristics also experience a loss when amino moldings are subjected to prolonged elevated temperatures. Some electrical characteristics are also affected adversely.

Finished products having an amino resin surface exhibit excellent resistance to moisture, greases, oils, and solvents; are tasteless and odorless; and resist scratching and marring. However, the melamine resins have better chemical, heat, and moisture resistance than the aminos.

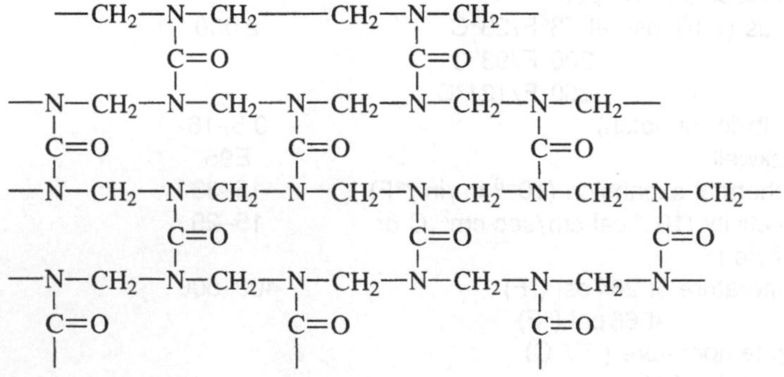

FIGURE 3.9 Chemical structure of urea formaldehyde.

TABLE 3.49 Physical and Mechanical Properties of Amino Compounds

Property	Alpha cellulose filled	Wood flour filled
Specific gravity	1.47–1.52	1.5
Water absorption (24 hr at 73°F/23°C) (%)	0.4–0.8	0.7
Dielectric strength, short-term (V/mil)	330–370	300–400
Tensile strength at break (psi)	5,500–13,000	8,500–10,000
Tensile modulus ($\times 10^3$ psi)	1,000–1,500	
Elongation at break (%)	< 1	
Compressive strength (psi)	25,000–45,000	25,000–35,000
Flexural strength (psi)	5,000–18,000	8,000–10,000
Compressive modulus ($\times 10^3$ psi)		
Flexural modulus ($\times 10^3$ psi) at 73°F/23°C	1,300–1,600	1,300–1,600
200°F/93°C		
250°F/121°C		
Izod impact (ft-lb/in. of notch)	0.25–0.4	0.25–0.35
Hardness, Rockwell	M110–120	E95
Coefficient of thermal expansion (10^{-6} in./in./°F)	22–30	30
Thermal conductivity (10^{-4} cal-cm/sec-cm^2 °C or Btu/hr/ft^2/°F/in.)	2–10	10.1
Deflection temperature at 264 psi (°F)	260–290	270
at 66 psi (°F)		
Max. operating temperature (°F/°C)	170/77	170/77
Limiting oxygen index (%)		
Flame spread		
Underwriters Lab. rating (Sub. 94)		

Amino resins are used in modifying other resins to increase their durability, specifically as modifiers for alkyds to increase hardness and accelerate cure.

Aminos are unsuitable for outdoor exposure.

3.15 ALLYLS

The allyls or diallyl phthalates are produced in several variations, but the most commonly used are diallyl phthalate (DAP) and diallyl isophthalate (DAIP). The primary difference between the two is that DAIP will withstand somewhat higher temperatures than DAP.

DAP and DAIP are seldom used as cast homopolymers, except for small electrical parts, because of their low tensile strength and impact resistance. When physical abuse is likely, the polymer is provided with a glass or mineral filling.

DAP finds application for the impregnation of ferrous and nonferrous castings because of its low viscosity, excellent sealing properties, low resin bleed out and ease of cleanup. It is also used to impregnate wood to reduce water absorption and increase impact, compressive, and shear strengths. Refer to Table 3.50 for the physical and mechanical properties of DAP.

DAP and DAIP glass laminates have high-temperature electrical properties superior to those of most other structural laminates. Cure cycles are shorter and little or no postcure is required to provide usable strength up to 482°F/250°C.

TABLE 3.50 Physical and Mechanical Properties of DAP

Property	Unfilled	Glass-filled	Mineral-filled
Specific gravity	1.3–1.4	1.7–1.98	1.65–1.85
Water absorption (24 hr at 73°F/23°C)(%)	0.2	0.12–0.35	0.2–0.5
Dielectric strength, short-term (V/mil)	380	400–450	400–450
Tensile strength at break (psi)	5,000–6,000	6,000–11,000	5,000–8,000
Tensile modulus ($\times 10^3$ psi)	300	1,400–2,200	1,200–2,200
Elongation at break (%)		3–5	3–5
Compressive strength (psi)	21,000–23,000	25,000–35,000	20,000–32,000
Flexural strength (psi)	6,000–13,000	9,000–20,000	8,500–11,000
Compressive modulus ($\times 10^3$ psi)	300		
Flexural modulus ($\times 10^3$ psi) at 73°F/23°C 200°F/93°C 250°F/121°C	350–300	1,200–1,500	1,000–1,450
Izod impact (ft-lb/in. of notch)	0.2–0.4	0.4–15	0.3–0.8
Hardness, Rockwell	M95–100	E80–87	E60–80
Coefficient of thermal expansion (10^6 in./in./°F)	81–143	10–36	10–42
Thermal conductivity (10^{-4} cal-cm/sec-cm^2 °C or Btu/hr/ft^2/°F/in.)	4.8–5.0	8.0–15.0	7.0–7.5
Deflection temperature at 264 psi (°F) at 66 psi (°F)	140–190	300–500	320–350
Max. operating temperature (°F/°C)			
Limiting oxygen index (%)			
Flame spread			
Underwriters Lab. rating (Sub. 94)			

TABLE 3.51 Chemical Resistance of DAP and DAIP

% gain in weight after 30 days immersion at 77°F/25°C

Corrodent	DAP	DAIP
Acetone	1.3	−0.03
Sodium hydroxide 1%	0.7	0.7
Sodium hydroxide 10%	0.5	0.6
Sulfuric acid 3%	0.8	0.7
Sulfuric acid 30%	0.4	0.4
Water	0.9	0.8

Allyl carbonate (diethylene glycol) is marketed as CR-39 by PPG Industries. It is used largely in optical castings in competition with glass or polymethyl methacrylate (PMMA) where its abrasion resistance and high heat distortion or impact resistance are required.

Refer to Table 3.51 for the chemical resistance of DAP and DAIP.

3.16 POLYBUTADIENES

Polybutadienes have essentially a hydrocarbon structure as follows:

$$\left[\begin{array}{c} CH-CH_2 \\ | \\ CH \\ | \\ CH_2 \end{array}\right]$$

The basic materials are monopolymers or copolymers that react through terminal hydroxyl groups, terminal carboxyl groups, vinyl groups, or a combination of these materials. Because of their high viscosity, they have limited use as casting compounds.

Polybutadiene polymers that have a 1,2 microstructure varying from 60 to 90% offer potential as moldings, laminating resins, coatings, and cast liquid and formed sheet products. Outstanding electrical and thermal stability results from the structure which is essentially pure hydrocarbon.

Peroxide catalysts used to cure the polybutadienes produce carbon–carbon double bonds in the vinyl groups. The final product is 100% hydrocarbon except where the starting polymer is the −OH or −COOH radical. Refer to Table 3.52 for the physical and mechanical properties of polybutadienes.

Poly BD marketed by Atochem North America Inc., is a family of hydroxyl-terminated butadiene homopolymers and copolymers with styrene or acrylonitrile. These materials are usually reacted with isocyanates to produce polyurethanes that have excellent resistance to boiling water.

TABLE 3.52 Physical and Mechanical Properties of Polybutadienes

Property	Unfilled	Mineral filled
Specific gravity	0.97	1.17
Water absorption (24 hr at 73°F/23°C)(%)	0.03	0.036
Dielectric strength, short-term (V/mil)	630	530
Tensile strength at break (psi)		3,000
Tensile modulus ($\times 10^3$ psi)	550	
Elongation at break (%)		12
Compressive strength (psi)		
Flexural strength (psi)	8,000–14,000	5,000
Compressive modulus ($\times 10^3$ psi)		
Flexural modulus ($\times 10^3$ psi) at 73°F/23°C		
200°F/93°C		
250°F/121°C		
Izod impact (ft-lb/in. of notch)		0.7–1.0
Hardness, Rockwell	R40	
Coefficient of thermal expansion (10^{-6} in./in./°F)		
Thermal conductivity (10^{-4} cal-cm/sec-cm^2 °C or Btu/hr/ft^2/°F/in.)		
Deflection temperature at 264 psi (°F)		
at 66 psi (°F)		
Max. operating temperature (°F/°C)		
Limiting oxygen index (%)		
Flame spread		
Underwriters Lab. rating (Sub. 94)		

B.F. Goodrich Chemical Company produces monopolymers and copolymers with acrylonitrile. These are known as the Hycar series. Acrylonitrile increases the viscosity and imparts oil resistance, adhesion, and compatibility with epoxy resins. Carboxy-terminated butadiene–acrylonitrile copolymer (CTBN) improves impact strength, low temperature shear strength, and crack resistance in epoxy compositions.

Amine-terminated butadiene–acrylonitrile copolymers (ATBN) are also used to modify epoxy resins. These are formulated on the amine hardener side of the mix.

Applications for these formulations include electrical potting compounds (with transformer oil), sealants, and moisture block compounds for telephone cables.

1. Corrosion Resistance Properties

The hydrocarbon structure is responsible for the excellent resistance shown by polybutadiene to chemicals and solvents and for the electrical properties that are good over a range of frequencies, temperatures, and humidity, and resistance to high temperature. Refer to Table 3.53 for the compatibility of polybutadiene with selected corrodents.

3.17 POLYIMIDES

Polyimide can be prepared as either thermoplastic or thermoset resins. There are two basic types of polyimides, condensation and addition resins. Condensation polyimides are available either as thermosets or thermoplastics. The addition polyimides are available only as thermosets.

The polyimides are heterocyclic polymers having an atom of nitrogen in the inside ring as shown below:

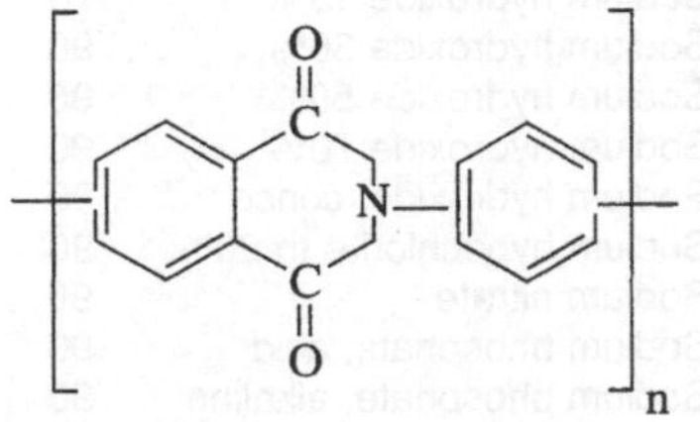

The fused rings provide stiffness which in turn provides high-temperature strength retention. The low concentration of hydrogen provides oxidative resistance by preventing thermal fracture of the chain.

Polyimides have continuous high-temperature stability up to 500–600°F/260–315°C and can withstand 900°F/482°C for short-term use. Their electrical and mechanical properties are relatively stable from low negative temperatures to high positive temperatures. They also exhibit dimensional stability (low cold flow) in most environments, excellent resistance to ionizing radiation, and very low outgassing in high vacuum. Their very low coefficient of friction can be improved even further by the addition of graphite or other fillers. Physical and mechanical properties are shown in Table 3.54.

Laminates of polyimide have flexural strengths approaching 50,000 psi with a tensile modulus of 3×10^6 psi. They resist burning without chemical modification and have a low dissipation factor (0.005) and dielectric constant (3.9).

TABLE 3.53 Compatibility of Polybutadiene with Selected Corrodents[a]

Chemical	Maximum temp. °F	Maximum temp. °C	Chemical	Maximum temp. °F	Maximum temp. °C
Alum	90	32	Nitric acid 10%	80	27
Alum ammonium	90	32	Nitric acid 20%	80	27
Alum ammonium sulfate	90	32	Nitric acid 30%	80	27
Alum chrome	90	32	Nitric acid 40%	x	x
Alum potassium	90	32	Nitric acid 50%	x	x
Aluminum chloride, aqueous	90	32	Nitric acid 70%	x	x
Aluminum sulfate	90	32	Nitric acid, anhydrous	x	x
Ammonia gas	90	32	Nitrous acid, concd	80	27
Ammonium chloride 10%	90	32	Ozone	x	x
Ammonium chloride 28%	90	32	Phenol	80	27
Ammonium chloride 50%	90	32	Sodium bicarbonate 20%	90	32
Ammonium chloride, sat.	90	32	Sodium bisulfate	80	27
Ammonium nitrate	90	32	Sodium bisulfite	90	32
Ammonium sulfate 10–40%	90	32	Sodium carbonate	90	32
Calcium chloride, sat.	80	27	Sodium chlorate	80	27
Calcium hypochlorite, sat.	90	32	Sodium hydroxide 10%	90	32
Carbon dioxide, wet	90	32	Sodium hydroxide 15%	90	32
Chlorine gas, wet	x	x	Sodium hydroxide 30%	90	32
Chrome alum	90	32	Sodium hydroxide 50%	90	32
Chromic acid 10%	x	x	Sodium hydroxide 70%	90	32
Chromic acid 30%	x	x	Sodium hydroxide, concd	90	32
Chromic acid 40%	x	x	Sodium hypochlorite to 20%	90	32
Chromic acid 50%	x	x	Sodium nitrate	90	32
Copper chloride	90	32	Sodium phosphate, acid	90	32
Copper sulfate	90	32	Sodium phosphate, alkaline	90	32
Fatty acids	90	32	Sodium phosphate, neutral	90	32
Ferrous chloride	90	32	Sodium silicate	90	32
Ferrous sulfate	90	32	Sodium sulfide to 50%	90	32
Hydrochloric acid, dil	80	27	Sodium sulfite 10%	90	32
Hydrochloric acid 20%	90	32	Sodium dioxide, dry	x	x
Hydrochloric acid 35%	90	32	Sulfur trioxide	90	32
Hydrochloric acid 38%	90	32	Sulfuric acid 10%	80	27
Hydrochloric acid 50%	90	32	Sulfuric acid 30%	80	27
Hydrochloric acid fumes	90	32	Sulfuric acid 50%	80	27
Hydrogen peroxide 90%	90	32	Sulfuric acid 60%	80	27
Hydrogen sulfide, dry	90	32	Sulfuric acid 70%	90	32
Nitric acid 5%	80	27	Toluene	x	x

[a] The chemicals listed are in the pure state or in a saturated solution unless otherwise indicated. Compatibility is shown to the maximum allowable temperature for which data are available. Incompatibility is shown by an x. A blank space indicates that the data are unavailable.

Source: Schweitzer, PA. Corrosion Resistance Tables, 4th ed., Vols. 1–3. New York: Marcel Dekker, 1995.

TABLE 3.54 Physical and Mechanical Properties of Polyimides

Property	Unfilled	50% glass fiber reinforced
Specific gravity	1.41–1.90	1.6–1.7
Water absorption (24 hr at 73°F/23°C) (%)	0.4–1.25	0.7
Dielectric strength, short-term (V/mil)	480–508	450
Tensile strength at break (psi)	4,300–22,900	6,400
Tensile modulus ($\times 10^3$ psi)	460–4,650	
Elongation at break (%)	1	1
Compressive strength (psi)	19,300–32,000	34,000
Flexural strength (psi)	6,500–50,000	21,300
Compressive modulus ($\times 10^3$ psi)	421	
Flexural modulus ($\times 10^3$ psi) at 73°F/23°C	422–3,000	1,980
200°F/93°C		
250°F/121°C		
Izod impact (ft-lb/in. of notch)	0.65–15	5.6
Hardness, Rockwell	M110–120	M118
Coefficient of thermal expansion (10^{-6} in./in./°F)	15–50	13
Thermal conductivity (10^{-4} cal-cm/sec-cm^2°C or Btu/hr/ft^2/°F/in.)	5.5–12	6.5
Deflection temperature at 264 psi (°F)	572–575	600
at 66 psi (°F)		
Max. operating temperature (°F/°C)	500–600/260–315	615/324
Limiting oxygen index (%)		
Flame spread		
Underwriters Lab. rating (Sub. 94)		

1. Corrosion Resistance Properties

Polyimides are sensitive to alkaline chemicals and will be dissolved by hot concentrated sodium hydroxide. They are also moisture sensitive, gaining 1% in weight after 1000 hours at 50% relative humidity and 72°F/23°C.

2. Typical Applications

Polyimides find use in various forms including laminates, moldings, films, coatings, and adhesives, specifically in areas of high temperature. Coatings are used in electrical applications as insulating varnishes and magnet wire enamels at high temperature. They also find application as a coating on cookware as an alternate to fluorocarbon coatings.

TABLE 3.55 Physical and Mechanical Properties of Cyanate Polymers

Composition by weight	A	B	C	D	E	F	G
AroCy L-10	100	100	50	50	50	75	35
AroCy M-50			50				
AroCy B-30				50			
AroCy F-40					50		
Epoxy resin A						25	
Epoxy resin B							25
Nonylphenol	6	2	2	2	2	2	2
Copper naphthenate 8% Cu	0.25						
Zinc Naphthenate 8% Zn		0.15	0.15	0.15	0.12		
Copper acetylacetonate 24% Cu						0.05	0.05
Properties							
Flexural strength (ksi)	27.2	23.5	20.3	22.8	22.1	24.8	11.7
Flexural modulus (msi)	0.44	0.40	0.44	0.40	0.43	0.44	0.40
Flexure strain (%)	8.1	7.7	5.0	6.9	5.6	6.9	3.1
% water absorption 64 hr at 100°C	1.47	1.67	1.62	1.88	1.26	1.46	1.28
Dielectric constant at 1 MHz, dry	2.99	2.98	2.91	2.96	2.92	3.06	3.18
Flammability (UL 94)	Burns	Burns	V-1	Burns	V-0	V-0	Burns
Density (g/cm³)	1.222	1.228	1.170	1.205	1.331	1.333	1.201

3.18   CYANATE ESTERS

Cyanate esters are a family of aryl dicyanate monomers that contain the reactive cyanate $(-O-C=N)$ functional group. When heated, this cyanate functionality undergoes an exothermic cyclotrimerization reaction to form triazine ring connecting units resulting in the formation of a thermoset polycyanate polymer.

The cyanate ester monomers are available from low-viscosity liquids to meltable solids. Ciba-Geigy produces a series of cyanate esters under the trademark AroCy L-10. Table 3.55 shows the cured properties of AroCy L-10 alone and in mixtures with other cyanate esters and epoxies.

Cyanate esters are used in manufacturing structural composites by filament winding, resin transfer molding, and pultrusion.

REFERENCES

1. Schweitzer PA. Corrosion Resistance Tables, 4th ed., Vols. 1–3. New York: Marcel Dekker, 1995
2. Schweitzer PA. Corrosion Resistant Piping Systems, New York: Marcel Dekker, 1994.

4

Elastomers

4.1 INTRODUCTION TO ELASTOMERS

An elastomer is generally considered to be any material, either natural or synthetic, that is elastic or resilient and in general resembles natural rubber in feeling and appearance. A more technical definition is provided by ASTM, which states, "An elastomer is a polymeric material which at room temperature can be stretched to at least twice its original length and upon immediate release of the stress will return quickly to its original length." These materials are sometimes referred to as rubbers.

Natural rubber is polymerized hydrocarbon whose commercial synthesis proved to be difficult. Synthetic rubbers now produced are similar to but not identical to natural rubber. Natural rubber has the hydrocarbon butadiene as its simplest unit. Butadiene, $CH_2=CH-CH=CH_2$, has two unsaturated linkages and is easly polymerized. It is produced commercially by cracking petroleum and also from ethyl alcohol. Natural rubber is a polymer of methyl butadiene (isoprene):

$$CH_2=\overset{\overset{\displaystyle CH_3}{|}}{C}-CH=CH_2$$

When butadiene or its derivatives become polymerized, the units link together to form long chains that each contain over 1000 units. In early attempts to develop a synthetic rubber, it was found that simple butadiene does not yield a good grade of rubber, apparently because the chains are too smooth and do not interlock sufficiently strongly. Better results are obtained by introducing side groups into the chain either by modifying butadiene or by making a copolymer of butadiene and some other compound.

As development work continued in the production of synthetic rubbers, other compounds were used as the parent material in place of butadiene. Two of them were isobutylene and ethylene.

Elastomers are composed primarily of large molecules that tend to form spiral threads, similar to a coiled spring, and are attached to each other at infrequent intervals. These coils tend to stretch or compress when a small stress is applied but exert an increasing resistance to the application of additional stresses. This phenomenon is illustrated by the reaction of rubber to the application of additional stress.

In the raw state, elastomers tend to be soft and sticky when hot, and hard and brittle when cold. Compounding increases the utility of rubber and synthetic elastomers. Vulcanization extends the temperature range within which they are flexible and elastic. In addition to vulcanizing agents, ingredients are added to make elastomers stronger, tougher, or harder, to make them age better, to color them, and in general to impart specific properties to meet specific application needs.

A.   The Importance of Compounding

The rubber chemist is able to optimize selected properties or fine-tune the formulation to meet a desired balance of properties. However, this does involve trade-offs. Improvement in one property can come at the expense of another property. Therefore it is important that these effects be taken into consideration. Table 4.1 illustrates some of these effects.

The following examples illustrate some of the important properties that are required of elastomers and the typical services that require these properties:

> Restance to abrasive wear: automobile tire treads, conveyor belt covers, soles and heels, cables, hose covers
>
> Resistance to tearing: tire treads, footwear, hot water bags, hose covers, belt covers, O-rings
>
> Resistance to flexing: auto tires, transmission belts, V-belts, mountings, footwear
>
> Resistance to high temperatures: auto tires, belts conveying hot materials, steam hose, steam packing, O-rings
>
> Resistance to cold: airplane parts, automotive parts, auto tires, refrigeration hose, O-rings
>
> Minimum heat buildup: auto tires, transmission belt, V-belts, mountings
>
> High resilience: sponge rubber, mountings, elastic bands, thread, sandblast hose, jar rings, O-rings

TABLE 4.1   Compounding Trade-Offs

An improvement in	Can be at the expense of
Tensile strength, hardness	Extensibility
	Dynamic properties
Dynamic properties	Thermal stability
	Compression set resistance
Balanced properties	Processability
Optimized properties	Higher cost
Compression set resistance	Flex fatigue strength
	Resilience
Oil resistance	Low-temperature flexibility
Abrasion resistance	Resilience
Dampening	Resilience
	Compression set resistance

High rigidity: packing, soles and heels, valve cups, suction hose, battery boxes

Long life: fire hose, transmission belts, tubing

Electrical resistivity: electricians' tape, switchboard mats, electricians' gloves, wire insulation

Electrical conductivity: hospital flooring, nonstatic hose, matting

Impermeability to gases: balloons, life rafts, gasoline hose, special diaphragms

Resistance to ozone: ignition distributor gaskets, ignition cables, windshield wipers

Resistance to sunlight: wearing apparel, hose covers, bathing caps, windshield wipers

Resistance to chemicals: tank linings, hoses for chemicals

Resistance to oils: gasoline hose, oil-suction hose, paint hose, creamery hose, packing house hose, special belts, tank linings, special footwear

Stickiness: cements, electricians' tape, adhesive tapes, pressure-sensitive tapes

Low specific gravity: airplane parts, forestry hose, balloons

Lack of odor or taste: milk tubing, brewery and winery hose, nipples, jar rings

Acceptance of color pigments: ponchos, life rafts, welding hose

Table 4.2 provides a comparison of the important properties of the most common elastomers. Specific values of each property will be found in the section dealing with each elastomer.

TABLE 4.2 Comparative Properties of Elastomers[a]

	Natural rubber (NR)	Isoprene (IR)	Neoprene (CR)	Butadiene–Styrene (Buna S)	Nitrile-NBR (Buna N)	Butyl (IIR)	Chlorobutyl (CIIR)	Hypalon (CSM)	Polybutadiene (BR)
Abrasion resistance	E	E	G	G	G	G	G	G	E
Acid resistance	P	P	GE	P	F	G	P	E	G
Chemical resistance									
Aliphatic hydrocarbons	P	P	G	G	E	P	E	G	P
Aromatic hydrocarbons	P	P	F	P	G	P	P	F	P
Oxygenated (ketones, etc.)	G	G	P	G	P	G	P	P	G
Oil and gasoline	P	P	FG	P	E	P	E	G	P
Animal and vegetable oils	PG	PG	G	PG	E	E	G	G	PG
Resistance to									
Water absorption	E	E	G	GE	FG	G	G	GE	G
Ozone	P	P	GE	P	P	E	E	E	P
Sunlight aging	P	P	E	P	P	G	G	E	P
Heat aging	P	G	G	F	G	G	G	E	G
Flame	P	P	G	P	P	P	P	G	P
Electrical properties	G	G	F	E	F	E	PF	G	E
Impermeability	G	G	G	GE	G	E	G	E	G
Compression set resistance	E	E	F	G	GE	P	F	G	G
Tear resistance	GE	GE	FG	P	FG	GE	G	G	G
Tensile strength	E	E	G	G	GE	G	G	GE	G
Water/steam resistance	E	E	F	E	FG	E	PF	G	E
Weather resistance	P	P	E	P	F	E	F	E	G
Adhesion to metals	E	E	E	E	E	G	G	E	E
Adhesion to fabrics	E	E	E	E	G	G	G	G	E
Rebound									
Cold	E	E	E	G	G	P	F	G	E
Hot	E	E	E	G	G	G	F	G	E

[a]E, excellent; GE, good to excellent; G, good; F, fair; P, poor; PF, poor to fair; PG, poor to good.

Ethylene–acrylic (EA)	Acrylate–butadiene (ABR)	Acrylic ester–acrylic halide (ACM)	Ethylene–propylene (EPDM)	Styrene–butadiene–styrene (SBS)	Styrene–ethylene–butylene–styrene (SEBS)	Polysulfide (ST)	Polysulfide (FA)	Urethane (AU)	Polyamides	Polyester (PE)	Thermoplastic (TPE)	Silicone (Si)	Fluorosilicone (FSI)	Vinylidene fluoride (HFP)	Fluoroelastomers (FKM)	Ethylene–tetrafluoroethylene (ETFE)	Ethylene–chlorotrifluoroethylene (ECTFE)	Perfluoroelastomers (FPM)
E	G	FG	GE	G	GE	PF	PF	E	E	E	G	P	P	G	G	G	E	G
PG	G	P	G	E	E	F	G	P	FG	F	G	FG	FG	E	E	E	E	E
G	E	E	P	E	P	E	E	E	E	G	E	G	E	G	E	E		
F	P	P	P	P	P	E	E	E	G	G	P	FG	G	G	E	E	G	E
P	P	P	GE	P	P	G	G	P	G	P	P	PF	G	E	G	FG		
G	E	E	P	F	FG	E	E	E	G	G	FG	P	G	E	E	E	E	
G	G	G	G	P	G			E	FG	G	G	G	G	G	E	E	E	
G	G	G	GE	E	E	G	G	G	G	G	E	G	G	G	GE	G	E	G
E	E	E	E	P	E	E	E	E	E	E	E	E	E	E	GE	E	E	E
E	E	E	P	E	E	E	P	P	G	G	E	E	E	GE	E	E	E	
G	E	G	E	G	E	P	G	G	E	E	E	E	E	E	E	E		
P	P	P	P	P	PG	P	P	P	F	G	P	F	G	E	G	G	E	
FG	G	PF	G	F	E	G	G	FG	G	FG	F	E	E	F	G	G	E	G
G	G	G	G	G	G	E	E	G	FG	G	P	P	G	G	E	G	G	E
G	G	FG	GE	G	G	P	P	F	G	F	G	GE	GE	GE	GE			G
G	G	G	GE	G	G	P	P	GE	G	G	E	G	P	P	F	F	G	P
G	G	G	GE	G	G	F	E	G	E	G	G	P	F	GE	GE	G	G	G
G	G	PF	E	FG	FG	E	G	P	G	FG	G	F	F	FG	G	G	G	E
G	E	F	E	F	G	E	E	E	G	E	G	E	E	E	E	E	E	
G	G	G	GE	G	E	E	G	E	G	G	G	GE	G	GE	P			
P	G	F	G	G	G	P	P	E	G	G	G	G	G	G				
F	G	F	G	G	G	P	P	E	G	E	G	G	G	E				

B. Physical Properties

Elastomers are organic materials and react in a completely different manner from metals. Consequently, engineering terms in conventional usage have different meanings when applied to elastomeric properties. Some of the more common definitions follow:

1. Tensile Strength

In elastomers, tensile strength refers to the force per unit of original cross section on elongating to rupture. Tensile strength as such is not an important property for elastomers since they are rarely used in tensile applications. However, there are other properties that correlate with tensile strength. For example, wear and tear resistance, resilience, cut resilience, stress relaxation, creep, flex fatigue, and in some elastomers, such as neoprene, better ozone resistance, all improve with tensile strength.

   The procedure for conducting tensile tests is standardized and described in ASTM D412. Dumbell-shaped specimens 4 or 5 in. long are die-cut from fleet sheets and marked in the narrow section with benchmarks 1 or 2 in. apart. The ends of the specimen are placed in the grips of a vertical testing machine. The lower grip is power driven at 2 in./min and stretches the specimen until it breaks. As the distance between benchmarks widens, measurements are taken between their centers to determine elongation.

   Tensile tests are frequently conducted before and after an exposure test to determine the relative resistance of a group of compounds to deterioration by such things as oil, sunlight, weathering, ozone, heat, oxygen, and chemicals. Even a small amount of deterioration results in appreciable changes in tensile properties.

2. Elongation

This is the maximum extension of rubber at the moment of rupture and is expressed as a percentage of the original distance between previously applied benchmarks. Ultimate elongation is the elongation at the moment of rupture. Usually an elastomer with less than 100% elongation will break if doubled over on itself.

3. Modulus

With elastomers, modulus refers to the force per unit of original cross section for a specific extension. Most modulus readings are taken at 300% elongation, but lower extensions can be used. Modulus is a ratio of the stress of elastomer to the tensile stress but differs from that of metal as it is not a Young's modulus stress–strain-type curve. Stress–strain values are extremely low for slight extensions but increase logarithmically with increased extension.

4. Hysteresis

Hysteresis denotes energy loss per loading cycle. The mechanical loss of energy is converted into heating of the elastomeric material and is capable of reaching destructive temperatures.

5. Heat Buildup

The temperature rise in an elastomeric product resulting from hysteresis is denoted by the term heat buildup. It can also mean the use of high frequencies on elastomers where the power factor is too high.

6. Permanent Set

This term refers to the deformation of an elastomer that remains after a given period of stress followed by the release of stress.

7. Stress Relaxation

This is the loss in stress that remains after the elastomer has been held at constant strain over a period of time.

8. Creep

Creep in elastomers refers to change in strain when the stress is held constant.

9. Abrasion Resistance

This term refers to the resistance of an elastomeric composition to wear. It is usually measured by the loss of material when an elastomeric part is brought into contact with a moving abrasive surface. It is measured as a percentage of volume lost of sample compared to a standard elastomeric composition. It is almost impossible to correlate these relative values with life expectancy. In addition, many formulations contain wax-type substances which exude to the surface that can make test results erroneous because of the lubricating effects of the waxes.

10. Flex Fatigue

This results from an elastomer being subjected to fluctuating stresses causing it to fracture.

11. Impact Resistance

This is the resistance of the elastomer to abrading or cutting when hit by a sharp object.

12. Tear Resistance

This is the measure of the stress required to continue rupturing an elastomeric sheet, after an initial cut.

13. Flame Resistance

Some elastomers will burn profusely when ignited while others are self-extinguishing when the flame source is removed. This is the relative flammability of an elastomer.

14. Low-Temperature Properties

These properties indicate a stiffening range and a brittle point of an elastomer. Of the two, stiffening range is the most useful. Brittle point has little meaning unless the deforming force and rate are known. Since some properties change as the temperature is lowered and held, time is also an important factor. Hardness, stress–strain rate, and modulus are examples. Many materials tend to crystallize, at which time the elastomer is brittle and will fracture easily.

15. Heat Resistance

No elastomer is completely heat resistant; time and temperature have their aging effects. Heat resistance is usually measured as change in tensile strength, elongation, and durometer readings from the original values, usually after a 72-hour period.

16. Aging

The properties of an elastomer can be destroyed only by additional chain growth and linkage, which would result in a hard rigid material, or a chain rupture, which would result in a plastic or resinous mass. The agents that contribute to this condition are sunlight, heat, oxygen, stress with atmospheric ozone, atmospheric moisture, and atomospheric nitrous oxide. Chain growth or cross-linkage will usually decrease elongation and increase hardness and tensile strength, while chain rupture will have the opposite effect. Some elastomers will continue to harden and some to soften, and some will show an initial hardening followed by softening. All effects are irreversible.

17. Radiation Resistance

The effects of radiation are similar and complementary to those of aging. Damage is dependent only on the amount of radiation energy absorbed, regardless of the form of radiation. The least heat resistant of the elastomers displays the most radiation resistance.

18. Hardness

Hardness, as applied to elastomeric products, is defined as relative resistance of the surface to indentation by a Shore A durometer. In this device, the indenter point projects upward from the flat bottom of the case and is held in the zone position by a spring. When pressed against a sample, the indenter point is pushed back into the case against the spring; this motion is translated through rack-and-pinion mechanisms into movement of the pointer on the durometer dial. Hardness numbers on a durometer scale for some typical products are as follows:

Faucet washer, flooring, typewriter platen	90 ± 5
Shoe sole	80 ± 5
Solid tire, heel	70 ± 5
Tire tread, hose cover, conveyer belt cover	60 ± 5
Inner tube, bathing cap	50 ± 5
Rubber band	40 ± 5

Erasers and printing rolls usually have hardness values below 30.

19. Compression Set

Compression set is permanent deformation that remains in the elastomer after a compression force has been removed. The area of the elastomer that has compression set is not only permanently deformed but is also less resilient than normal. The possibility of compression set occurring increases with increasing temperature, compression force, and length of time that the force is applied. Each type of elastomer has a different resistance to compression set. As with any material, each elastomer also has a limiting temperature range within which it may be used. Table 4.3 shows the allowable operating temperature range for each of the common elastomers.

C. Similarities of Elastomers and Thermoplastic Polymers

All thermoplastic polymers and elastomers, with the exception of silicones, are carbon-based. They are made up from the linking of one or more monomers into long molecular chains. Many of the same monomers are found in both thermoplastic and elastomeric polymers. Typical examples include styrene, acrylonitrile, ethylene, propylene, and acrylic acid and its esters. An elastomer is in a thermoplastic state prior to vulcanization.

D. Differences Between Elastomers and Thermoplasts

At room temperature, an uncured elastomer can be a soft, pliable gum or a leathery, flexible solid, while an engineering thermoplastic is a rigid solid. Molecular mobility is the fundamental property that accounts for the differences

TABLE 4.3 Operating Temperature Ranges of Common Elastomers

Elastomer	Temperature range			
	°F		°C	
	Min	Max	Min	Max
Natural rubber NR	−59	175	−50	80
Isoprene rubber IR	−59	175	−50	80
Neoprene rubber CR	−13	203	−25	95
Buna-S SBR	−66	175	−56	80
Nitrile Rubber Buna-NBR	−40	250	−40	105
Butyl rubber IIR	−30	300	−34	149
Chlorobutyl rubber CIIR	−30	300	−34	149
Hypalon CSM	−20	250	−30	105
Polybutadiene rubber BR	−150	200	−101	93
Ethylene–acrylic rubber EA	−40	340	−40	170
Acrylate–butadiene rubber ABR	−40	340	−40	170
Ethylene–propylene EPDM	−65	300	−54	149
Styrene–butadiene–styrene SBS		150		65
Styrene–ethylene–butylene–styrene SEBS	−102	220	−75	105
Polysulfide ST	−50	212	−45	100
Polysulfide FA	−30	250	−35	121
Polyurethane AU	−65	250	−54	121
Polyamides	−40	300	−40	149
Polyesters PE	−40	302	−40	150
Thermoplastic elastomers TPE	−40	277	−40	136
Silicone SI	−60	450	−51	232
Fluorosilicone FSI	−100	375	−73	190
Vinylidene fluoride HEP	−40	450	−40	232
Fluoroelastomers FKM	−10	400	−18	204
Ethylene–tetrafluoroethylene elastomer ETFE	−370	300	−223	150
Ethylene–chlorotrifluoroethylene elastomer ECTFE	−105	340	−76	171
Perfluoroelastomers FPM	−58	600	−50	316

among elastomers and the differences between elastomers and engineering thermoplasts.

This distinction refers to a modulus of the material—a ratio between an applied force and the amount of resulting deformation. When the molecular resistance to motion is the least, deformation is the greatest. The factors that influence the resistance to motion include various intermolecular attractions, crystallinity, the presence of side chains, and physical entanglements of the molecular strands. The glass-transition temperature of the polymer is determined by the cumulative effect of these factors. Below this temperature, a thermoplast or elastomeric polymer is a supercooled liquid that behaves in many ways like a rigid solid. Above this temperature, a cross-linked elastomer will display rubber-like properties. The behavior of a thermoplast above its glass-transition temperature will depend on its level of crystallinity. A noncrystalline (amorphous) polymer will display a large decrease in modulus at the glass-transition temperature. Between the glass-transition temperature and the melting temperature, the modulus is relatively insensitive to temperature. A partially crystalline thermoplast will display a relatively small modulus change at the glass-transition temperature, followed by a steadily decreasing modulus as the temperature increases.

Engineering thermoploastics have glass-transition temperatures that fall in a wide temperature range, extending well above and below room temperature. Elastomers in the uncured state have glass-transition temperatures well below room temperature.

The major physical differences between a cured elastomer and an engineering thermoplastic is the presence of cross-links. The cross-links join the elastomer's molecular chains together at a cross-link density such that many molecular units exist between cross-link sites. This increases only slightly the resistance of deformation but increases the resilience of the elastomer, causing it to spring back to its original shape when the deforming stress is removed. In principle, most amorphous thermoplastic materials could be converted into elastomers in some appropriate temperature range by the controlled addition of a limited number of cross-links.

E. Causes of Failure

Elastomeric materials can fail as the result of chemical action and/or mechanical damage. Chemical deterioration occurs as the result of a chemical reaction between the elastomer and the medium or by the absorption of the medium into the elastomer. This attack results in a swelling of the elastomer and a reduction in its tensile strength.

The degree of deterioration is a function of the temperature and concentration of the corrodent. In general, the higher the temperature and the higher the concentration of the corrodent, the greater will be the chemical attack. Elastomers, unlike metals, absorb varying quantities of the material they are in contact with, especially organic liquids. This can result in swelling, cracking, and penetration to the substrate in an elastomer-lined vessel. Swelling can cause softening of the elastomer and in a lined vessel introduces high stresses and failure of the bond. If an elastomeric lining has high absorption, permeation will probably result. Some elastomers such as the fluorocarbons are easily permeated but have very little absorption. An approximation of the expected permeation and/or absorption of an elastomer can be based on the absorption of water. These data are usually available.

Permeation is a factor closely related to absorption but is a function of other physical effects, such as diffusion and temperature. All materials are somewhat permeable to chemical molecules, but the permeability rate of elastomers tends to be an order of magnitude greater than that of metals. This permeation has been a factor in elastomer-lined vessels where corrodents have permeated the rubber and formed bubbles between the rubber lining and the steel substrate. Permeation and absorption can result in

1. Bond failure and blistering. These are caused by an accumulation of fluids at the bond when the substrate is less permeable than the lining or from the formation of corrosion or reaction products if the substrate is attacked by the corrodent.
2. Failure of the substrate due to corrosive attack.
3. Loss of contents through lining and substrate as the result of eventual failure of the substrate.

The lining thickness is a factor affecting permeation. For general corrosion resistance, thicknesses of 0.010–0.020 in. are usually satisfactory, depending upon the combination of elastomeric materials and specific corrodent. When mechanical factors such as thinning due to cold flow, mechanical abuse, and permeation rates are a consideration, thicker linings may be required.

Increasing the lining thickness will normally decrease permeation by the square of the thickness. Although this would appear to be the approach to follow to control permeation, there are disadvantages. First, as the thickness increases, the thermal stresses on the boundary increase, which can result in bond failure. Temperature changes and large differences in coefficients of thermal expansion are the most common causes of bond failure. Thickness and modulus of elasticity of the elastomer are two of the factors that would influence these stresses. Second, as the thickness of the lining increases, installation becomes more difficult, with a resulting increase in labor costs.

The rate of permeation is also affected by temperature and temperature gradient in the lining. Lowering these will reduce the rate of permeation. Lined vessels that are used under ambient conditions, such as storage tanks, provide the best service.

In unbonded linings, it is important that the space between the liner and the support member be vented to the atmosphere, not only to allow the escape of minute quantities of permeant vapors, but also to prevent the expansion of entrapped air from collapsing the liner.

Although elastomers can be damaged by mechanical means alone, this is not usually the case. When in good physical condition, an elastomer will exhibit abrasion resistance superior to that of metal. The actual size, shape, and hardness of the particles and their velocity are the determining factors in how well a

TABLE 4.4 Ozone Resistance of Selected Elastomers

Excellent resistance

EPDM	Chlorosulfonated polyethylene
Epichlorhydrin	Polyacrylate
ETFE	Polysulfide
ECTFE	Polyamides
Fluorosilicones	Polyesters
Ethylene acrylate	Perfluoroelastomers
Ethylene vinyl acetate	Urethanes
Chlorinated polyethylene	Vinylidene fluoride

Good Resistance

Polychloroprene	Butyl

Fair Resistance
Nitrile

Poor Resistance

Buna S	Polybutadiene
Isoprene	SBR
Natural rubber	SBS

particular rubber resists mechanical damage from the medium. Hard, sharp objects, including those foreign to the normal medium, may cut or gouge the elastomer. Most mechanical damage occurs as a result of chemical deterioration of the elastomer. When the elastomer is in a deteriorated condition, the material is weakened, and consequently it is more susceptible to mechanical damage from flowing or agitated media.

Elastomers in outdoor use can be subject to degradation as a result of the action of ozone, oxygen, and sunlight. These three weathering agents can greatly affect the properties and appearance of a large number of elastomeric materials. Surface cracking, discoloration of colored stocks, and serious loss of tensile strength, elongation, and other rubberlike properties are the result of this attack. Table 4.4 shows the ozone resistance of selected elastomers. The low-temperature properties must also be taken into account for outdoor use. With many elastomers, crystallization takes place at low temperatures, at which time the elastomer is brittle and will fracture easily. Table 4.5 gives the relative low-temperature flexibility of the more common elastomers. Table 4.6 gives the brittle points of the common elastomers.

F. Selecting an Elastomer

Many factors must be taken into account when selecting an elastomer for a specific application. First and foremost is the compatibility of the elastomer with the medium at the temperature and concentration to which it will be exposed.

TABLE 4.5 Relative Low-Temperature Flexibility of the Common Elastomers

Elastomer	Relative flexibility[a]
Natural rubber NR	G–E
Butyl rubber IIR	F
Ethylene–propylene rubber EPDM	E
Ethylene–acrylic rubber EA	P–F
Fluoroelastomer FKM	F–G
Fluorosilicone FSI	E
Chlorosulfonated polyethylene (Hypalon) CSM	F–G
Polychloroprene (Neoprene) CR	F–G
Nitrile rubber (Buna-N) NBR	F–G
Polybutadiene BR	G–E
Polyisoprene IR	G
Polysulfide T rubber	F–G
Butadiene–styrene rubber (Buna-S) SBR	G
Silicone rubber SI	E
Polyurethane rubber AU	G
Polyether–urethane rubber EU	G

[a]E = excellent; G = good; F = fair; P = poor.

It should also be remembered that each of the materials can be formulated to improve certain of its properties. However, the improvement in one property may have an adverse effect on another property, such as corrosion resistance. Consequently, specifications of an elastomer should include the specific properties required for the application, such as relience, hysteresis, static or dynamic shear and compression modulus, flex fatigue and cracking, creep resistance to oils and chemicals, permeability, and brittle point, all in the temperature range to be encountered in service. This must also be accompanied by a complete listing of the concentrations of all media to be encountered. Providing this information will permit a competent manufacturer to supply an elastomer that will give years of satisfactory service. Because of the ability to change the formulations of many of these elastomers, the wisest policy is to permit a competent manufacturer to make the selection of the elastomer to satisfy the application. In addition to being able to change the formulation of each elastomer, it is also a common practice to blend two or more elastomers to produce a compound having specific properties. By so doing, the advantageous properties of each elastomer may be used.

Fabrics are very often used as a reinforcing member in conjunction with elastomers. Cotton, because of its ease of processing, availability in a wide range of weaves, and high adhesive strength, is the most widely used. It is also priced relatively low in comparison with the synthetic fibers. The disadvantages of

TABLE 4.6 Brittle Points of Common Elastomers

Elastomer	°F	°C
NR; IR	−68	−56
CR	−40	−40
SBR	−76	−60
NBR	−32 to −40	−1 to −40
CSM	−40 to −80	−40 to −62
BR	−60	−56
EA	−75	−60
EPDM; EPT	−90	−68
SEBS	−58 to −148	−50 to −100
Polysulfide SF	−60	−51
Polysulfide FA	−30	−35
AU	−85 to −100	−65 to −73
Polyamide No. 11	−94	−70
PE	−94	−70
TPE	−67 to −70	−55 to −60
SI	−75	−60
FSI	−75	−60
HEP	−80	−62
FKM	−25 to −75	−32 to −59
ETFE	−150	−101
ECTFE	−105	−70
FPM	−9 to −58	−23 to −50

cotton are its poor heat resistance and the need for bulk in order to obtain the proper strength.

When operating temperatures of reinforced elastomeric products are in the range of 200–250°F (93–120°C), DuPont's Dacron polyester fiber is used to provide good service life. In addition to better heat-resisting qualities than cotton, Dacron has strength comparable to that of cotton, with considerably less bulk. On the negative side, Dacron is more difficult to process than cotton, has lower adhesive strength, and is initially more expensive.

G. Applications

Elastomeric or rubber materials find a wide range of applications. One of the major areas of application is that of linings for vessels. Both natural and synthetic materials are used for this purpose. These linings have provided many years of service in the protection of steel vessels from corrosion. They are sheet applied and bonded to a steel substrate.

These materials are also used extensively as membranes in acid-brick-lined vessels to protect the steel shell from corrosive attack. The acid-brick lining in turn protects the elastomer from abrasion and excessive temperature. Another major use is as an impermeable lining for settling ponds and basins. These materials are employed to prevent pond contaminants from seeping into the soil and causing pollution of groundwater and contamination of the soil.

Natural rubber and most of the synthetic elastomers are unsaturated compounds that oxidize and deteriorate rapidly when exposed to air in thin films. These materials can be saturated by reacting with chlorine under the proper conditions, producing compounds that are clear, odorless, nontoxic, and noninflammable. They may be dissolved and blended with varnishes to impart high resistance to moisture and to the action of alkalies. This makes these products particularly useful in paints for concrete, where the combination of moisture and alkali causes the disintegration of ordinary paints and varnishes. These materials also resist mildew and are used to impart flame resistance and waterproofing properties to canvas. Application of these paints to steel will provide a high degree of protection against corrosion.

Large quantities of elastomeric materials are used to produce a myriad of products such as hoses, cable insulation, O-rings, seals and gaskets, belting, vibration mounts, flexible couplings, expansion joints, automotive and airplane parts, electrical parts and accessories. With such a wide variety of applications requiring very diverse properties, it is essential that an understanding of the properties of each elastomer be acquired so that proper choices can be made.

H. Elastomer Designations

Many of the elastomers have common names or trade names as well as the chemical name. For example, polychloroprene also goes under the name of neoprene (DuPont's tradename) or Bayprene (Mobay Corp.'s tradename). In addition, they also have an ASTM designation, which for polychloroprene is CR. Following on page 275 is an elastomer cross reference which gives the generic name, ASTM designation, and common or trade name, along with the companies who manufacture each elastomer.

4.2 NATURAL RUBBER

Natural rubber of the best quality is prepared by coagulating the latex of the *Hevea brasiliensis* tree, which is cultivated primarily in the Far East. However, there are other sources such as the wild rubbers of the same tree growing in Central America, guyayule rubber coming from shrubs grown mostly in Mexico, and balata. Balata is a resinous material and cannot be tapped like the *Hevea* tree sap. The balata tree must be cut down and boiled to extract balata which cures to a hard, tough product used as golf ball covers.

Elastomer Cross Reference

Generic name	Designation	Manufacturers'[a] common or trade names
Natural rubber	NR	26–31
Isoprene	IR	
Polychloroprene	CR	26–31, neoprene (1), Bayprene (2)
Butadiene–styrene	SBR	26–30, Buna-S, GR-S
Butadiene–acrylonitrile	NBR	16, 26–31, nitrile rubber, Buna-N, Perbunan (2), Nytek (21)
Butyl rubber	IIR	Gr-1, 26–30, Kalar (19)
Chlorobutyl rubber	CIIR	26–30
Carboxylic-acrylonitrile–butadiene	NBR	16, 26–31
Chlorosulfonated polyethylene	CSM	26–28, 30, 31, Hypalon (1)
Polybutadiene	BR	26–28, 30, 31, Buna-85, Buna-DB (2)
Ethylene–acrylic	EA	13, 28 Vamac (1)
Acrylate–butadiene	ABR	13, 28
Acrylic ester–acrylic halide	ACM	13, 28
Ethylene–propylene	EPDM	26–31
	EPT	Nordel (1), Royalene, EPDM (8), Dutral (9)
Styrene–butadiene–styrene	SBS	Kraton G (3)
Styrene–ethylene–butylene–styrene	SEBS	Kraton G (3)
Polysulfide	ST	27, 28, 30, Thiokol (4)
	FA	Blak-Stretchy (14), Blak-Tufy (14), Gra-Tufy (14)
Urethane	AU	16, 27, 30, 38, 31, Adiprene (1), Baytec (2), Futrathane (11), Conathane (16), Texion (2), Urane (23), Pellethane (22), pure CMC (14)
Polyamides	Nylon	Nylon (1), Rilsan (12), Vydyne (18), Plaskin (25)
Polyester	PE	Hytrel (1), Kodar (20)
Thermoplastic elastomers	TPE	Duracryn (1), Flexsorb (17), Geolast (18), Kodapak (20), Santoprene (18), Zurcon (24)
Silicone	SI	27–29, 32, Cohrplastic (15), Green-Sil (14), Parshield (13), Baysilone (2), Blue-Sil (14)
Fluorosilicone	FSI	Parshield (13)
Vinylidene fluoride	HFP	Kynar (7), Foraflon (5)
Fluoroelastomers	FKM	24, 26, 28–31, Viton (1), Fluorel (6), Technoflon (9)
Ethylene–tetrafluoroethylene	ETFE	Tefzel (1), Halon ET (9)
Ethylene–chlorotrifluoroethylene	ECTFE	Halar (9)
Perfluoroelastomers	FPM	Kalrez (1), Chemraz (10), Kel-F (6)

[a]List of manufacturers:
1. E. I. du Pont
2. Mobay Corp.
3. Shell Chemical Co.
4. Morton Thiokol Inc.
5. Atochem Inc.
6. 3-M Corp
7. Pennwalt Corporation
8. Uniroyal
9. Ausimont
10. Greene, Tweed & Co., Inc.
11. Futura Coatings Inc.
12. Atochem Inc.
13. Parker Seal Group
14. The Perma-Flex Mold Co.
15. CHR Industries
16. Conap Inc

17. Polymer Corp.
18. Monsanto Co.
19. Hardman Inc.
20. Eastman Chemical Products Inc.
21. Edmont Div. of Becton, Dickinson & Co.
22. Dow Chemical USA
23. Krebs Engineers
24. W. S. Shamban & Co.
25. Allied Signal
26. General Rubber Co.
27. Hecht Rubber Co.
28. Minor Rubber Co.
29. Newco Holz Rubber Co.
30. Alvan Rubber Co.
31. Burke Rubber Co.
32. Unaflex

Another source of rubber is the planation leaf gutta-percha. This material is produced from the leaves of trees grown in bush formation. The leaves are picked and the rubber is boiled out as with the balata. Gutta-percha has been used successfully for submarine-cable insulation for more than 40 years.

Chemically, natural rubber is a polymer of methyl butadiene (isoprene):

$$CH_2{=}C{-}CH{=}CH_2 \atop \ \ \ \ \ \overset{|}{CH_3}$$

When polymerized, the units link together forming long chains that each contain over 1000 units. Simple butadiene does not yield a good grade of rubber, apparently because the chains are too smooth and do not form a strong enough interlock. Synthetic rubbers are produced by introducing side groups into the chain either by modifying butadiene or by making a copolymer of butadiene and some other compound.

Purified raw rubber becomes sticky in hot weather and brittle in cold weather. Its valuable properties become apparent after vulcanization.

Depending upon the degree of curing, natural rubber is classified as soft, semihard, or hard rubber. Only soft rubber meets the ASTM definition of an elastomer, and therefore the information that follows pertain only to soft rubber. The properties of semihard and hard rubber differ somewhat, particularly in the area of corrosion resistance.

Most rubber is made to combine with sulfur or sulfur-bearing organic compounds or with other cross-linking chemical agents in a process known as vulcanization, which was invented by Charles Goodyear in 1839 and forms the basis of all later developments in the rubber industry.

When properly carried out, vulcanization improves mechanical properties, eliminates tackiness, renders the rubber less susceptible to temperature changes, and makes it insoluble in all known solvents. Other materials are added for various purposes as follows:

Carbon blacks, precipitated pigments, and organic vulcanization accelerators are added to increase tensile strength and resistance to abrasion.

Whiting, barite, talc, silica, silicates, clays, and fibrous materials are added to cheapen and stiffen.

Bituminous substances, coal tar and its products, vegetable and mineral oils, paraffin, petrolatum, petroleum, oils, and asphalt are added to soften (for purposes of processing or for final properties).

Condensation amines and waxes are added as protective agents against natural aging, sunlight, heat, and flexing.

Pigments are added to provide coloration.

A. Physical and Mechanical Properties

The physical and mechanical properties of natural rubber are shown in Table 4.7. It is these properties that are responsible for the many varied applications of natural rubber. Many of these properties are modified somewhat through the process of vulcanization. Freshly cut or torn raw rubber possesses the power of self-adhesion. This property is for all intents and purposes absent in vulcanized rubber.

Dry heat up to 120°F (49°C) has little deteriorating effect on natural rubber. At temperatures of 300–400°F (148–205°C), rubber begins to melt and becomes sticky: at higher temperatures, it becomes entirely carbonized. Natural rubber possesses good electrical insulation properties but has poor flame resistance.

TABLE 4.7 Physical and Mechanical Properties of Natural Rubber[a]

Specific gravity	0.92
Refractive index	1.52
Specific heat (cal/g)	0.452
Swelling (% by volume)	
in kerosene at 77°F (25°C)	200
in benzene at 77°F (25°C)	200
in acetone at 77°F (25°C)	25
in mineral oil at 100°F (70°C)	120
Brittle point	−68°F (−56°C)
Relative permeability to hydrogen	50
Relative permeability to air	11
Insulation resistance (ohms/cm)	10^{17}
Resilience (%)	90
Tear resistance (psi)	1,640
Coefficient of linear expansion at 32–140°F (in./in.-°F)	0.000036
Coefficient of heat conduction K (Btu/ft^2-in.-°F)	1.07
Tensile strength (psi)	3,000–4,500
Elongation (% at break)	775–780
Hardness, Shore A	40–100
Abrasion resistance	Excellent
Maximum temperature, continuous use	175°F (80°C)
Impact resistance	Excellent
Compression set	Good
Machine qualities	Can be ground
Effect of sunlight	Deteriorates
Effect of aging	Moderately resistant
Effect of heat	Softens

[a] These are representative values, and they may be altered by compounding.

Vulcanization has the greatest effect on the mechanical properties of natural rubber. Unvulcanized rubber can be stretched to approximnately ten times its length and at this point will bear a load of 10 tons/in.^{2}. It can be compressed to one-third of its thickness thousands of times without injury. When most types of vulcanized rubbers are stretched, their resistance increases in greater proportion than their extension. Even when stretched just short of the point of rupture, they recover almost all of their original dimensions on being released and then gradually recover a portion of the residual distortion. The outstanding property of natural rubber in comparison to the synthetic rubbers is its resilience. It has excellent rebound properties, either hot or cold.

B. Resistance to Sun, Weather, and Ozone

Cold water preserves natural rubber, but if exposed to the air, particularly in sunlight, rubber tends to become hard and brittle. It has only fair resistance to ozone. Unlike the synthetic elastomers, natural rubber softens and reverts with aging to sunlight. In general, it has relatively poor weathering and aging properties.

C. Chemical Resistance

Natural rubber offers excellent resistance to most inorganic salt solutions, alkalies, and nonoxidizing acids. Hydrochloric acid will react with soft rubber to form rubber hydrochloride, and therefore it is not recommended that natural rubber be used for items that will come into contact with that acid. Strong oxidizing media such as nitric acid, concentrated sulfuric acid, permanganates, dichromates, chlorine dioxide, and sodium hypochlorite will severely attack rubber. Mineral and vegetable oils, gasoline, benzene, toluene, and chlorinated hydrocarbons also affect rubber. Cold water tends to preserve natural rubber. Natural rubber offers good resistance to radiation and alcohols.

Unvulcanized rubber is soluble in gasoline, naphtha, carbon bisulfide, benzene, petroleum ether, turpentine, and other liquids.

Refer to Table 4.8 and to Appendix 1 for the compatibility of natural rubber with selected corrodents.

D. Applications

Natural rubber finds its major use in the manufacture of pneumatic tires and tubes, power transmission belts, conveyer belts, gaskets, mountings, hose, chemical tank linings, printing press platens, sound and/or shock absorbers, and seals against air, moisture, sound, and dirt.

Rubber has been used for many years as a lining material for steel tanks, particularly for protection against corrosion by inorganic salt solutions, especially

TABLE 4.8 Compatibility of Natural Rubber with Selected Corrodents[a]

Chemical	Maximum temp.		Chemical	Maximum temp.	
	°F	°C		°F	°C
Acetaldehyde	x	x	Barium hydroxide	140	60
Acetamide	x	x	Barium sulfate	140	60
Acetic acid 10%	150	66	Barium sulfide	140	60
Acetic acid 50%	x	x	Benzaldehyde	x	x
Acetic acid 80%	x	x	Benzene	x	x
Acetic acid, glacial	x	x	Benzenesulfonic acid 10%	x	x
Acetic anhydride	x	x	Benzoic acid	140	60
Acetone	140	60	Benzyl alcohol	x	x
Acetyl chloride	x	x	Benzyl chloride	x	x
Acrylic acid			Borax	140	60
Acrylonitrile			Boric acid	140	60
Adipic acid			Bromine gas, dry		
Allyl alcohol			Bromine gas, moist		
Allyl chloride			Bromine, liquid		
Alum	140	60	Butadiene		
Aluminum acetate			Butyl acetate	x	x
Aluminum chloride, aqueous	140	60	Butyl alcohol	140	60
			n-Butylamine		
Aluminum chloride, dry	160	71	Butyric acid	x	x
Aluminum fluoride	x	x	Calcium bisulfide		
Aluminum hydroxide			Calcium bisulfite	140	60
Aluminum nitrate	x	x	Calcium carbonate	140	60
Aluminum oxychloride			Calcium chlorate	140	60
Aluminum sulfate	140	60	Calcium chloride	140	60
Ammonia gas			Calcium hydroxide 10%	140	60
Ammonium bifluoride			Calcium hydroxide, sat.	140	60
Ammonium carbonate	140	60	Calcium hypochlorite	x	x
Ammonium chloride 10%	140	60	Calcium nitrate	x	x
Ammonium chloride 50%	140	60	Calcium oxide	140	60
Ammonium chloride, sat.	140	60	Calcium sulfate	140	60
Ammonium fluoride 10%	x	x	Caprylic acid		
Ammonium fluoride 25%	x	x	Carbon bisulfide	x	x
Ammonium hydroxide 25%	140	60	Carbon dioxide, dry		
Ammonium hydroxide, sat.	140	60	Carbon dioxide, wet		
Ammonium nitrate	140	60	Carbon disulfide	x	x
Ammonium persulfate			Carbon monoxide	x	x
Ammonium phosphate	140	60	Carbon tetrachloride	x	x
Ammonium sulfate 10–40%	140	60	Carbonic acid	140	60
Ammonium sulfide	140	60	Cellosolve	x	x
Ammonium sulfite			Chloroacetic acid, 50% water	x	x
Amyl acetate	x	x			
Amyl alcohol	140	60	Chloroacetic acid	x	x
Amyl chloride	x	x	Chlorine gas, dry	x	x
Aniline	x	x	Chlorine gas, wet	x	x
Antimony trichloride			Chlorine, liquid	x	x
Aqua regia 3 : 1	x	x	Chlorobenzene	x	x
Barium carbonate	140	60	Chloroform	x	x
Barium chloride	140	60		*(Continued)*	

TABLE 4.8 Continued

Chemical	Maximum temp. °F	Maximum temp. °C	Chemical	Maximum temp. °F	Maximum temp. °C
Chlorosulfonic acid	x	x	Malic acid	x	x
Chromic acid 10%	x	x	Manganese chloride		
Chromic acid 50%	x	x	Methyl chloride	x	x
Chromyl chloride			Methyl ethyl ketone	x	x
Citric acid 15%	140	60	Methyl isobutyl ketone	x	x
Citric acid, concd	x	x	Muriatic acid	140	60
Copper acetate			Nitric acid 5%	x	x
Copper carbonate	x	x	Nitric acid 20%	x	x
Copper chloride	x	x	Nitric acid 70%	x	x
Copper cyanide	140	60	Nitric acid, anhydrous	x	x
Copper sulfate	140	60	Nitrous acid, concd	x	x
Cresol	x	x	Oleum		
Cupric chloride 5%	x	x	Perchloric acid 10%		
Cupric chloride 50%	x	x	Perchloric acid 70%		
Cyclohexane	x	x	Phenol	x	x
Cyclohexanol			Phosphoric acid 50–80%	140	60
Dibutyl phthalate			Picric acid		
Dichloroacetic acid			Potassium bromide 30%	140	60
Dichloroethane (ethylene dichloride)	x	x	Salicylic acid		
			Silver bromide 10%		
Ethylene glycol	140	60	Sodium carbonate	140	60
Ferric chloride	140	60	Sodium chloride	140	60
Ferric chloride 50% in water	140	60	Sodium hydroxide 10%	140	60
Ferric nitrate 10–50%	x	x	Sodium hydroxide 50%	x	x
Ferrous chloride	140	60	Sodium hydroxide, concd	x	x
Ferrous nitrate	x	x	Sodium hypochlorite 20%	x	x
Fluorine gas, dry	x	x	Sodium hypochlorite, concd	x	x
Fluorine gas, moist			Sodium sulfide to 50%	140	60
Hydrobromic acid, dil	140	60	Stannic chloride	140	60
Hydrobromic acid 20%	140	60	Stannous chloride	140	60
Hydrobromic acid 50%	140	60	Sulfuric acid 10%	140	60
Hydrochloric acid 20%	x	x	Sulfuric acid 50%	x	x
Hydrochloric acid 38%	140	60	Sulfuric acid 70%	x	x
Hydrocyanic acid 10%			Sulfuric acid 90%	x	x
Hydrofluoric acid 30%	x	x	Sulfuric acid 98%	x	x
Hydrofluoric acid 70%	x	x	Sulfuric acid 100%	x	x
Hydrofluoric acid 100%	x	x	Sulfuric acid, fuming	x	x
Hypochlorous acid			Sulfurous acid	x	x
Iodine solution 10%			Thionyl chloride		
Ketones, general			Toluene		
Lactic acid 25%	x	x	Trichloroacetic acid		
Lactic acid, concd	x	x	White liquor		
Magnesium chloride	140	60	Zinc chloride	140	60

[a]The chemicals listed are in the pure state or in a saturated solution unless otherwise indicated. Compatibility is shown to the maximum allowable temperature for which data are available. Incompatibility is shown by an x. A blank space indicates that data are unavailable.
Source: Schweitzer PA. Corrosion Resistance Tables, 4th ed., Vols. 1–3. New York: Marcel Dekker, 1995.

brine, alkalies, and nonoxidizing acids. These linings have the advantage of being readily repaired in place. Natural rubber is also used for lining pipelines used to convey these types of materials. Some of these applications have been replaced by synthetic rubbers that have been developed over the years.

4.3 ISOPRENE RUBBER (IR)

Chemically, natural rubber is natural *cis*-polyisoprene. The synthetic form of natural rubber, synthetic *cis*-polyisoprene, is called isoprene rubber. The physical and mechanical properties of isoprene rubber are similar to the physical and mechanical properties of natural rubber, the one major difference being that isoprene does not have an odor. This feature permits the use of isoprene rubber in certain food-handling applications.

Isoprene rubber can be compounded, processed, and used in the same manner as natural rubber. Other than the lack of odor, isoprene rubber has no advantages over natural rubber.

4.4 NEOPRENE (CR)

Neoprene is one of the oldest and most versatile of the synthetic rubbers. Chemically is is polychloroprene. Its basic unit is a chlorinated butadiene whose formula is

$$CH_2-\underset{\underset{Cl}{|}}{C}-CH=CH_2$$

The raw material is acetylene, which makes this product more expensive than some of the other elastomeric materials.

Neoprene was introduced commercially by DuPont in 1932 as an oil-resistant substitute for natural rubber. Its dynamic properties are very similar to those of natural rubber, but its range of chemical resistance overcomes many of the shortcomings of natural rubber.

As with other elastomeric materials, neoprene is available in a variety of formulations. Depending on the compounding procedure, material can be produced to impart specific properties to meet specific application needs.

Neoprene is also available in a variety of forms. In addition to a neoprene latex that is similar to natural rubber latex, neoprene is produced in a "fluid" form as either a compounded latex dispersion or a solvent solution. Once these materials have solidified or cured, they have the same physical and chemical properties as the solid or cellular forms.

A. Physical and Mechanical Properties

The properties discussed here are attainable with neoprene but may not necessarily be incorporated into every neoprene product. Nor will every neoprene product perform the same in all environments. The reason for this variation is compounding. By selective addition and/or deletion of specific ingredients during compounding, specific properties can be enhanced or reduced to provide the neoprene formulation best suited for the application. A neoprene compound can be produced that will provide whichever of the properties discussed are desired. When the hardness of neoprene is above 55 Shore A, its resilience exceeds that of natural rubber by approximately 5%. At hardnesses below 50 Shore A, its resilience is not as good as that of natural rubber, even though its resilience is measured at 75%, which is a high value. Because of its high resilience, neoprene products have low hysteresis and a minimum heat buildup during dynamic operations.

Solid neoprene products can be ignited by an open flame but will stop burning when the flame is removed. Because of its chlorine content, neoprene is more resistant to burning than exclusively hydrocarbon elastomers. Natural rubber and many of the other synthetic elastomers will continue to burn once ignited, even if the flame is removed. In an actual fire situation, neoprene will burn. Although compounding can improve the flame resistance of neoprene, it cannot make it immune to burning.

Compared to natural rubber, neoprene is relatively impermeable to gases. Table 4.9 lists typical permeability constants. Because of this impermeability, neoprene can be used to seal against Freon blowing agents, propane, butane, and other gases.

Neoprene is used in many electrical applications, although its dielectric characteristics limit its use as an insulation to low voltage (600 V) and low frequency (60 Hz). Because of its high degree of resistance to indoor and outdoor aging and its resistance to weathering, neoprene is often used as a protective outer jacket to insulation at all voltages. It is also immune to high-voltage corona discharge effects that cause severe surface cutting in many types of elastomers.

At the maximum operating temperature of 200°F (93°C), neoprene continues to maintain good physical properties and has excellent resistance to long-term heat degradation. Unlike other elastomers, neoprene does not soften or melt when heated, regardless of the degree of heat. Heat failure results from the hardening of the product and lack of resilience. Neoprene products display little change in performance characteristics down to approximately 0°F (−18°C). As temperatures decrease further, the material stiffens until the brittle point is reached. Although the brittle point for standard neoprene products is −40°F (−40°C), special compounding can produce materials that can be used at temperatures as low as −67°F (−55°C).

TABLE 4.9 Physical and Mechanical Properties of Neoprene (CR)[a]

Specific gravity	1.4
Refractive index	1.56
Specific heat (cal/g)	0.40
Volumetric coefficient of thermal expansion	
at 77°F	403×10^{-6}°F
at 25°F	725×10^{-6}°C
Thermal conductivity	
(Btu/hr-ft^2-in.°F or	1.45
g-cal/hr-cm^{2}cm°C)	1.80
Brittle point	−40°F (−40°C)
dc resistivity (ohm-cm)	2×10^{13}
Dielectric strength (V/mil)	600
Permeability (cm^3/cm^2-cm-sec-atm) at 77°F (25°C)	
to nitrogen	1×10^{-8}
to methane	2×10^{-8}
to oxygen	3×10^{-8}
to helium	10×10^{-8}
to carbon dioxide	19×10^{-8}
Tensile strength (psi)	1,000–2,500
Elongation (% at break)	200–600
Hardness, Shore A	40–95
Abrasion resistance	Excellent
Maximum temperature, continuous use	180–200°F
	(82–93°C)
Impact resistance	Excellent
Compression set (%)	15–35
Machine qualities	Can be ground
Resistance to sunlight	Excellent
Effect of aging	Little effect
Resistance to heat	Good

[a] These are representative values, and they may be altered by compounding.

Neoprene will form an extremely strong mechanical bond with cotton fabric. If suitable treatments of additives are provided, it can also be made to adhere to such man-made fibers as glass, nylon, rayon, acrylic, and polyester fibers. It can also be molded in contact with metals, particularly carbon and alloy steels, stainless steeels, aluminum and aluminum alloys, brass, and copper, using any one of the commercially available bonding agents.

Neoprene provides no nourishment for microorganisms, but it will not deter them from consuming other ingredients in the compound. Consequently, products containing metabolizable compounding ingredients require the inclusion of fungicide, bactericide, or pesticide in the formulation to provide protection.

Pigmentation of neoprene products is a simple matter since the elastomer readily accepts color additives. However, the lighter shades, such as tones of yellow, red, blue, and other bright colors, will eventually discolor with prolonged exposure to sunlight and ultraviolet light. Because of this, products intended for prolonged outdoor service are usually produced in shades of grey, maroon, brown, or black. The lighter shades are used for products having limited exposure to sunlight, such as rainwear and appliance parts.

Most neoprene products have tensile strengths ranging from 1000 to 2500 psi with elongation at break running from 200 to 600% and hardness of 40–95 Shore A. These hardnesses are equivalent to those of a rubber band or a typewriter roller. By making use of cellular neoprene, softer products can be produced. Although harder compositions can be produced, their application is limited. Products made from elastomeric materials can be subjected to abrasion, continual flexing and twisting, and impact. Neoprene provides extremely good resistance to this type of mechanical abuse. Its low degree of heat buildup during continual flexing guards against fatigue from dynamic operations. Protection from flex crack inititation and cut growth is provided by its toughness. High resistance to impact, abrasion, and tearing can be provided by proper compounding. In general, the innate properties of neoprene plus the ability to compound it in various formulations can provide a product having the desired mechanical properties. Table 4.9 lists the mechanical and physical properties of neoprene.

In the design of many products, the recovery of a component after it has been held under a load for a long period of time is an important consideration. The degree of permanent deformation or compression set of neoprene is relatively low (see Table 4.9). Through selective compounding, products with lower compression sets than those shown are available.

The opposite of compression set is *permanent elongation set*, which is the permanent increase in the length of a sample that has been stretched and held in tension for a specified period of time. The average elongation set of neoprene products is 5%.

B. Resistance to Sun, Weather, and Ozone

Neoprene displays excellent resistance to sun, weather, and ozone. Because of its low rate of oxidation, products made of neoprene have high resistance to both outdoor and indoor aging. Over prolonged periods of time in an outdoor environment, the physical properties of neoprene display insignificant change. If neoprene is properly compounded, ozone in atmospheric concentrations has little effect on the product. When severe ozone exposure is expected, as for example around electrical equipment, compositions of neoprene can be provided to resist thousands of parts per million of ozone for hours without surface cracking. Natural rubber will crack within minutes when subjected to ozone concentrations of only 50 ppm.

C. Chemical Resistance

Neoprene's resistance to attack from solvents, waxes, fats, oils, greases, and many other petroleum-based products is one of its outstanding properties. Excellent service is also experienced when it is in contact with aliphatic compounds (methyl and ethyl alcohols, ethylene glycols, etc.), aliphatic hydrocarbons, and most Freon refrigerants. A minimum amount of swelling and relatively little loss of strength occur when neoprene is in contact with these fluids.

When exposed to dilute mineral acids, inorganic salt solutions, or alkalies, neoprene products show little if any change in appearance or change in properties.

Chlorinated and aromatic hydrocarbons, organic esters, aromatic hydroxy compounds, and certain ketones have an adverse effect on neoprene, and consequently only limited serviceability can be expected with them. Highly oxidizing acid and salt solutions also cause surface deterioration and loss of strength. Included in this category are nitric acid and concentrated sulfuric acid.

Neoprene formulations can be produced that provide products with outstanding resistance to water absorption. These products can be used in continuous or periodic immersion in either freshwater or saltwater without any loss of properties.

Properly compounded neoprene can be buried underground successfully, since moisture, bacteria, and soil chemicals usually found in the earth have little effect on its properties. It is unaffected by soils saturated with seawater, chemicals, oils, gasolines, wastes, and other industrial byproducts. Refer to Table 4.10 and Appendix 1 for the compatibilities of neoprene with selected corrodents.

D. Applications

Neoprene products are available in three basic forms:

1. Conventional solid rubber parts
2. Highly compressed cellular materials
3. Free-flowing liquids

Each form has certain specific properties that can be of advantage to final products.

Solid products can be produced by molding, extruding, or calendering. Molding can be accomplished by compression, transfer, injection, blow, vacuum, or wrapped-mandrel methods. Typical products produced by these methods are instrument seals, shoe soles and heels, auto spark plug boots, radiator hose,

TABLE 4.10 Compatibility of Neoprene with Selected Corrodents[a]

Chemical	Maximum temp.		Chemical	Maximum temp.	
	°F	°C		°F	°C
Acetaldehyde	200	93	Barium hydroxide	230	110
Acetamide	200	93	Barium sulfate	200	93
Acetic acid 10%	160	71	Barium sulfide	200	93
Acetic acid 50%	160	71	Benzaldehyde	x	x
Acetic acid 80%	160	71	Benzene	x	x
Acetic acid, glacial	x	x	Benzenesulfonic acid 10%	100	38
Acetic anhydride	x	x	Benzoic acid	150	66
Acetone	x	x	Benzyl alcohol	x	x
Acetyl chloride	x	x	Benzyl chloride	x	x
Acrylic acid	x	x	Borax	200	93
Acrylonitrile	140	60	Boric acid	150	66
Adipic acid	160	71	Bromine gas, dry	x	x
Allyl alcohol	120	49	Bromine gas, moist	x	x
Allyl chloride	x	x	Bromine, liquid	x	x
Alum	200	93	Butadiene	140	66
Aluminum acetate			Butyl acetate	60	16
Aluminum chloride, aqueous	150	66	Butyl alcohol	200	93
			n-Butylamine		
Aluminum chloride, dry			Butyric acid	x	x
Aluminum fluoride	200	93	Calcium bisulfide		
Aluminum hydroxide	180	82	Calcium bisulfite	x	x
Aluminum nitrate	200	93	Calcium carbonate	200	93
Aluminum oxychloride			Calcium chlorate	200	93
Aluminum sulfate	200	93	Calcium chloride	150	66
Ammonia gas	140	60	Calcium hydroxide 10%	230	110
Ammonium bifluoride	x	x	Calcium hydroxide, sat.	230	110
Ammonium carbonate	200	93	Calcium hypochlorite	x	x
Ammonium chloride 10%	150	66	Calcium nitrate	150	66
Ammonium chloride 50%	150	66	Calcium oxide	200	93
Ammonium chloride, sat.	150	66	Calcium sulfate	150	66
Ammonium fluoride 10%	200	93	Caprylic acid		
Ammonium fluoride 25%	200	93	Carbon bisulfide	x	x
Ammonium hydroxide 25%	200	93	Carbon dioxide, dry	200	93
Ammonium hydroxide, sat.	200	93	Carbon dioxide, wet	200	93
Ammonium nitrate	200	93	Carbon disulfide	x	x
Ammonium persulfate	200	93	Carbon monoxide	x	x
Ammonium phosphate	150	66	Carbon tetrachloride	x	x
Ammonium sulfate 10–40%	150	66	Carbonic acid	150	66
Ammonium sulfide	160	71	Cellosolve	x	x
Ammonium sulfite			Chloroacetic acid, 50% water	x	x
Amyl acetate	x	x			
Amyl alcohol	200	93	Chloroacetic acid	x	x
Amyl chloride	x	x	Chlorine gas, dry	x	x
Aniline	x	x	Chlorine gas, wet	x	x
Antimony trichloride	140	60	Chlorine, liquid	x	x
Aqua regia 3 : 1	x	x	Chlorobenzene	x	x
Barium carbonate	150	66	Chloroform	x	x
Barium chloride	150	66	Chlorosulfonic acid	x	x

TABLE 4.10 Continued

Chemical	Maximum temp. °F	Maximum temp. °C	Chemical	Maximum temp. °F	Maximum temp. °C
Chromic acid 10%	140	60	Manganese chloride	200	93
Chromic acid 50%	100	38	Methyl chloride	x	x
Chromyl chloride			Methyl ethyl ketone	x	x
Citric acid 15%	150	66	Methyl isobutyl ketone	x	x
Citric acid, concd	150	66	Muriatic acid	x	x
Copper acetate	160	71	Nitric acid 5%	x	x
Copper carbonate			Nitric acid 20%	x	x
Copper chloride	200	93	Nitric acid 70%	x	x
Copper cyanide	160	71	Nitric acid, anhydrous	x	x
Copper sulfate	200	93	Nitrous acid, concd	x	x
Cresol	x	x	Oleum	x	x
Cupric chloride 5%	200	93	Perchloric acid 10%		
Cupric chloride 50%	160	71	Perchloric acid 70%	x	x
Cyclohexane	x	x	Phenol	x	x
Cyclohexanol	x	x	Phosphoric acid 50–80%	150	66
Dibutyl phthalate			Picric acid	200	93
Dichloroacetic acid	x	x	Potassium bromide 30%	160	71
Dichloroethane (ethylene dichloride)	x	x	Salicylic acid		
			Silver bromide 10%		
Ethylene glycol	100	38	Sodium carbonate	200	93
Ferric chloride	160	71	Sodium chloride	200	93
Ferric chloride 50% in water	160	71	Sodium hydroxide 10%	230	110
Ferric nitrate 10–50%	200	93	Sodium hydroxide 50%	230	110
Ferrous chloride	90	32	Sodium hydroxide, concd	230	110
Ferrous nitrate	200	93	Sodium hypochlorite 20%	x	x
Fluorine gas, dry	x	x	Sodium hypochlorite, concd	x	x
Fluorine gas, moist	x	x	Sodium sulfide to 50%	200	93
Hydrobromic acid, dil	x	x	Stanic chloride	200	93
Hydrobromic acid 20%	x	x	Stannous chloride	x	x
Hydrobromic acid 50%	x	x	Sulfuric acid 10%	150	66
Hydrochloric acid 20%	x	x	Sulfuric acid 50%	100	38
Hydrochloric acid 38%	x	x	Sulfiric acid 70%	x	x
Hydrocyanic acid 10%	x	x	Sulfuric acid 90%	x	x
Hydrofluoric acid 30%	x	x	Sulfuric acid 98%	x	x
Hydrofluoric acid 70%	x	x	Sulfuric acid 100%	x	x
Hydrofluoric acid 100%	x	x	Sulfiric acid, fuming	x	x
Hypochlorous acid	x	x	Sulfurous acid	100	38
Iodine solution 10%	80	27	Thionyl chloride	x	x
Ketones, general	x	x	Toluene	x	x
Lactic acid 25%	140	60	Trichloroacetic acid	x	x
Lactic acid, concd	90	32	White liquor	140	60
Magnesium chloride	200	93	Zinc chloride	160	71
Malic acid					

[a]The chemicals listed are in the pure state or in a saturated solution unless otherwise indicated. Compatibility is shown to the maximum allowable temperature for which data are available. Incompatibility is shown by an x. A blank space indicates that data are unavailable.
Source: Schweitzer PA. Corrosion Resistance Tables, 4th ed., Vols. 1–3. New York: Marcel Dekker, 1995.

boating accessories, appliance parts, O-rings, and other miscellaneous components.

Extrusion processes provide means to economically and uniformly mass produce products quickly. Neoprene products manufactured by these processes include tubing, sealing strips, wire jacketing, filaments, rods, and many types of hose.

Calendered products include sheet stock, belting, and friction and coated fabrics. A large proportion of sheet stock is later die-cut into finished products such as gaskets, pads, and diaphragms.

Cellular forms of neoprene are used primarily for gasketing, insulation, cushioning, and sound and vibration damping. This material provides compressibility not found in solid rubber but still retains the advantageous properties of neoprene. It is available as an open-cell sponge, a closed-cell neoprene, and a foam neoprene.

Open-cell neoprene is a compressible, absorbent material whose cells are uniform and interconnected. This material is particularly useful for gasketing and dustproofing applications where exposure to fluids is not expected.

Closed-cell neoprene is a resilient complex of individual, nonconnecting cells that impart the added advantage of nonabsorbency. This property makes closed-cell neoprene especially suitable for sealing applications where fluid contact is expected, for products such as wet suits for divers, shoe soles, automotive deck lid seals, and for other appplications where a compressible nonabsorbent weather-resistant material is required.

Foam neoprene is similar to open-cell neoprene in that it is a compressible material with interconnecting cells. Its main area of application is for cushioning, for example, in mattresses, seating, and carpet underlay. Because of the good heat and oil resistance of neoprene, it has also found application as a railroad car lubricator. This absorbent open-cell structure provides a wicking action to deliver oil to the journal bearings.

Fluid forms of neoprene are important in the manufacture of many products because of their versatility. "Fluid" neoprene is the primary component in such products as adhesives, coatings and paints, sealants, caulks, and fiber binders. It is available in two forms—as a neoprene latex or as a solvent solution. Neoprene latex is an elastomer–water dispersion. It is used primarily in the manufacture of dipped products such as tool grips, household and industrial gloves, meteorological balloons, sealed fractional-horsepower motors, and a variety of rubber-coated metal parts. Other applications include use as a binder for curled animal hair in resilient furniture cushioning, transportation seating, acoustical filtering and packaging. It is also used extensively in latex-based adhesives, foams, protective coatings, and knife-coated fabrics, as a binder for cellulose and asbestos, and as an elasticizing additive for concrete, mortar, and asphalt. These products produced from neoprene latex possess the same properties as those associated with solid neoprene, including resistance to oil, chemicals, ozone, weather, and flame.

Neoprene solvent solutions are prepared by dissolving neoprene in standard rubber solvents. These solutions can be formulated in a range of viscosities suitable for application by brush, spray, or roller. Major areas of application include coatings for storage tanks, industrial equipment, and chemical processing equipment. These coatings protect the vessels from corrosion by acids, oils, alkalies, and most hydrocarbons.

Neoprene roofing applied in liquid form is used to protect concrete, plywood, and metal decks. The solvent solution can be readily applied and will cure into a roofing membrane that is tough, elastic, and weather resistant.

Solvent-based neoprene adhesives develop quick initial bonds and remain strong and flexible most indefinitely. They can be used to join a wide variety of rigid and flexible materials.

Collapsible nylon containers coated with neoprene are used for transporting and/or storing liquids, pastes, and flowable dry solds. Containers have been designed to hold oils, fuels, molasses, and various bulk-shipped products.

Neoprene in its many forms has proved to be a reliable, indispensable substitute for natural rubber, possessing many of the advantageous properties of natural rubber while also overcoming many of its shortcomings.

4.5 BUTADIENE–STYRENE RUBBER (SBR, BUNA-S, GR-S)

During World War II, a shortage of natural rubber was created when Japan occupied the Far Eastern nations from which natural rubber was obtained. Because of the great need for rubber, the U.S. government developed what was originally known as Government Rubber Styrene-Type (GR-S) because it was the most practical to put into rapid production on a wartime scale. It was later designated GR-S.

The rubber is produced by copolymerizing butadiene and styrene. As with natural rubber and the other synthetic elastomers, compounding with other ingredients will improve certain properties. Continued development since World War II has improved its properties considerably over what was initially produced by either Germany or the United States.

A. Physical and Mechanical Properties

In general, Buna-S is very similar to natural rubber although some of its physical and mechanical properties are inferior. It is lacking in tensile strength, elongation, resilience, hot tear, and hysteresis. These disadvantages are offset somewhat by its low cost, cleanliness, slightly better heat-aging properties, slightly better wear than natural rubber for passenger tires, and availability at a stable price. The electrical properties of SBR are generally good but are not outstanding in any one area.

Because of these shortcomings, there are many variations and blends of NR-SBR. Basic is the ratio of butadiene to styrene, temperature of polymerization, and type of chemicals used during polymerization. The NR-SBR blends must be properly formulated to achieve the correct balance of properties. SBR is superior to NR in several of its processing properties. It also has excellent abrasion resistance and resistance to hydraulic brake fluid.

Molded SBR parts have a strong tendency to bloom due to the presence of added stabilizers. The bloom creates an unsightly appearance and can cause migration staining in contact with paint. Consequently, SBR is not generally used in readily visible locations or in contact with painted surfaces.

Buna-S has a maximum operating temperature of 175°F (80°C), which is not exceptional. At reduced temperatures below 0°F, Buna-S products are more flexible than those produced from natural rubber.

Butadiene–styrene rubber has poor flame resistance and will support combustion. Table 4.11 lists the physical and mechanical properties of Buna-S.

TABLE 4.11 Physical and Mechanical Properties of Butadiene–Styrene Rubber (SBR, Buna-S, GR-S)[a]

Specific gravity	0.94
Refractive index	1.53
Specific heat (cal/g)	0.454
Brittle point	$-76°F$ ($-60°C$)
Insulation resistance (ohms/cm)	10^{15}
Dielectric constant at 50 Hz	2.9
Swelling (% by volume)	
in kerosene at 77°F (25°C)	100
in benzene at 77°F (25°C)	200
in acetone at 77°F (25°C)	30
in mineral oil at 100°F (70°C)	150
Tear resistance (psi)	550
Creep at 70°C (%)	14.6
Tensile strength (psi)	1,600–3,700
Elongation (% at break)	650
Hardness, Shore A	35–90
Abrasion resistance	Excellent
Maximum temperature, continuous use	175°F (80°C)
Resistance to compression set	Poor
Machine qualities	Can be ground
Resistance to sunlight	Deteriorates
Effect of aging	Little effect
Effect of heat	Stiffens

[a]These are representative values, and they may be altered by compounding.

B. Resistance to Sun, Weather, and Ozone

Butadiene–styrene rubber has poor weathering and aging properties. Sunlight will cause it to deteriorate. However, it does have better water resistance than natural rubber.

C. Chemical Resistance

The chemical resistance of Buna-S is similar to that of natural rubber. It is resistant to water and exhibits fair to good resistance to dilute acids, alkalies, and alcohols. It is not resistant to oils, gasoline, hydrocarbons, or oxidizing agents. Refer to Table 4.12 for the compatibility of SBR with selected corrodents.

D. Applications

The major use of Buna-S is in the manufacture of automobile tires, although Buna-S materials are also used to manufacture conveyor belts, hose, gaskets, and seals against air, moisture, sound, and dirt.

4.6 NITRILE RUBBER (NBR, BUNA-N)

Nitrile rubbers are an outgrowth of German Buna-N or Perbunan. They are copolymers of butadiene and acrylonitrile ($CH_2=CH-C=N$) and are one of the four most wildely used elastomers. XNBR is a carboxylic–acrylonitrile–buta-diene–nitrile rubber with somewhat improved abrasion resistance over that of the standard NBR nitrile rubbers. The main advantages of nitrile rubbers are their low cost, good oil and abrasion resistance, and good low-temperature and swell characteristics. Their greater resistance to oils, fuel, and solvents compared to that of neoprene is their primary advantage. As with other elastomers, appropriate compounding will improve certain properties.

A. Physical and Mechanical Properties

The physical and mechanical properties of nitrile rubbers are very similar to those of natural rubber (see Table 4.13). These rubbers have the exceptional ability to retain both their strength and elasticity at extremely low temperatures. It is this property that makes them valuable for use as hose used in operating the hydraulic controls of airplanes.

Buna-N does not have exceptional heat resistance. It has a maximum operating temperature of 250°F (120°C), but it has a tendency to harden at high temperatures. Nitrile rubbers will support combustion and burn. Their electrical properties are relatively poor, and consequently they do not find wide use in the electrical field since there are so many other elastomeric materials with far superior electrical properties.

TABLE 4.12 Compatibility of SBR with Selected Corrodents[a]

Chemical	Maximum temp.		Chemical	Maximum temp.	
	°F	°C		°F	°C
Acetic acid 10%	x	x	Formaldehyde 50%	200	93
Acetic acid 50%	x	x	Glycerine	200	93
Acetic acid 80%	x	x	Hydrochloric acid, dil	x	x
Acetic acid, glacial	x	x	Hydrochloric acid 20%	x	x
Acetone	200	93	Hydrochloric acid 35%	x	x
Ammonium chloride 10%	200	93	Hydrochloric acid 38%	x	x
Ammonium chloride 28%	200	93	Hydrochloric acid 50%	x	x
Ammonium chloride 50%	200	93	Hydrofluoric acid, dil	x	x
Ammonium chloride, sat.	200	93	Hydrofluoric acid 30%	x	x
Ammonium sulfate 10–40%			Hydrofluoric acid 40%	x	x
Aniline	x	x	Hydrofluoric acid 50%	x	x
Benzene	x	x	Hydrofluoric acid 70%	x	x
Benzoic acid	x	x	Hydrofluoric acid 100%	x	x
Butane	x	x	Hydrogen peroxide, dil, all	200	93
Butyl acetate	x	x	concn		
Butyl alcohol	200	93	Lactic acid all concn	200	93
Calcium chloride, sat	200	93	Methyl ethyl ketone	x	x
Calcium hydroxide 10%	200	93	Methyl isobutyl ketone	x	x
Calcium hydroxide 20%	200	93	Muriatic acid	x	x
Calcium hydroxide 30%	200	93	Nitric acid	x	x
Calcium hydroxide sat	200	93	Phenol	x	x
Carbon bisulfide	x	x	Phosphoric acid 10%	200	93
Carbon tetrachloride	x	x	Potassium hydroxide to	x	x
Chloroform	x	x	50%		
Chromic acid 10%	x	x	Propane	x	x
Chromic acid 30%	x	x	Sodium chloride	200	93
Chromic acid 40%	x	x	Sodium hydroxide, all concn	x	x
Chromic acid 50%	x	x	Sulfuric acid, all concn	x	x
Copper sulfate	200	93	Toluene	x	x
Corn oil	x	x	Trichloroethylene	x	x
Ethyl acetate	x	x	Water, demineralized	210	99
Ethyl alcohol	200	93	Water, distilled	200	93
Ethylene glycol	200	93	Water, salt	200	93
Formaldehyde, dil	200	93	Water, sea	200	93
Formaldehyde 37%	200	93			

[a] The chemicals listed are in the pure state or in a saturated solution unless otherwise indicated. Compatibility is shown to the maximum allowable temperature for which data are available. Incompatibility is shown by an x. A blank space indicates that data are unavailable. *Source:* Schweitzer PA. Corrosion Resistance Tables, 4th ed., Vols. 1–3. New York: Marcel Dekker, 1995.

TABLE 4.13 Physical and Mechanical Properties of Nitrile Rubber (NBR, Buna-N)[a]

Specific gravity	0.99
Refractive index	1.54
Brittle point	−32°F to −40°F
	(−1°C to −40°C)
Swelling (% by volume)	
in kerosene at 77°F (25°C)	9–10
in benzene at 77°F (25°C)	120
in acetone at 77°F (25°C)	60–50
in mineral oil at 100°F (70°C)	2–10
in air at 77°F (25°C)	30–50
Tensile strength (psi)	500–4,000
Elongation (% at break)	400
Hardness, Shore A	40–95
Abrasion resistance	Excellent
Maximum temperature, continuous use	250°F (120°C)
Compression set	Good
Tear resistance	Excellent
Resilience (%)	63–74
Machining qualities	Can be ground
Resistance to sunlight	Fair
Effect of aging	Highly resistant
Resistance to heat	Stiffens

[a]These are representative values, and they may be altered by compounding.

NBR has good compression set recovery from deformation and good abrasion resistance and tensile strength.

NBR is available in several grades with acrylonitrile content varying between 15 and 20%. Resistance to hydrocarbons, tensile strength, and heat resistance increase as the nitrile content increases. Conversely, resilience, low temperature flexibility, and set resistance all decrease.

By the addition of protective additives, resistance to heat, oxidation, and ozone can be increased. The addition of polyvinyl chloride (PVC) improves processibibility and causes a dramatic improvement in ozone resistance, as well as increasing tensile strength. One caution. PVC is capable of releasing small amounts of chloride ion, probably as hydrochloric acid. In vapor tubing applications that are in sequence with delicate electronic sensors, PVC/NBR blends should be avoided since the chloride ion will foul the sensor.

Hydrogenated nitrile rubber (HNBR) is produced from nitrile rubber polymer that is further modified in a catalytic hydrogenating process. The unsaturation in the polymer backbone is selectively hydrogenated while leaving the acrylonitrile group unaffected. HNBR retains the useful properties of NBR with the additional advantages of ozone and weathering resistance, plus an upper temperature limit in the range of 300°F/149°C. An HNBR polymer has greater resilience than an NBR polymer with the same nitrile content.

After hydrogenation, the residual unsaturation is low. Consequently, HNBR has good resistance to sulfur and sulfur-bearing compounds. When peroxide cured, it also has temperature set resistance. HNBR is suitable for fuel and oil seals, hoses, and belts.

Carbonated nitrile rubber (XNBR) incorporates up to 10% of a third comonomer with organic acid functionality. When compared to NBR, XNBR has improved abrasion resistance and strength. XNBR can be difficult to process, and it requires special formulation to prevent sticking to mixer surfaces and premature vulcanization.

B. Resistance to Sun, Weather, and Ozone

Nitrile rubbers offer poor resistance to sunlight and ozone, and their weathering qualities are not good.

C. Chemical Resistance

Nitrile rubbers exhibit good resistance to solvents, oil, water, and hydraulic fluids. Very slight swelling occurs in the presence of aliphatic hydrocarbons, fatty acids, alcohols, and glycols. The deterioration of physical properties as a result of this swelling is small, making NBR suitable for gasoline- and oil-resistant applications. NBR has excellent resistance to water. The use of highly polar solvents such as acetone and methyl ethyl ketone, chlorinated hydrocarbons, ozone, nitro hydrocarbons, ether, or esters should be avoided, since these materials will attack the nitrile rubbers.

XNBR rubbers are used primarily in nonalkaline service. Refer to Table 4.14 and Appendix 1 for the compatibility of NBR with selected corrodents.

D. Applications

Because of its exceptional resistance to fuels and hydraulic fluids, Buna-N's major area of application is in the manufacture of aircraft hose, gasoline and oil hose, and self-sealing fuel tanks. Other applications include carburetor diaphragms, gaskets, cables, printing rolls, and machinery mountings.

TABLE 4.14 Compatibility of Nitrile Rubber with Selected Corrodents[a]

Chemical	Maximum temp.		Chemical	Maximum temp.	
	°F	°C		°F	°C
Acetaldehyde	x	x	Boric acid	150	66
Acetamide	180	82	Calcium hypochlorite	x	x
Acetic acid 10%	x	x	Carbonic acid	100	38
Acetic acid 50%	x	x	Ethylene glycol	100	38
Acetic acid 80%	x	x	Ferric chloride	150	66
Acetic acid, glacial	x	x	Ferric nitrate 10–50%	150	66
Acetic anhydride	x	x	Hydrofluoric acid 70%	x	x
Acetone	x	x	Hydrofluoric acid 100%	x	x
Acetyl chloride	x	x	Hypochlorous acid		
Acrylic acid	x	x	Nitric acid 20%	x	x
Acrylonitrile	x	x	Nitric acid 70%	x	x
Adipic acid	180	82	Nitric acid, anhydrous	x	x
Allyl alcohol	180	82	Phenol	x	x
Allyl chloride	x	x	Phosphoric acid 50–80%	150	66
Alum	150	66	Sodium carbonate	125	52
Aluminum chloride,			Sodium chloride	200	93
aqueous	150	66	Sodium hydroxide 10%	150	66
Aluminum hydroxide	180	82	Sodium hypochlorite 20%	x	x
Aluminum nitrate	190	88	Sodium hypochlorite, concd	x	x
Aluminum sulfate	200	93	Stannic chloride	150	66
Ammonia gas	190	88	Sulfuric acid 10%	150	66
Ammonium carbonate	x	x	Sulfuric acid 50%	150	66
Ammonium nitrate	150	66	Sulfuric acid 70%	x	x
Ammonium phosphate	150	66	Sulfuric acid 90%	x	x
Amyl alcohol	150	66	Sulfuric acid 98%	x	x
Aniline	x	x	Sulfuric acid 100%	x	x
Barium chloride	125	52	Sulfuric acid, fuming	x	x
Benzene	150	66	Zinc chloride	150	66
Benzoic acid	150	66			

[a] The chemicals listed are in the pure state or in a saturated solution unless otherwise indicated. Compatibility is shown to the maximum allowable temperature for which data are available. Incompatibility is shown by an x. A blank space indicates that data are unavailable.
Source: Schweitzer PA. Corrosion Resistance Tables, 4th ed., Vols. 1–3. New York: Marcel Dekker, 1995.

4.7 BUTYL RUBBER (IIR) AND CHLOROBUTYL RUBBER (CIIR)

Butyl rubber contains isobutylene

$$\begin{array}{c} CH_3 \\ | \\ C-CH_2 \\ | \\ CH_3 \end{array}$$

as its parent material with small proportions of butadiene or isoprene added. Commercial butyl rubber may contain 5% butadiene as a copolymer. It is a general-purpose synthetic rubber whose outstanding physical properties are low permeability to air (approximately one-fifth that of natural rubber) and high energy absorption.

Chlorobutyl rubber is chlorinated isobutylene–isoprene. It has the same general properties as butyl rubber but with slightly higher allowable operating temperatures.

A. Physical and Mechanical Properties

The single outstanding physical property of butyl rubber is its impermeability. It does not permit gases like hydrogen or air to diffuse through it nearly as rapidly as ordinary rubber does, and it has excellent resistance to the aging action of air. These properties make butyl rubber valuable in the production of life jackets (inflatable type), life rafts, and inner tubes for tires.

At room temperature, the resiliency of butyl rubber is poor, but as the temperature increases, the resiliency increases. At elevated temperatures, butyl rubber exhibits good resiliency. Its abrasion resistance, tear resistance, tensile strength, and adhesion to fabrics and metals is good. Butyl rubber has a maximum continuous service temperature of 250–350°F (120–150°C) with good resistance to heat aging. Its electrical properties are generally good but not outstanding in any one category. The flame resistance of butyl rubber is poor.

The polymer has excellent flexibility but very low resilience due to its molecular structure. As a result, it absorbs a great deal of any mechanical energy that is put into it. This energy absorption is responsible for butyl rubber's vibration-damping properties, which are maintained over a broad temperature range. Consequently, butyl rubber is an excellent choice where high vibration damping is required. This property is useful in automobile body mounts and suspension bumpers.

Table 4.15 lists the physical and mechanical properties of butyl rubber.

There are two commercially available halogenated butyl rubber derivatives bromobutyl (BIIR) and chlorobutyl (CIIR). The halogen atoms are incorporated into the polymer on the isoprene units. These compounds have chemical properties that permit the polymers to be covulcanized with other elastomers more readily than IIR. The end-use properties and applications are similar to those for IIR. A typical application is the cover for air conditioning hoses.

Chlorobutyl (CIIR) rubbers have a maximum operating temperature of 300°F (150°C) and can be operated as low as −30°F (−34°C). The other physical and mechanical properties are similar to those of butyl rubber.

TABLE 4.15 Physical and Mechanical Properties of Butyl Rubber (IIR)[a]

Specific gravity	0.91
Dielectric strength (V/mm)	25,000
Tensile strength (psi)	500–3,000
Hardness, Shore A	15–90
Abrasion resistance	Excellent
Maximum temperature, continuous use	250–300°F (120–148°C)
Machining qualities	Can be ground
Resistance to sunlight	Excellent
Effect of aging	Highly resistant
Resistance to heat	Stiffens slightly

[a] These are representative values, and they may be altered by compounding.

B. Resistance to Sun, Weather, and Ozone

Butyl rubber has excellent resistance to sun, weather, and ozone. Its weathering qualities are outstanding, as is its resistance to water absorption.

C. Chemical Resistance

Butyl rubber is very nonpolar. It has exceptional resistance to dilute mineral acids, alkalies, phosphate ester oils, acetaone, ethylene, ethylene glycol, and water. Resistance to concentrated acids, except nitric and sulfuric, is good. Unlikle natural rubber, it is very resistant to swelling by vegetable and animal oils. It has poor resistance to petroleum oils, gasoline, and most solvents (except oxygenated solvents).

CIIR has the same general resistance as natural rubber but can be used at higher temperatures. Unlike butyl rubber, CIIR cannot be used with hydrochloric acid.

Refer to Table 4.16 and Appendix 1 for the compatibility of butyl rubber with selected corrodents and to Table 4.17 for the compatibility of chlorobutyl rubber with selected corrodents.

D. Applications

Because of its impermeability, butyl rubber finds many uses in the manufacture of inflatable items such as life jackets, life boats, balloons, and inner tubes. The excellent resistance it exhibits in the presence of water and steam makes it suitable for hoses and diaphragms. Applications are also found as flexible

TABLE 4.16 Compatibility of Butyl Rubber with Selected Corrodents[a]

Chemical	Maximum temp. °F	°C	Chemical	Maximum temp. °F	°C
Acetaldehyde	80	27	Borax	190	88
Acetic acid 10%	150	66	Boric acid	150	66
Acetic acid 50%	110	43	Butyl acetate	x	x
Acetic acid 80%	110	43	Butyl alcohol	140	60
Acetic acid, glacial	x	x	Butyric acid	x	x
Acetic anhydride	x	x	Calcium bisulfite	120	49
Acetone	100	38	Calcium carbonate	150	66
Acrylonitrile	x	x	Calcium chlorate	190	88
Adipic acid	x	x	Calcium chloride	190	88
Allyl alcohol	190	88	Calcium hydroxide 10%	190	88
Allyl chloride	x	x	Calcium hydroxide, sat.	190	88
Alum	200	93	Calcium hypochlorite	x	x
Aluminum acetate	200	93	Calcium nitrate	190	88
Aluminum chloride, aqueous	200	93	Calcium sulfate	100	38
			Carbon dioxide, dry	190	88
Aluminum chloride, dry	200	93	Carbon dioxide, wet	190	88
Aluminum fluoride	180	82	Carbon disulfide	190	88
Aluminum hydroxide	100	38	Carbon monoxide	x	x
Aluminum nitrate	100	38	Carbon tetrachloride	90	32
Aluminum sulfate	200	93	Carbonic acid	150	66
Ammonium bifluoride	x	x	Cellosolve	150	66
Ammonium carbonate	190	88	Chloroacetic acid, 50% water	150	66
Ammonium chloride 10%	200	93			
Ammonium chloride 50%	200	93	Chloroacetic acid	100	38
Ammonium chloride, sat.	200	93	Chlorine gas, dry	x	x
Ammonium fluoride 10%	150	66	Chlorine, liquid	x	x
Ammonium fluoride 25%	150	66	Chlorobenzene	x	x
Ammonium hydroxide 25%	190	88	Chloroform	x	x
Ammonium hydroxide, sat.	190	88	Chlorosulfonic acid	x	x
Ammonium nitrate	200	93	Chromic acid 10%	x	x
Ammonium persulfate	190	88	Chromic acid 50%	x	x
Ammonium phosphate	150	66	Citric acid 15%	190	88
Ammonium sulfate 10–40%	150	66	Citric acid, concd	190	88
Amyl acetate	x	x	Copper chloride	150	66
Amyl alcohol	150	66	Copper sulfate	190	88
Aniline	150	66	Cresol	x	x
Antimony trichloride	150	66	Cupric chloride 5%	150	66
Barium chloride	150	66	Cupric chloride 50%	150	66
Barium hydroxide	190	88	Cyclohexane	x	x
Barium sulfide	190	88	Dichloroethane (ethylene dichloride)	x	x
Benzaldehyde	90	32			
Benzene	x	x	Ethylene glycol	200	93
Benzenesulfonic acid 10%	90	32	Ferric chloride	175	79
Benzoic acid	150	66	Ferric chloride 50% in water	160	71
Benzyl alcohol	190	88	Ferric nitrate 10–50%	190	88
Benzyl chloride	x	x	Ferrous chloride	175	79

TABLE 4.16 Continued

Chemical	Maximum temp. °F	Maximum temp. °C	Chemical	Maximum temp. °F	Maximum temp. °C
Ferrous nitrate	190	88	Perchloric acid 10%	150	66
Fluorine gas, dry	x	x	Phenol	150	66
Hydrobromic acid, dil	125	52	Phosphoric acid 50–80%	150	66
Hydrobromic acid 20%	125	52	Salicylic acid	80	27
Hydrobromic acid 50%	125	52	Sodium chloride	200	93
Hydrochloric acid 20%	125	52	Sodium hydroxide 10%	150	66
Hydrochloric acid 38%	125	52	Sodium hydroxide 50%	150	66
Hydrocyanic acid 10%	140	60	Sodium hydroxide, concd	150	66
Hydrofluoric acid 30%	150	66	Sodium hypochlorite 20%	x	x
Hydrofluoric acid 70%	150	66	Sodium hypochlorite, concd	x	x
Hydrofluoric acid 100%	150	66	Sodium sulfide to 50%	150	66
Hypochlorous acid	x	x	Stannic chloride	150	66
Lactic acid 25%	125	52	Stannous chloride	150	66
Lactic acid, concd	125	52	Sulfuric acid 10%	200	93
Magnesium chloride	200	93	Sulfuric acid 50%	150	66
Malic acid	x	x	Sulfuric acid 70%	x	x
Methyl chloride	90	32	Sulfuric acid 90%	x	x
Methyl ethyl ketone	100	38	Sulfuric acid 98%	x	x
Methyl isobutyl ketone	80	27	Sulfuric acid 100%	x	x
Muriatic acid	x	x	Sulfuric acid, fuming	x	x
Nitric acid 5%	200	93	Sulfurous acid	200	93
Nitric acid 20%	150	66	Thionyl chloride	x	x
Nitric acid 70%	x	x	Toluene	x	x
Nitric acid, anhydrous	x	x	Trichloroacetic acid	x	x
Nitrous acid, concd	125	52	Zinc chloride	200	93
Oleum	x	x			

[a] The chemicals listed are in the pure state or in a saturated solution unless otherwise indicated. Compatibility is shown to the maximum allowable temperature for which data are available. Incompatibility is shown by an x. A blank space indicates that data are unavailable.
Source: Schweitzer PA. Corrosion Resistance Tables, 4th ed., Vols. 1–3. New York: Marcel Dekker, 1995.

electrical insulation, shock and vibration absorbers, curing bags for tire vulcanization, and molding.

4.8 CHLOROSULFONATED POLYETHYLENE RUBBER (HYPALON)

Chlorosulfonated polyethylene synthetic rubber (CSM) is manufactured by DuPont under the trade name Hypalon. In many respects it is similar to neoprene, but it does possess some advantages over neoprene in certain types of service. It

TABLE 4.17 Compatibility of Chlorobutyl Rubber with Selected Corrodents[a]

Chemical	Maximum temp.		Chemical	Maximum temp.	
	°F	°C		°F	°C
Acetic acid 10%	150	60	Cupric chloride 5%	150	66
Acetic acid 50%	150	60	Cupric chloride 50%	150	66
Acetic acid 80%	150	60	Ethylene glycol	200	93
Acetic acid, glacial	x	x	Ferric chloride	175	79
Acetic anhydride	x	x	Ferric chloride 50% in water	100	38
Acetone	100	38	Ferric nitrate 10–50%	160	71
Alum	200	93	Ferrous chloride	175	79
Aluminum chloride,			Hydrobromic acid, dil	125	52
aqueous	200	93	Hydrobromic acid 20%	125	52
Aluminum nitrate	190	88	Hydrobromic acid 50%	125	52
Aluminum sulfate	200	93	Hydrochloric acid 20%	x	x
Ammonium carbonate	200	93	Hydrochloric acid 38%	x	x
Ammonium chloride 10%	200	93	Hydrofluoric acid 70%	x	x
Ammonium chloride 50%	200	93	Hydrofluoric acid 100%	x	x
Ammonium chloride, sat.	200	93	Lactic acid 25%	125	52
Ammonium nitrate	200	93	Lactic acid, concd	125	52
Ammonium phosphate	150	66	Magnesium chloride	200	93
Ammonium sulfate 10–40%	150	66	Nitric acid 5%	200	93
Amyl alcohol	150	66	Nitric acid 20%	150	66
Aniline	150	66	Nitric acid 70%	x	x
Antimony trichloride	150	66	Nitric acid, anhydrous	x	x
Barium chloride	150	66	Nitrous acid, concd	125	52
Benzoic acid	150	66	Phenol	150	66
Boric acid	150	66	Phosphoric acid 50–80%	150	66
Calcium chloride	160	71	Sodium chloride	200	93
Calcium nitrate	160	71	Sodium hydroxide 10%	150	66
Calcium sulfate	160	71	Sodium sulfide to 50%	150	66
Carbon monoxide	100	38	Sulfuric acid 10%	200	93
Carbonic acid	150	66	Sulfuric acid 70%	x	x
Chloroacetic acid	100	38	Sulfuric acid 90%	x	x
Chromic acid 10%	x	x	Sulfuric acid 98%	x	x
Chromic acid 50%	x	x	Sulfuric acid 100%	x	x
Citric acid 15%	90	32	Sulfuric acid, fuming	x	x
Copper chloride	150	66	Sulfurous acid	200	93
Copper cyanide	160	71	Zinc chloride	200	93
Copper sulfate	160	71			

[a]The chemicals listed are in the pure state or in a saturated solution unless otherwise indicated. Compatibility is shown to the maximum allowable temperature for which data are available. Incompatibility is shown by an x. A blank space indicates that data are unavailable. *Source:* Schweitzer PA. Corrosion Resistance Tables, 4th ed., Vols. 1–3. New York: Marcel Dekker, 1995.

has better heat and ozone resistance, better electrical properties, better color stability, and better chemical resistance.

Hypalon, when properly compounded, also exhibits good resistance to wear and abrasion, good flex life, high impact resistance, and good resistance to permanent deformation under heavy loading.

A. Physical and Mechanical Properties

The ability of Hypalon to retain its electrical properties after long-term exposure to heat, water immersion, and weathering is outstanding. These properties make the elastomer useful as insulation for low-voltage applications (less than 600 V), particularly as a covering for power and control cable, mine trailing cable, locomotive wire, nuclear power station cable, and motor lead wire. Because of Hypalon's outstanding weathering resistance, it is used as an outer protective jacket in high-voltage applications. The elastomer also exhibits excellent resistance to corona discharge.

Another property of Hypalon that is important in electrical applications is its ability to be colored and not discolor or fade when exposed to sunlight and ultraviolet light for long periods of time. The white raw polymer will accept any color, including light pastels, without impairing the true brilliance or hue. Because of the polymer's natural ozone resistance, it is not necessary to add antiozonates during compounding. The antiozonates are strong discoloring agents and when added to elastomers will cause colors to fade and become unstable.

When coloring agents are added to most elastomers, it is usually necessary to sacrifice some physical properties. This is not the case with Hypalon. Except in cases where the elastomer is being specially compounded for exceptionally high heat resistance or set characteristics, its physical properties will be unaffected by the addition of coloring agents. In these special cases a black material must be used if the maximum performance is to be gotten from the elastomer.

Hypalon will burn in an actual fire situation but is classified as self-extinguishing. If the flame is removed, the elastomer will stop burning. This phenomenon is due to its chlorine content, which makes it more resistant to burning than exclusively hydrocarbon polymers.

Hypalon's resistance to abrasion is superior to that of natural rubber and many other elastomers by as much as 2 to 1. It also possesses high resistance to fatigue cracking and cut growth from constant flexing. These latter properties make Hypalon suitable for products intended for dynamic operation. Good resistance to impact, crushing, cutting, gouging, and other types of physical abuse are also present in rubber parts produced from this elastomer.

The chlorine content of the elastomer protects it against the attack of microorganisms, and it will not promote the growth of mold, mildew, fungus, or bacteria. This feature is important when the elastomer is to be used in coating fabrics. To maintain this property, it is important that proper compounding procedures be followed. The addition of wax and those plasticizers that provide food for microorganisms should be avoided if the maximum resistance to mold, mildew, and fungus is to be maintained.

Fabrics coated with Hypalon are highly resistant to soiling and staining from atmospheric deposits and from abrasive contact with soiling agents. Most deposits left on the elastomeric surface can be removed by the application of soap and water. Stubborn deposits can be removed with detergents, dry cleaning fluids, bleaches, and other cleaning agents without causing damage to the elastomers. Table 4.18 lists the physical and mechanical properties of Hypalon.

Hypalon has a broad range of service temperatures with excellent thermal properties. General-purpose compounds can operate continuously at temperatures of 248–275°F (120–135°C). Special compounds can be formulated that can be used intermittently up to 302°F (150°C).

On the low-temperature side, conventional compounds can be used continuously down to 0 to −20°F (−18 to −28°C). Special compounds can be produced that will retain their flexibility down to −40°F (−40°C), but to produce such a compound, it is necessary to sacrifice performance of some of the other properties.

TABLE 4.18 Physical and Mechanical Properties of Chlorosulfonated Polyethylene (Hypalon; CSM)[a]

Specific gravity	1.08–1.28
Brittle point	−40°F to −80°F
	(−40°C to −62°C)
Dielectric strength, V/mil	500
Dielectric constant at 1,000 Hz	8^{-10}
Dissipation factor at 1,000 Hz	0.05–0.07
Tensile strength (psi)	2,500
Elongation (% at break)	430–540
Hardness, Shore A	60
Abrasion resistance	Excellent
Maximum temperature, continuous use	250°F (121°C)
Impact resistance	Good
Compression set (%)	
at 158°F (70°C)	16
at 212°F (100°C)	25
at 250°F (121°C)	44
Resistance to sunlight	Excellent
Effect of aging	None
Resistance to heat	Good

[a] These are representative values, and they may be altered by compounding.

Heat aging does not have any effect on the tensile strength of Hypalon, since it acts as additional heat curing. However, the elongation at break does not decrease as the temperature increases.

Hypalon exhibits good recovery from deformation after being subjected to a heavy load or a prolonged deflection. Refer to Table 4.18 for compression set values. Refer to Table 4.19 and Appendix 1 for the compatibility of Hypalon with selected corrodents.

B. Resistance to Sun, Weather, and Ozone

Hypalon is one of the most weather-resistance elastomers available. Oxidation takes place at a very slow rate. Sunlight and ultraviolet light have little, if any, adverse effect on its physical properties. It is also inherently resistant to ozone attack without the need for the addition of special antioxidants or antiozonants to the formulation.

Many elastomers are degraded by ozone concentrations of less than 1 part per million parts of air. Hypalon, however, is unaffected by concentrations as high as 1 part per 100 parts of air.

C. Chemical Resistance

When properly compounded, Hypalon is highly resistant to attack by hydrocarbon oils and fuels, even at elevated temperatures. It is also resistant to such oxidizing chemicals as sodium hypochlorite, sodium peroxide, ferric chloride, and sulfuric, chromic, and hydrofluoric acids. Concentrated hydrochloric acid (37%) at elevated temperatures above 158°F (70°C) will attack Hypalon but can be handled without adverse effect at all concentrations below this temperature. Nitric acid at room temperature and up to 60% concentration can also be handled without adverse effects.

Hypalon is also resistant to salt solutions, alcohols, and both weak and concentrated alkalies and is generally unaffected by soil chemicals, moisture, and other deteriorating factors associated with burial in the earth. Long-term contact with water has little or no effect on Hypalon. It is also resistant to radiation.

Hypalon has poor resistance to aliphatic, aromatic, and chlorinated hydrocarbons, aldehydes, and ketones.

D. Applications

Hypalon finds useful applications in many industries and many fields. Because of its outstanding resistance to oxidizing acids, it has found widespread use as an acid transfer hose. For the same reason, it is used to line railroad tanks cars and other tanks containing acids and other oxidizing chemicals. Its physical and

TABLE 4.19 Compatibility of Hypalon with Selected Corrodents[a]

Chemical	Maximum temp. °F	Maximum temp. °C	Chemical	Maximum temp. °F	Maximum temp. °C
Acetaldehyde	60	16	Borax	200	93
Acetamide	x	x	Boric acid	200	93
Acetic acid 10%	200	93	Bromine gas, dry	60	16
Acetic acid 50%	200	93	Bromine gas, moist	60	16
Acetic acid 80%	200	93	Bromine, liquid	60	16
Acetic acid, glacial	x	x	Butadiene	x	x
Acetic anhydride	200	93	Butyl acetate	60	16
Acetone	x	x	Butyl alcohol	200	93
Acetyl chloride	x	x	Butyric acid	x	x
Acrylonitrile	140	60	Calcium bisulfide	200	93
Adipic acid	140	60	Calcium carbonate	90	32
Allyl alcohol	200	93	Calcium chlorate	90	32
Aluminum fluoride	200	93	Calcium chloride	200	93
Aluminum hydroxide	200	93	Calcium hydroxide 10%	200	93
Aluminum nitrate	200	93	Calcium hydroxide, sat.	200	93
Aluminum sulfate	180	82	Calcium hypochlorite	200	93
Ammonia gas	90	32	Calcium nitrate	100	38
Ammonium carbonate	140	60	Calcium oxide	200	93
Ammonium chloride 10%	190	88	Calcium sulfate	200	93
Ammonium chloride 50%	190	88	Caprylic acid	x	x
Ammonium chloride, sat.	190	88	Carbon dioxide, dry	200	93
Ammonium fluoride 10%	200	93	Carbon dioxide, wet	200	93
Ammonium hydroxide 25%	200	93	Carbon disulfide	200	93
Ammonium hydroxide, sat.	200	93	Carbon monoxide	x	x
Ammonium nitrate	200	93	Carbon tetrachloride	200	93
Ammonium persulfate	80	27	Carbonic acid	x	x
Ammonium phosphate	140	60	Chloroacetic acid	x	x
Ammonium sulfate 10–40%	200	93	Chlorine gas, dry	x	x
			Chlorine gas, wet	90	32
Ammonium sulfide	200	93	Chlorobenzene	x	x
Amyl acetate	60	16	Chloroform	x	x
Amyl alcohol	200	93	Chlorosulfonic acid	x	x
Amyl chloride	x	x	Chromic acid 10%	150	66
Aniline	140	60	Chromic acid 50%	150	66
Antimony trichloride	140	60	Chromyl chloride		
Barium carbonate	200	93	Citric acid 15%	200	93
Barium chloride	200	93	Citric acid, concd	200	93
Barium hydroxide	200	93	Copper acetate	x	x
Barium sulfate	200	93	Copper chloride	200	93
Barium sulfide	200	93	Copper cyanide	200	93
Benzaldehyde	x	x	Copper sulfate	200	93
Benzene	x	x	Cresol	x	x
Benzenesulfonic acid 10%	x	x	Cupric chloride 5%	200	93
Benzoic acid	200	93	Cupric chloride 50%	200	93
Benzyl alcohol	140	60	Cyclohexane	x	x
Benzyl chloride	x	x	Cyclohexanol	x	x

TABLE 4.19 Continued

Chemical	Maximum temp.		Chemical	Maximum temp.	
	°F	°C		°F	°C
Dichloroethane (ethylene dichloride)	x	x	Nitric acid 70%	x	x
			Nitric acid, anhydrous	x	x
Ethylene glycol	200	93	Oleum	x	x
Ferric chloride	200	93	Perchloric acid 10%	100	38
Ferric chloride 50% in water	200	93	Perchloric acid 70%	90	32
Ferric nitrate 10–50%	200	93	Phenol	x	x
Ferrous chloride	200	93	Phosphoric acid 50–80%	200	93
Fluorine gas, dry	140	60	Picric acid	80	27
Hydrobromic acid, dil	90	32	Potassium bromide 30%	200	93
Hydrobromic acid 20%	100	38	Sodium carbonate	200	93
Hydrobromic acid 50%	100	38	Sodium chloride	200	93
Hydrochloric acid 20%	160	71	Sodium hydroxide 10%	200	93
Hydrochloric acid 38%	140	60	Sodium hydroxide 50%	200	93
Hydrocyanic acid 10%	90	32	Sodium hydroxide, concd	200	93
Hydrofluoric acid 30%	90	32	Sodium hypochlorite 20%	200	93
Hydrofluoric acid 70%	90	32	Sodium hypochlorite, concd		
Hydrofluoric acid 100%	90	32	Sodium sulfide to 50%	200	93
Hypochlorous acid	x	x	Stannic chloride	90	32
Ketones, general	x	x	Stannous chloride	200	93
Lactic acid 25%	140	60	Sulfuric acid 10%	200	93
Lactic acid, concd	80	27	Sulfuric acid 50%	200	93
Magnesium chloride	200	93	Sulfuric acid 70%	160	71
Manganese chloride	180	82	Sulfuric acid 90%	x	x
Methyl chloride	x	x	Sulfuric acid 98%	x	x
Methyl ethyl ketone	x	x	Sulfuric acid 100%	x	x
Methyl isobutyl ketone	x	x	Sulfurous acid	160	71
Muriatic acid	140	60	Toluene	x	x
Nitric acid 5%	100	38	Zinc chloride	200	93
Nitric acid 20%	100	38			

[a] The chemicals listed are in the pure state or in a saturated solution unless otherwise indicated. Compatibility is shown to the maximum allowable temperature for which data are available. Incompatibility is shown by an x. A blank space indicates that data are unavailable. *Source:* Schweitzer PA. Corrosion Resistance Tables, 4th ed., Vols. 1–3. New York: Marcel Dekker, 1995.

mechanical properties make it suitable for use in hoses undergoing continuous flexing and/or those carrying hot water or steam.

The electrical industry makes use of Hypalon to cover automotive ignition and primary wire, nuclear power station cable, control cable, and welding cable. As an added protection from storms at sea, power and lighting cable on offshore oil platforms are sheathed with Hypalon. Because of its heat and radiation resistance, it is also used as a jacketing material on heating cable embedded in roadways to melt ice and on X-ray machine cable leads. It is also used in applicance cords, insulating hoods and blankets, and many other electrical accessories.

In the automotive industry, advantage is taken of Hypalon's color stability and good weathering properties by using the elastomer for exterior parts on cars, trucks, and other commercial vehicles. Its resistance to heat, ozone, oil, and grease makes it useful for application under the hood for such components as emission control hose, tubing, ignition wire jacketing, spark plug boots, and air-conditioning and power-steering hoses. The ability to remain soil-free and easily cleanable makes it suitable for tire whitewalls.

When combined with cork, Hypalon provides a compressible yet set-resistant gasket suitable for automobile crankcase and rocker pans. Hypalon protects the cork from oxidation at elevated temperatures and also provides excellent resistance to oil, grease, and fuels.

The construction industry has made use of Hypalon for sheet roofing, pond liners, reservoir covers, curtain-wall gaskets, floor tiles, escalator rails, and decorative and maintenance coatings. In these applications the properties of color stability, excellent weatherability, abrasion resistance, useful temperature range, light weight, flexibility, and good aging characteristics are of importance.

Application is also found in the coating of fabrics that are used for inflatable structures, flexible fuel tanks, tarpaulins, and hatch and boat covers. These produces offer the advantages of being lightweight and colorful. Consumer items such as awnings, boating garb, convertible tops, and other products also make use of fabrics coated with this elastomer.

4.9 POLYBUTADIENE RUBBER (BR)

Butadiene ($CH_2=CH-CH=CH_2$) has two unsaturated linkages and can be polymerized readily. When butadiene or its derivatives becomes polymerized, the units link together to form long chains that each contains over 1,000 units. Simple butadiene does not yield a good grade of rubber, apparently because the chains are too smooth. Better results are obtained by introducing side groups into the chain either by modifying butadiene or by making a copolymer of butadiene and some other compound.

A. Physical and Mechanical Properties

Polybutadiene, designated as BR, is very similar to butadiene–styrene rubbers but is extremely difficult to process. As a result, it is widely used as an admixture with Buna-S and other elastomers. It is rarely used in an amount larger than 75% of the total polymer in a compound.

It has outstanding properties of resilience and hysteresis almost equivalent to that of natural rubber, excellent abrasion resistance, and good resistance to

TABLE 4.20   Physical and Mechanical Properties of Polybutadiene (BR)[a]

Specific gravity	0.94
Specific heat (cal/g)	0.45
Brittle point	−68°F (−56°C)
Insulation resistance (ohms/cm)	10^{17}
Coefficient of linear expansion at 32–140°F (in./in.-°F)	0.000036
Dielectric constant at 50 Hz	2.9
Power factor at 50 Hz	7×10^{-4}
Swelling (% by volume)	
in kerosene at 77°F (25°C)	200
in benzene at 77°F (25°C)	200
in acetone at 77°F (25°C)	25
in mineral oil at 100°F (70°C)	120
Tensile strength (psi)	2,000–3,000
Elongation (% at break)	700–750
Hardness, Shore A	45–80
Abrasion resistance	Excellent
Maximum temperature, continuous use	200°F (93°C)
Impact resistance	Excellent
Compression set	Good
Resilience (%)	90
Tear resistance (psi)	1,600
Machining qualities	Can be ground
Resistance to sunlight	Deteriorates
Resistance to heat	Poor

[a] These are representative values, and they may be altered by compounding.

water absorption and heat aging and possesses good electrical properties. Its tensile strength, tear resistance, and impermeability are all good. Polybutadiene has an operating temperature range only slightly greater than that of natural rubber, ranging from −150 to 200°F (−101 to 93°C).

Polybutadiene will burn and has poor flame resistance. Its physical and mechanical properties are given in Table 4.20.

B. Resistance to Sun, Weather, and Ozone

Although polybutadiene has good weather resistance, it will deteriorate when exposed to sunlight for prolonged periods of time. It also exhibits poor resistance to ozone.

TABLE 4.21 Compatibility of Polybutadiene (BR) with Selected Corrodents[a]

Chemical	Maximum temp.		Chemical	Maximum temp.	
	°F	°C		°F	°C
Alum	90	32	Nitric acid 10%	80	27
Alum ammonium	90	32	Nitric acid 20%	80	27
Alum ammonium sulphate	90	32	Nitric acid 30%	80	27
Alum chrome	90	32	Nitric acid 40%	x	x
Alum potassium	90	32	Nitric acid 50%	x	x
Alum chloride, aqueous	90	32	Nitric acid 70%	x	x
Aluminum sulfate	90	32	Nitric acid, anhydrous	x	x
Ammonia gas	90	32	Nitrous acid, concd	80	27
Ammonium chloride 10%	90	32	Ozone	x	x
Ammonium chloride 28%	90	32	Phenol	80	27
Ammonium chloride 50%	90	32	Sodium bicarbonate to 20%	90	32
Ammonium chloride, sat.	90	32	Sodium bisulfate	80	27
Ammonium nitrate	90	32	Sodium bisulfite	90	32
Ammonium sulfate 10–40%	90	32	Sodium carbonate	90	32
Calcium chloride, sat.	80	27	Sodium chlorate	80	27
Calcium hypochlorite, sat.	90	32	Sodium hydroxide 10%	90	32
Carbon dioxide, wet	90	32	Sodium hydroxide 15%	90	32
Chlorine gas, wet	x	x	Sodium hydroxide 30%	90	32
Chrome alum	90	32	Sodium hydroxide 50%	90	32
Chromic acid 10%	x	x	Sodium hydroxide 70%	90	32
Chromic acid 30%	x	x	Sodium hydroxide, concd	90	32
Chromic acid 40%	x	x	Sodium hypochlorite to 20%	90	32
Chromic acid 50%	x	x	Sodium nitrate	90	32
Copper chloride	90	32	Sodium phosphae, acid	90	32
Copper sulfate	90	32	Sodium phosphate, alkaline	90	32
Fatty acids	90	32	Sodium phosphate, neutral	90	32
Ferrous chloride	90	32	Sodium silicate	90	32
Ferrous sulfate	90	32	Sodium sulfide to 50%	90	32
Hydrochloric acid, dil	80	27	Sodium sulfite 10%	90	32
Hydrochloric acid 20%	90	32	Sodium dioxide, dry	x	x
Hydrochloric acid 35%	90	32	Sulfur trioxide	90	32
Hydrochloric acid 38%	90	32	Sulfuric acid 10%	80	27
Hydrochloric acid 50%	90	32	Sulfuric acid 30%	80	27
Hydrochloric acid fumes	90	32	Sulfuric acid 50%	80	27
Hydrogen peroxide 90%	90	32	Sulfuric acid 60%	80	27
Hydrogen sulfide, dry	90	32	Sulfuric acid 70%	90	32
Nitric acid 5%	80	27	Toluene	x	x

[a]The chemicals listed are in the pure state or in a saturated solution unless otherwise indicated. Compatibility is shown to the maximum allowable temperature for which data are available. Incompatibility is shown by an x. A blank space indicates that data are unavailable.
Source: Schweitzer PA. Corrosion Resistance Tables, 4th ed., Vols. 1–3. New York: Marcel Dekker, 1995.

C. Chemical Resistance

The chemical resistance of polybutadiene is similar to that of natural rubber. It shows poor resistance to aliphatic and aromatic hydrocarbons, oil, and gasoline but displays fair to good resistance in the presence of mineral acids and oxygenated compounds. Refer to Table 4.21 for the compatibility of polybutadiene rubber with selected corrodents.

D. Applications

Very rarely is polybutadiene used by itself. It is generally used as a blend with other elastomers to impart better resiliency, abrasion resistance, and/or low-temperature properties, particularly in the manufacture of automobile tire trends, shoe heels and soles, gaskets, seals, and belting.

4.10 ETHYLENE–ACRYLIC (EA) RUBBER

Ethylene–acrylic rubber is produced from ethylene and acrylic acid. As with other synthetic elastomers, the properties of the EA rubbers can be altered by compounding. Basically EA is a cost-effective, hot-oil-resistant rubber with good low-temperature properties.

A. Physical and Mechanical Properties

Ethylene–acrylic elastomers have good tear strength and tensile strength and high elongation at break. In addition, exceptionally low compression set values are an added advantage, making the product suitable for many hose, sealing, and cut gasket applications. A unique feature is their practically constant damping characteristics over broad ranges of temperature, frequency, and amplitude. Very little change in damping values takes place between −4 and 320°F (−20 and 160°C). This property, which shows up as a poor rebound in resiliency tests, is actually a design advantage. Combined with EA's heat and chemical resistance, it allows the use of EA in vibration-damping applications. This elastomer provides heat resistance surpassed by only the most expensive polymers such as the fluorocarbon of fluorosilicone elastomers. In measurements of dry heat resistance, EA outlasts other moderately priced oil-resistant rubbers. Parts retain elasticity and remain functional after continuous air-oven exposures from 18 months at 250°F (121°C) to 7 days at 400°F (204°C). Parts fabricated of EA will perform at least as long as parts made of Hypalon or general-purpose nitrile rubber but at exposure temperatures 50–100°F (27°C) higher.

The low-temperature performance of EA is inherently superior to that of most other heat- and oil-resistant rubbers, including standard fluoroelastomers, chlorosulfonated polyethylene, polyacrylates, and polyepichlorhydrin. Typical

TABLE 4.22 Physical and Mechanical Properties of Ethylene–
Acrylic (EA) Rubber[a]

Specific gravity	1.08–1.12
Hardness, Shore A	40–95
Tensile strength (psi)	2,500
Elongation (% at break)	650
Compression set (%)	Good
Tear resistance	Good
Maximum temperature, continuous use	340°F (170°C)
Brittle point	−75°F (−60°C)
Water absorption (%/24 hr)	Very low
Abrasion resistance	Excellent
Resistance to sunlight	Excellent
Effect of aging	Nil
Resistance to heat	Excellent
Dielectric strength	Good
Electrical insulation	Fair to good
Permeability to gases	Very low

[a]These are representative values, and they may be altered by
compounding.

unplasticized compounds are flexible to −20°F (−29°C) and have brittle points
as low as −75°F (−60°C). Compounding EA with ester plasticizers will extend
its low-temperature flexibility limits to −50°F (−46°C). When exposed to flame,
EA has poor flame resistance but does have low smoke emission.

The physical and mechanical properties of ethylene-acrylic rubber are given
in Table 4.22.

B. Resistance to Sun, Weather, and Ozone

EA elastomers have extremely good resistance to sun, weather, and ozone. Long-
term exposures have no effect on these rubbers.

C. Chemical Resistance

Ethylene–acrylic elastomers exhibit very good resistance to hot oils and hydro-
carbon- or glycol-based proprietary lubricants and to transmission and power
steering fluids. The swelling characteristics of EA will be retained better than
those of the silicone rubbers after oil immersion.

Ethylene–acrylic rubber also has outstanding resistance to hot water. Its
resistance to water absorption is very good. Good resistance is also displayed to
dilute acids, aliphatic hydrocarbons, gasoline, and animal and vegetable oils.

Ethylene–acrylic rubber is not recommended for immersion in esters, ketones, highly aromatic hydrocarbons, or concentrated acids. Neither should it be used in applications calling for long-term exposure to high-pressure steam.

D. Applications

Ethylene–acrylic rubber is used in such products as gaskets, hoses, seals, boots, damping components, low-smoke floor tiling, and cable jackets for offshore oil platforms, ships, and building plenum installations. Ethylene–acrylic rubber in engine parts provides good resistance to heat, fluids, and wear, as well as good low-temperature sealing ability.

4.11 ACRYLATE–BUTADIENE RUBBER (ABR) AND ACRYLIC ESTER–ACRYLIC HALIDE (ACM) RUBBERS

Acrylate–butadiene and acrylic ester–acrylic halide rubbers are similar to ethylene–acrylic rubbers.

A. Physical and Mechanical Properties

The ABRs and ACM rubbers exhibit good resilience and tear resistance but poor impact resistance. Abrasion resistance and compression set are good. The maximum temperatures rating is 340°F (170°C), the same as for the EA rubbers.

Table 4.23 lists the physical and mechanical properties of ACM rubbers. The polymer backbone of an ACM rubber contains alkyl or alkoxy esters of acrylic acid plus a small amount of a reactive comonomer that introduces sites for

TABLE 4.23 Physical and Mechanical Properties of Acrylic Ester–Acrylic Halide (ACM) Rubbers[a]

Specific gravity	1.1
Hardness range, Shore A	45–90
Tensile strength (psi)	2,175
Elongation (% at break)	400
Compression set (%)	Fair
Tear resistance	Good
Maximum temperature, continuous use	340°F (170°C)
Electrical properties	Poor to fair
Abrasion resistance	Fair to good
Permeability to gases	Low
Resistance to sunlight	Good
Resistance to heat	Excellent

[a] These are representative values, and they may be altered by compounding.

cross-linking. Because of its fully saturated backbone, the polymer has excellent resistance to heat, oxidation, and ozone. It is one of the few elastomers with higher heat resistance than EPDM.

B. Resistance to Sun, Weather, and Ozone

Acrylate-butadiene and acrylic ester–acrylic halide rubbers exhibit good resistance to sun, weather, and ozone.

C. Chemical Resistance

Acrylate–butadiene and ACM rubbers have excellent resistance to aliphatic hycrocarbons (gasoline, kerosene) and offer good resistance to water, acids, synthetic lubricants, and silicate hydraulic fluids. They are unsatisfactory for use in contact with alkali, aromatic hydrocarbons (benzene, toluene), halogenated hydrocarbons, alcohol, and phosphate hydraulic fluids.

D. Applications

These rubbers are used where resistance to atmospheric conditions and heat are required.

4.12   ETHYLENE–PROPYLENE RUBBERS (EPDM AND EPT)

Ethylene–propylene rubber is a synthetic hydrocarbon-based rubber made either from ethylene–propylene diene monomer or ethylene–propylene terpolymer. These monomers are combined in such a manner as to produce an elastomer with a completely saturated backbone and pendant unsaturation for sulfur vulcanization. As a result of this configuration, vulcanizates of EPDM elastomers are extremely resistant to attack by ozone, oxygen, and weather.

Ethylene–propylene rubber possesses many properties superior to those of natural rubber and conventional general-purpose elastomers. In some applications, it will perform better than other materials, while in other applications it will last longer or require less maintenance and may even cost less.

EPDM has exceptional heat resistance, being able to operate at temperatures of 300–350°F (148–176°C), while also finding application at temperatures as low as −70°F (−56°C). Experience has shown that EPDM has exceptional resistance to steam. Hoses manufactured from EPDM have had lives several times longer than that of hoses manufactured from other elastomers.

The dynamic properties of EPDM remain constant over a wide temperature range, making this elastomer suitable for a variety of applications. It also has very high resistance to sunlight, aging, and weather, excellent electrical properties, good mechanical properties, and good chemical resistance. However, being hydrocarbon based, it is not resistant to petroleum-based oils or flame.

This material may be processed and vulcanized by the same techniques and with the same equipment as those used for processing other general-purpose elastomers. As with other elastomers, compounding plays an important part in tailoring the properties of EPDM to meet the needs of a specific application. Each of the properties of the elastomer can be enhanced or reduced by the addition or deletion of chemicals and fillers. Because of this, the properties discussed must be considered in general terms.

Ethylene–propylene terpolymer (EPT) is a synthetic hydrocarbon-based rubber produced from an ethylene–propylene terpolymer. It is very similar in physical and mechanical properties to EPDM.

A. Physical and Mechanical Properties

EPDM is available in several grades. The different grades are produced by varying the ethylene–propylene ratio and the type and amount of diene monomer. The three diene monomers which are used are dicyclopentadiene, norbornadiene, and hexadiene. These factors can affect the processibility of the compound as well as the performance properties of the vulcanized elastomer. EPDM can accept a much greater loading of extending oils and mineral fillers than any other elastomer. This offers opportunities to manufacture compounds in which the polymer content is only a small fraction of the total formulation. Thus EPDM is susceptible to the production of off-specification material resulting from excessive incorporation of inexpensive fillers. Excess filler content will reduce the mechanical properties of the compound and lead to inferior set resistance.

It is possible to compound EPDM to provide either high resilience or high damping. When compounded to provide high resilience, the products are similar to natural rubber in liveliness and minimum hysteresis values. The energy-absorbing compounds have low resilience values approaching those of specialty elastomers used for shock and vibration damping. Whether the compound has been compounded for resilience or high damping, its properties remain relatively constant over a wide temperature range. As can be seen from Table 4.24 a temperature variation of 200°F (110°C) has little effect on the resilience of the compound. The isolation efficiency (based on the percentage of disturbing force transmitted) over a temperature range of 0–180°F (−18 to 82°C) varies by only 10%. This property is particularly important in vibration-isolation applications such as in automotive body mounts.

EPDM exhibits little sensitivity to changes in load. When properly compounded, it also has excellent resistance to creep under both static and dynamic conditions.

TABLE 4.24 Physical and Mechanical Properties of Ethylene–
Propylene Rubber (EDPM)[a]

Specific gravity	0.85
Specific heat (cal/g)	0.56
Brittle point	−90°F (−68°C)
Resilience (%)	
at 212°F (100°C)	78
at 75°F (24°C)	77
at 14°F (−10°C)	63
Dielectric strength (V/mil)	800
Insulation resistance (megohms/1,000 ft)	25,500
Insulation resistance, constant K, (megohms/1,000 ft)	76,400
Permeability to air at 86°F (30°C), (cm^3/cm^2-cm-sec-atm)	8.5×10^{-8}
Tensile strength (psi)	To 3,500
Elongation (% at break)	560
Hardness, Shore A	30–90
Abrasion resistance	Good
Maximum temperature, continuous use	300°F (148°C)
Impact resistance	Good
Compression set (%)	
at 158°F (70°C)	8–10
at 212°F (100°C)	12–26
Resistance to sunlight	Excellent
Effect of aging	Nil
Resistance to heat	Excellent
Tear resistance	Good

[a] These are representative values, and they may be altered by compounding.

Flexibility at low temperature is another advantage of this elastomer. Standard compounds have brittle points of −90°F (−68°C) or below. Special compounding can supply material with stiffness values of −90°F (−68°C) and brittle points below −100°F (−73°C).

The electrical properties of EPDM are excellent, particularly for high-voltage insulation. The properties are also stable after long periods of immersion in water. Excellent resistance is also provided against cutting caused by high-voltage corona discharge.

Ethylene–propylene rubber can be produced in any color including white and pastel shades. The color stability is excellent, with aging characteristics that are available in other elastomers only in black. Since techniques have been developed whereby the material can be painted with permanent waterproof colors, the elastomer can be produced in a black stock providing the maximum physical properties.

Ethylene–propylene rubber has relatively high resistance to heat. Standard formulations can be used continuously at temperatures of 250–300°F (121–148°C) in air. In the absence of air, such as in a steam hose lining or cable insulation covered with an outer jacket, higher temperatures can be tolerated. It is also possible by special compounding to produce material that can be used in services up to 350°F (176°C). Standard compounds can be used in intermittent service at 350°F (176°C).

Other advantageous mechanical properties of EPDM include good resistance to impact, tearing, abrasion, and cut growth over a wide temperature range. These properties make the elastomer suitable for applications that involve continuous flexing or twisting during operation.

Ethylene–propylene rubber exhibits a low degree of permanent deformation. Table 4.24 gives examples of these values. The ranges shown are for both standard and special compounds. In addition, special compounds can be supplied that will have a permanent deformation of only 26% after compression at 350°F (176°C).

B. Resistance to Sun, Weather, and Ozone

Ethylene–propylene rubber is particularly resistant to sun, weather, and ozone attack. Excellent weather resistance is obtained whether the material is formulated in color, white, or black. The elastomer remains free of surface crazing and retains a high percentage of its properties after years of exposure. Ozone resistance is inherent in the polymer, and for all practical purposes it can be considered immune to ozone attack. It is not necessary to add any special compounding ingredients to produce this resistance.

C. Chemical Resistance

Ethylene–propylene rubber resists attack from oxygenated solvents, such as acetone, methyl ethyl ketone, ethyl acetate, weak acids and alkalies, detergents, phosphate esters, alcohols, and glycols. It exhibits exceptional resistance to hot water and high-pressure steam. The elastomer, being hydrocarbon-based, is not resistant to hydrocarbon solvents and oils, chlorinated hydrocarbons, or turpentine. However, by proper compounding, its resistance to oil can be improved to provide adequate service life in many applications where such resistance is required. Ethylene–propylene terpolymer rubbers, in general, are resistant to most of the same corrodents as EPDM but do not have as broad a resistance to mineral acids and some organics.

Refer to Table 4.25 and Appendix 1 for the compatibility of EPDM with selected corrodents and to Table 4.26 and Appendix 1 for the compatibility of EPT with selected corrodents.

TABLE 4.25 Compatibility of EDPM Rubber with Selected Corrodents[a]

Chemical	Maximum temp. °F	°C	Chemical	Maximum temp. °F	°C
Acetaldehyde	200	93	Benzenesulfonic acid 10%	x	x
Acetamide	200	93	Benzoic acid	x	x
Acetic acid 10%	140	60	Benzyl alcohol	x	x
Acetic acid 50%	140	60	Benzyl chloride	x	x
Acetic acid 80%	140	60	Borax	200	93
Acetic acid, glacial	140	60	Boric acid	190	88
Acetic anhydride	x	x	Bromine gas, dry	x	x
Acetone	200	93	Bromine gas, moist	x	x
Acetyl chloride	x	x	Bromine, liquid	x	x
Acrylonitrile	140	60	Butadiene	x	x
Adipic acid	200	93	Butyl acetate	140	60
Allyl alcohol	200	93	Butyl alcohol	200	93
Allyl chloride	x	x	Butyric acid	140	60
Alum	200	93	Calcium bisulfite	x	x
Aluminum fluoride	190	88	Calcium carbonate	200	93
Aluminum hydroxide	200	93	Calcium chlorate	140	60
Aluminum nitrate	200	93	Calcium chlorite	200	93
Aluminum sulfate	190	88	Calcium hydroxide 10%	200	93
Ammonia gas	200	93	Calcium hydroxide, sat.	200	93
Ammonium bifluoride	200	93	Calcium hypochlorite	200	93
Ammonium carbonate	200	93	Calcium nitrate	200	93
Ammonium chloride 10%	200	93	Calcium oxide	200	93
Ammonium chloride 50%	200	93	Calcium sulfate	200	93
Ammonium chloride, sat.	200	93	Carbon bisulfide	x	x
Ammonium fluoride 10%	200	93	Carbon dioxide, dry	200	93
Ammonium fluoride 25%	200	93	Carbon dioxide, wet	200	93
Ammonium hydroxide 25%	100	38	Carbon disulfide	200	93
Ammonium hydroxide, sat.	100	38	Carbon monoxide	x	x
Ammonium nitrate	200	93	Carbon tetrachloride	200	93
Ammonium persulfate	200	93	Carbonic acid	x	x
Ammonium phosphate	200	93	Cellosolve	200	93
Ammonium sulfate 10–40%	200	93	Chloroacetic acid	160	71
Ammonium sulfide	200	93	Chlorine gas, dry	x	x
Amyl acetate	200	93	Chlorine gas, wet	x	x
Amyl alcohol	200	93	Chlorine, liquid	x	x
Amyl chloride	x	x	Chlorobenzene	x	x
Aniline	140	60	Chloroform	x	x
Antimony trichloride	200	93	Chlorosulfonic acid	x	x
Aqua regia 3:1	x	x	Chromic acid 50%	x	x
Barium carbonate	200	93	Citric acid 15%	200	93
Barium chloride	200	93	Citric acid, concd	200	93
Barium hydroxide	200	93	Copper acetate	100	38
Barium sulfate	200	93	Copper carbonate	200	93
Barium sulfide	140	60	Copper chloride	200	93
Benzaldehyde	150	66	Copper cyanide	200	93
Benzene	x	x	Copper sulfate	200	93

TABLE 4.25 Continued

Chemical	Maximum temp. °F	°C	Chemical	Maximum temp. °F	°C
Cresol	x	x	Methyl isobutyl ketone	60	16
Cupric chloride 5%	200	93	Nitric acid 5%	60	16
Cupric chloride 50%	200	93	Nitric acid 20%	60	16
Cyclohexane	x	x	Nitric acid 70%	x	x
Cyclohexanol	x	x	Nitric acid, anhydrous	x	x
Dichloroethane (ethylene dichloride)	x	x	Oleum	x	x
			Perchloric acid 10%	140	60
Ethylene glycol	200	93	Phosphoric acid 50–80%	140	60
Ferric chloride	200	93	Picric acid	200	93
Ferric chloride 50% in water	200	93	Potassium bromide 30%	200	93
Ferric nitrate 10–50%	200	93	Salicylic acid	200	93
Ferrous chloride	200	93	Sodium carbonate	200	93
Ferrous nitrate	200	93	Sodium chloride	140	60
Fluorine gas, moist	60	16	Sodium hydroxide 10%	200	93
Hydrobromic acid, dil	90	32	Sodium hydroxide 50%	180	82
Hydrobromic acid 20%	140	60	Sodium hydroxide, concd	180	82
Hydrobromic acid 50%	140	60	Sodium hypochlorite 20%	200	93
Hydrochloric acid 20%	100	38	Sodium hypochlorite, concd	200	93
Hydrochloric acid 38%	90	32	Sodium sulfide to 50%	200	93
Hydrocyanic acid 10%	200	93	Stannic chloride	200	93
Hydrofluoric acid 30%	60	16	Stannous chloride	200	93
Hydrofluoric acid 70%	x	x	Sulfuric acid 10%	150	66
Hydrofluoric acid 100%	x	x	Sulfuric acid 50%	150	66
Hypochlorous acid	200	93	Sulfuric acid 70%	140	66
Iodine solution 10%	140	60	Sulfuric acid 90%	x	x
Ketones, general	x	x	Sulfuric acid 98%	x	x
Lactic acid 25%	140	60	Sulfuric acid 100%	x	x
Lactic acid, concd			Sulfuric acid, fuming	x	x
Magnesium chloride	200	93	Toluene	x	x
Malic acid	x	x	Trichloroacetic acid	80	27
Methyl chloride	x	x	White liquor	200	93
Methyl ethyl ketone	80	27	Zinc chloride	200	93

[a]The chemicals listed are in the pure state or in a saturated solution unless otherwise indicated. Compatibility is shown to the maximum allowable temperature for which data are available. Incompatibility is shown by an x. A blank space indicates that data are unavailable.
Source: Schweitzer PA. Corrosion Resistance Tables, 4th ed., Vols. 1–3. New York: Marcel Dekker, 1995.

TABLE 4.26 Compatibility of EPT with Selected Corrodents[a]

Chemical	Maximum temp.		Chemical	Maximum temp.	
	°F	°C		°F	°C
Acetamide	200	93	Citric acid, concd	180	82
Acetic acid 10%	x	x	Copper sulfate	180	82
Acetic acid 50%	x	x	Corn oil	x	x
Acetic acid 80%	x	x	Cottonseed oil	80	27
Acetic acid, glacial	x	x	Cresol	100	38
Acetone	x	x	Diacetone alcohol	210	99
Ammonia gas	140	60	Dibutyl phthalate	100	38
Ammonium chloride 10%	180	82	Ethers, general		
Ammonium chloride 28%	180	82	Ethyl acetate	x	x
Ammonium chloride 50%	180	82	Ethyl alcohol	180	82
Ammonium chloride, sat.	180	82	Ethylene chloride	x	x
Ammonium hydroxide 10%	140	60	Ethylene glycol	180	82
Ammonium hydroxide 25%	140	60	Formaldehyde, dil	180	82
Ammonium hydroxide, sat.	140	60	Formaldehyde 37%	180	82
Ammonium sulfate 10–40%	180	82	Formaldehyde 50%	140	60
Amyl alcohol	180	82	Glycerine	180	82
Aniline	x	x	Hydrochloric acid, dil	210	99
Benzene	x	x	Hydrochloric acid 20%	x	x
Benzoic acid	140	60	Hydrochloric acid 35%	x	x
Bromine water, dil	x	x	Hydrochloric acid 38%	x	x
Bromine water, sat.	x	x	Hydrochloric acid 50%	x	x
Butane	x	x	Hydrofluoric acid dil	210	99
Butyl acetate	x	x	Hydrofluoric acid 30%	140	60
Butyl alcohol	180	82	Hydrofluoric acid 40%	x	x
Calcium chloride dil	180	82	Hydrofluoric acid 50%	x	x
Calcium chloride sat	180	82	Hydrofluoric acid 70%	x	x
Calcium hydroxide 10%	180	82	Hydrofluoric acid 100%	x	x
Calcium hydroxide 20%	180	82	Hydrogen peroxide,	x	x
Calcium hydroxide 30%	180	82	all concn		
Calcium hydroxide, sat.	210	99	Lactic acid, all concn	210	99
Cane sugar liquors	180	82	Methyl ethyl ketone	x	x
Carbon bisulfide	x	x	Methyl isobutyl ketone	x	x
Carbon tetrachloride	x	x	Monochlorobenzene	x	x
Carbonic acid	180	82	Muriatic acid	x	x
Castor oil	160	71	Nitric acid	x	x
Cellosolve	x	x	Oxalic acid	140	60
Chlorine water, sat.	80	27	Phenol	80	27
Chlorobenzene	x	x	Phosphoric acid, all concn	180	82
Chloroform	x	x	Potassium hydroxide to	210	99
Chromic acid 10%	x	x	50%		
Chromic acid 30%	x	x	Potassium sulfate 10%	210	99
Chromic acid 40%	x	x	Propane	x	x
Chromic acid 50%	x	x	Silicone oil	200	93
Citric acid 5%	180	82	Sodium carbonate	180	82
Citric acid 10%	180	82	Sodium chloride	180	82
Citric acid 15%	180	82	Sodium hydroxide to 70%	200	93

TABLE 4.26   Continued

Chemical	Maximum temp. °F	Maximum temp. °C	Chemical	Maximum temp. °F	Maximum temp. °C
Sodium hypochlorite, all concn	x	x	Water, demineralized	210	99
			Water, distilled	210	99
Sulfuric acid to 70%	200	93	Water, salt	210	99
Sulfuric acid 93%	x	x	Water, sea	210	99
Toluene	x	x	Whiskey	180	82
Trichloroethylene	x	x	Zinc chloride	180	82

[a] The chemicals listed are in the pure state or in a saturated solution unless otherwise indicated. Compatibility is shown to the maximum allowable temperature for which data are available. Incompatibility is shown by an x. A blank space indicates that data are unavailable.
Source: Schweitzer PA. Corrosion Resistance Tables, 4th ed., Vols. 1–3. New York: Marcel Dekker, 1995.

D. Applications

Extensive use is made of ethylene–propylene rubber in the automotive industry. Because of its paintability, this elastomer is used as the gap-filling panel between the grills and bumper, which provides a durable and elastic element. Under hood components such as radiator hose, ignition wire insulation, overflow tubing, window washer tubing, exhaust emission control tubing, and various other items make use of EPDM because of its resistance to heat, chemicals, and ozone. Other automotive applications include body mounts, spring mounting pads, miscellaneous body seals, floor mats, and pedal pads. Each application takes advantage of one or more specific properties of the elastomer.

Appliance manufacturers, especially washer manufacturers, have also found wide use for EPDM. Its heat and chemical resistance combined with its physical properties make it ideal for such appliances as door seals and cushions, drain and water-circulating hoses, bleach tubing, inlet nozzles, boots, seals, gaskets, diaphragms, vibration isolators, and a variety of grommets. This elastomer is also used in dishwashers, refrigerators, ovens, and a variety of small appliances.

Ethylene–propylene rubber finds application in the electrical industry and in the manufacture of electrical equipment. One of the primary applications is as an insulating material. It is used for medium-voltage (up to 35 kV) and secondary network power cable, coverings for line and multiplex distribution wire, jacketing and insulation for types S and SJ flexible cords, and insulation for automotive ignition cable.

Accessory items such as molded terminal covers, plugs, transformer connectors, line-tap and switching devices, splices, and insulating and semiconductor tape are also produced from EPDM.

Medium- and high-voltage underground power distribution cable insulated with EPDM offers many advantages. It provides excellent resistance to tearing and failure caused by high voltage contaminants and stress. Its excellent electrical properties make it suitable for high-voltage cable insulation. It withstands heavy corona discharge without sustaining damage.

Manufacturers of other industrial products take advantage of the heat and chemical resistance, physical durability, ozone resistance, and dynamic properties of EPDM. Typical applications include high-pressure steam hose, high-temperature conveyor belting, water and chemical hose, hydraulic hose for phosphate-type fluids, vibration mounts, industrial tires, tugboat and dock bumpers, tanks and pump linings, O-rings, gaskets, and a variety of molded products. Standard formulations of EPDM are also used for such consumer items as garden hose, bicycle tires, sporting goods, marine accessories, and tires for garden equipment.

4.13   STYRENE–BUTADIENE–STYRENE (SBS) RUBBER

Styrene–butadiene–styrene (SBS) rubbers are either pure or oil-modified block copolymers. They are most suitable as performance modifiers in blends with thermoplastics or as a base rubber for adhesive, sealant, or coating formulations. SBS compounds are formulations containing block copolymer rubber and other suitable ingredients. These compounds have a wide range of properties and provide the benefits of rubberiness and easy processing on standard thermoplastic processing equipment.

A.   Physical and Mechanical Properties

Since the physical and mechanical properties vary greatly depending upon the formulation, this discussion of these properties is based on the fact that proper formulation will provide a material having the desired combination of properties.

The degree of hardness will determine the flexibility of the final product. With a Short A hardness range from 37 to 74, SBS rubbers offer excellent impact resistance and low-temperature flexibility. Their maximum service temperature is 150°F (65°C). They also exhibit good abrasion resistance and good resistance to water absorption and heat aging. Their resistance to compression set and tear is likewise good, as is their tensile strength.

The electrical properties of SBS rubber are only fair, and resistance to flame is poor.

The physical and mechanical properties of SBS rubber are given in Table 4.27.

TABLE 4.27 Physical and Mechanical Properties of Styrene–Butadiene–Styrene (SBS) Rubber[a]

Specific gravity	0.92–1.09
Tear resistance	Good
Tensile strength (psi)	625–4,600
Elongation (% at break)	500–1,400
Hardness, Shore A	37–74
Abrasion resistance	Good
Maximum temperature, continuous use	150°F (65°C)
Impact resistance	Good
Compression set at 74°F (23°C) (%)	10–15
Machining qualities	Can be ground
Resistance to sunlight	Poor
Effect of aging	Little
Resistance to heat	Fair

[a] These are representative values, and they may be altered by compounding.

B. Resistance to Sun, Weather, and Ozone

SBS rubbers are not resistant to ozone, particularly when they are in a stressed condition. Neither are they resistant to prolonged exposure to sun or weather.

C. Chemical Resistance

The chemical resistance of SBS rubbers is similar to that of natural rubber. They have excellent resistance to water, acids, and bases. Prolonged exposure to hydrocarbon solvents and oils will cause deterioration; however, short exposures can be tolerated.

D. Applications

The specific formulation will determine the applicability of various products. Applications include a wide variety of general-purpose rubber items and use in the footwear industry. These rubbers are used primarily in blends with other thermoplastic materials and as performance modifiers.

4.14 STYRENE–ETHYLENE–BUTYLENE–STYRENE (SEBS) RUBBER

Styrene–ethylene–butylene–styrene (SEBS) rubbers are either pure or oil-modified block copolymer rubbers. These rubbers are used as performance modifiers in blends with thermoplastics or as the base rubber for adhesive, sealant, or coating formulations. Formulations of SEBS compounds provide a wide range of properties with the benefits of rubberiness and easy processing on standard thermoplastic processing equipment.

A. Physical and Mechanical Properties

The SEBS rubbers offer excellent impact resistance and low-temperature flexibility, are highly resistant to oxidation and ozone, and do not require vulcanization. Their degree of flexibility is a function of hardness, with their Shore A

TABLE 4.28 Physical and Mechanical Properties of Styrene–Ethylene–Butylene–Styrene (SEBS) Rubber[a]

Specific gravity	0.885–1.17
Brittle point	−58 to −148°F
	(−50 to −100°C)
Tear strength (psi)	275–470
Moisture absorption (mg/in.^{2})	1.2–3.3
Dielectric strength at 77°F (25°C)	
at 60 Hz	2.1–2.8
at 1 kHz	2.1–2.8
at 1 MHz	2.1–2.35
Dissipation factor at 77°F (25°C)	
at 60 Hz	0.0001–0.002
at 1 kHz	0.0001–0.003
at 1 MHz	0.0001–0.01
Dielectric strength (V/mil)	625–925
Volume resistivity (ohm/cm)	$9 \times 10^5 - 9.1 \times 10^{16}$
Surface resistivity (ohm)	$2 \times 10^{16} - 9.5 \times 10^{16}$
Insulation resistance constant at 60°F (15.6°C) and 500 V dc	$6.8 \times 10^4 - 2.5 \times 10^6$
Insulation resistance at 60°F (15.6°C), (megohms/1,000 ft)	$2.1 \times 10^4 - 1 \times 10^6$
Tensile strength (psi)	1,600–2,700
Elongation (% at break)	500–675
Hardness, Shore A	65–95
Abrasion resistance	Good
Maximum temperature, continuous use	220°F (105°C)
Impact resistance	Good
Resistance to sunlight	Good
Effect of aging	Small
Resistance to heat	Good

[a] These are representative values, and they may be altered by compounding.

hardness ranging from 37 to 95. These rubbers are serviceable from -120 to $220°F$ (-85 to $105°C$). They have excellent resistance to very low temperature impact and bending. Their thermal life and aging properties are excellent, as is their abrasion resistance.

The electrical properties of SEBS rubbers are extremely good. Although SEBS rubbers have poor flame resistance, compounding can improve their flame retardancy. This compounding reduces the operating temperatures slightly.

The physical and mechanical properties of SEBS rubbers are given in Table 4.28.

B. Resistance to Sun, Weather, and Ozone

SEBS rubbers and compounds exhibit excellent resistance to ozone. For prolonged outdoor exposure, the addition of an ultraviolet absorber or carbon black pigment or both is recommended.

C. Chemical Resistance

The chemical resistance of SEBS rubbers is similar to that of natural rubber. They have excellent resistance to water, acids, and bases. Soaking in hydrocarbon solvents and oils will deteriorate the rubber, but short exposures can be tolerated.

D. Applications

SEBS rubbers find applications for a wide variety of general-purpose rubber items as well as in automotive, sporting goods, and other products. Many applications are found in the electrical industry for such items as flexible cords, welding and booster cables, flame-resistant appliance wiring materials, and automotive primary wire insulation.

4.15 POLYSULFIDE RUBBERS (ST AND FA)

Polysulfide rubbers are manufactured by combining ethylene ($CH_2{=}CH_2$) with an alkaline polysulfide. The sulfur forms part of the polymerized molecule. They are also known as Thiokol rubbers. In general these elastomers do not have great elasticity, but they do have good resistance to heat and are resistant to most solvents. Compared to nitrile rubber, they have poor tensile strength, a pungent odor, poor rebound, high creep under strain, and poor abrasion resistance.

Modified organic polysulfides are made by substituting other unsaturated compounds for ethylene, which results in compounds that have little objectionable odor.

A. Physical and Mechanical Properties

ST polysulfide rubber is prepared from bis(2-chloroethyl)formal and sodium polysulfide. Products made from these rubbers have good low-temperature properties and exhibit outstanding resistance to oils and solvents, to gas permeation, ozone, and weathering. Pure gum vulcanizates possess poor physical properties, and as a result all practical compounds contain reinforcing fillers, usually carbon black.

Compounds can be used continuously at temperatures of 212°F (100°C) and intermittently at temperatures up to 300°F (140°C). The polymer does not melt or soften at elevated temperatures, but a gradual shortening in elongation and a drop in tensile strength do occur. As the temperature increases, the effect becomes more pronounced. With the addition of plasticizers, ST compounds can be made to remain flexible at tempertures below −60°F (−51°C).

Compression set values for ST elastomers are shown in Table 4.29. The use of ST rubber in compressive applications should be limited to those applications whose service temperature does not exceed 200°F (93°C).

ST polysulfide rubber is blended with nitrile rubber (NBR) and neoprene to obtain a balance of properties unattainable with either polymer alone. High ratios of ST to NBR or neoprene W decrease swelling from aromatics, fuels, ketones, and esters. The low-temperature flexibility is also improved. Higher ratios of NBR or neoprene W to ST result in improvements in physical properties, tear strength, and compressive set resistance before and after heat aging.

The electrical insulating properties of polysulfide ST are poor, as is its flame resistance.

Table 4.29 lists the physical and mechanical properties of ST polysulfide.

FA polysulfide is prepared by reacting a mixture of bis(2-chloroethyl)formal and ethylene dichloride with sodium polysulfide. Cured compounds are particularly noted for outstanding volume swell resistance to aliphatic and aromatic solvents and to alcohols, ketones, and esters; exceptionally low permeability to gases, water, and organic liquids; and excellent low-temperature flexibility.

FA polysulfide rubbers that contain no plasticizers are flexible at temperature as low as −50°F (−45°C) and will retain their excellent flexing characteristics at subnormal temperatures even when oil and solvents are present. With the addition of plasticizers or reinforcing pigments, a wide range of hardnesses can be achieved. Depending on the hardness, tensile strengths up to 1500 psi can be developed. When the materials are immersed in solvents, they retain a very high percentage of this tensile strength. FA polysulfide rubbers have a wider operating temperature range than ST elastomers. The FA series remains serviceable over a temperature range of −50 to 250°F (−45 to 121°C).

TABLE 4.29 Physical and Mechanical Properties of Polysulfide ST Rubber[a]

Specific gravity		1.27	
Brittle point		−60°F (−51°C)	
Permeability			
to helium gas (cm^3/sec-in.^{2}, 0.1-in. thick film)		1.5	
to solvents (fl oz/in.-24 h-ft^2)	75°F (24°C)		180°F (82°C)
methanol	0.005		0.140
carbon tetrachloride	0.042		0.210
ethyl acetate	0.150		0.510
benzene	0.540		1.800
diisobutylene	0.0		0.001
methyl ethyl ketone	0.28		0.65
Tensile strength (psi)		500–1,750	
Elongation (% at break)		230–450	
Hardness, Shore A		30–90	
Abrasion resistance		Good	
Maximum temperature, continuous use		212°F (100°C)	
Compression set, (% after 22 h)			
at 158°F (70°C)		15	
at 212°F (100°C)		75	
Machining qualities		Excellent	
Resistance to sunlight		Excellent	
Effect of aging		None	
Resistance to heat		Fair	

[a]These are representative values, and they may be altered by compounding.

The electrical properties of the FA polysulfide rubbers are good, but their flame resistance is poor. Table 4.30 lists the physical and mechanical properties of FA polysulfide rubber.

B. Resistance to Sun, Weather, and Ozone

FA polysulfide rubber compounds display excellent resistance to ozone, weathering, and exposure to ultraviolet light. Their resistance is superior to that of ST polysulfide ribbers. If high concentrations of ozone are to be present, the use of 0.5 part of nickel dibutyldithiocarbamate (NBC) per 100 parts of FA polysulfide rubber will improve the ozone resistance.

ST polysulfide rubber compounded with carbon black is resistant to ultraviolet light and sunlight. Its resistance to ozone is good but can be improved by the addition of NBC; however, this addition can degrade the material's

TABLE 4.30 Physical and Mechanical Properties of Polysulfide FA Rubber[a]

Specific gravity	1.34
Refractive index	1.65
Brittle point	$-30°F$ ($-35°C$)
Dielectric constant	
at 1 kHz	7.3
at 1 MHz	6.8
Dissipation factor	
at 1 kHz	5.3×10^{-3}
at 1 MHz	5.2×10^{13}
Volume resistivity (ohm-cm)	
at 73°F (23°C)	5×10^{13}
at 140°F (60°C)	2×10^{12}
Surface resistivity (ohms)	
at 73°F (23°C)	7×10^{14}
at 140°F (60°C)	2×10^{14}
Permeability at 75°F (24°C)[b]	
to methanol	0.001
to carbon tetrachloride	0.01
to ethyl acetate	0.04
to benzene	0.14
Permeability at room temperature[c]	
to hydrogen	14.8×10^{-6}
to helium	9.4×10^{-6}
Swelling (% by volume)	
in kerosene at 77°F (25°C)	4
in benzene at 77°F (25°C)	50
in acetone at 77°F (25°C)	25
in mineral oil at 158°F (70°C)	1
Tensile strength (psi)	150–1,200
Elongation (% at break)	210–700
Hardness, Shore A	25–90
Abrasion resistance	Fair
Maximum temperature, continuous use	250°F (121°C)
Machining qualities	Excellent
Resistance to sunlight	Excellent
Effect of aging	None

[a] These are representative values, and they may be altered by compounding.
[b] $\frac{1}{16}$-in. (1.6 mm) sheet, (in.-oz.-in.^{2} (24 h-ft.)
[c] 0.25 mm thick sheet in cm^3/cm^2-min.

compression set. ST polysulfide rubber also possesses satisfactory weather resistance.

C. Chemical Resistance

Polysulfide rubbers posses outstanding resistance to solvents. They exhibit excellent resistance to oils, gasoline, and aliphatic and aromatic hydrocarbon solvents, very good water resistance, goold alkali resistance, and fair acid resistance. FA polysulfide rubbers are somewhat more resistant to solvents than ST rubbers. Compounding of FA polymers with NBR will provide high resistance to aromatic solvents and improve the physical properties of the blend. For high resistance to esters and keteones, neoprene W is compounded with FA polysulfide rubber to produce improved physical properties.

ST polysulfide rubbers exhibit better resistance to chlorinated organics than FA polysulfide rubbers. Contact of either rubber with strong, concentrated inorganic acids such as sulfuric, nitric or hydrochloric acid should be avoided.

Refer to Table 4.31 for the compatibility of polysulfide ST rubber with selected corrodents.

D. Applications

FA polysulfide rubber is one of the elastomeric materials commonly used for the fabrication of rubber rollers for printing and coating equipment. The major reason for this is its high degree of resistance to many types of solvents, including ketones, esters, aromatic hydrocarbons, and plasticizers, that are used as vehicles for various printing inks and coatings.

Applications are also found in the fabrication of hose and hose liners for the handling of aromatic solvents, esters, ketones, oils, fuels, gasoline, paints, lacquers, and thinners. Large amounts of material are also used to produce caulking compounds, cements, paint can gaskets, seals, and flexible mountings.

The impermeability of the polysuldife rubbers to air and gas has promoted the use of these materials for inflatable products such as life jackets, life rafts, balloons, and other inflatable items.

4.16 URETHANE RUBBERS (AU)

The first commercial thermoplastic elastomers (TPE) were the thermoplastic urethanes (TPU). Their general structure is A—B—A—B, where A represents a hard crystalline block derived by chain extension of a diisocyanate with a glycol. The soft block is represented by B and can be derived from either a polyester or a polyether. Figure 4.1 shows typical TPU structures, both polyester and polyether types.

TABLE 4.31 Compatibility of Polysulfide ST Rubber with Selected Corrodents[a]

Chemical	Maximum temp.		Chemical	Maximum temp.	
	°F	°C		°F	°C
Acetamide	x	x	Citric acid, concd	x	x
Acetic acid 10%	80	27	Copper sulfate	x	x
Acetic acid 50%	80	27	Corn oil	90	32
Acetic acid 80%	80	27	Cottonseed oil	90	32
Acetic acid, glacial	80	27	Cresol	x	x
Acetone	80	27	Diacetone alcohol	80	27
Ammonia gas	x	x	Dibutyl phthalate	80	27
Ammonium chloride 10%	150	66	Ethers, general	90	32
Ammonium chloride 28%	150	66	Ethyl acetate	80	27
Ammonium chloride 50%	150	66	Ethyl alcohol	80	27
Ammonium chloride, sat.	90	32	Ethylene chloride	80	27
Ammonium hydroxide 10%	x	x	Ethylene glycol	150	66
Ammonium hydroxide 25%	x	x	Formaldehyde, dil	80	27
Ammonium hydroxide, sat.	x	x	Formaldehyde 37%	80	27
Ammonium sulfate 10–40%	x	x	Formaldehyde 50%	80	27
Amyl alcohol	80	27	Glycerine	80	27
Aniline	x	x	Hydrochloric acid, dil	x	x
Benzene	x	x	Hydrochloric acid 20%	x	x
Benzoic acid	150	66	Hydrochloric acid 35%	x	x
Bromine water, dilute	80	27	Hydrochloric acid 38%	x	x
Bromine water, sat.	80	27	Hydrochloric acid 50%	x	x
Butane	150	66	Hydrdofluoric acid, dil	x	x
Butyl acetate	80	27	Hydrdofluoric acid, 30%	x	x
Butyl alcohol	80	27	Hydrofluoric acid 40%	x	x
Calcium chloride, dil	150	66	Hydrofluoric acid 50%	x	x
Calcium chloride, sat.	150	66	Hydrofluoric acid 70%	x	x
Calcium hydroxide 10%	x	x	Hydrofluoric acid 100%	x	x
Calcium hydroxide 20%	x	x	Hydrogen peroxide, all concn	x	x
Calcium hydroxide 30%	x	x			
Calcium hydroxide, sat.	x	x	Lactic acid, all concn	x	x
Cane sugar liquors	x	x	Methyl ethyl ketone	150	66
Carbon bisulfide	x	x	Methyl isobutyl ketone	80	27
Carbon tetrachloride	x	x	Monochlorobenzene	x	x
Carbonic acid	150	66	Muriatic acid	x	x
Castor oil	80	27	Nitric acid, all concn	x	x
Cellosolve	80	27	Oxalic acid, all concn	x	x
Chlorine water, sat.	x	x	Phenol, all concn	x	x
Chlorobenzene	x	x	Phosphoric acid, all concn	x	x
Chloroform	x	x	Potassium hydroxide to 50%	80	27
Chromic acid 10%	x	x			
Chromic acid 30%	x	x	Potassium sulfate 10%	90	32
Chromic acid 40%	x	x	Propane	150	66
Chromic acid 50%	x	x	Silicone oil	x	x
Citric acid 5%	x	x	Sodium carbonate	x	x
Citric acid 10%	x	x	Sodium chloride	80	27
Citric acid 15%	x	x	Sodium hydroxide, all concn	x	x

TABLE 4.31 Continued

Chemical	Maximum temp. °F	°C	Chemical	Maximum temp. °F	°C
Sodium hypochlorite, all concn	x	x	Water, distilled	80	27
			Water, salt	80	27
Sulfuric acid, all concn	x	x	Water, sea	80	27
Toluene	x	x	Whiskey	x	x
Trichloroethylene	x	x	Zinc chloride	x	x
Water, demineralized	80	27			

[a] The chemicals listed are in the pure state or in a saturated solution unless otherwise indicated. Compatibility is shown to the maximum allowable temperature for which data are available. Incompatibility is shown by an x. A blank space indicates that data are unavailable.
Source: Schweitzer PA. Corrosion Resistance Tables, 4th ed., Vols. 1–3. New York: Marcel Dekker, 1995.

The urethane linkages in the hard blocks are capable of a high degree of inter- and intramolecular hydrdogen bonding. Such bonding increases the crystallinity of the hard phase and can influence the mechanical properties—hardness, modulus, tear strength—of the TPU.

As with other block copolymers, the nature of the soft segments determines the elastic behavior and low-temperature performance. TPUs based on polyester soft blocks have excellent resistance to nonpolar fluids and high tear strength and abrasion resistance. Those based on polyether soft blocks have excellent

$$R = \left[CH_2 - CH_2 - CH_2 - CH_2 \right]$$

or

$$\left[CH_2 - CH \atop CH_3 \right]$$

or

$$\left[CH_2 - CH_2 - O - \underset{\underset{O}{\|}}{C} - CH_2 - CH_2 - CH_2 - CH_2 - \underset{\underset{O}{\|}}{C} \right]$$

n = 30 to 120   m = 8 to 50

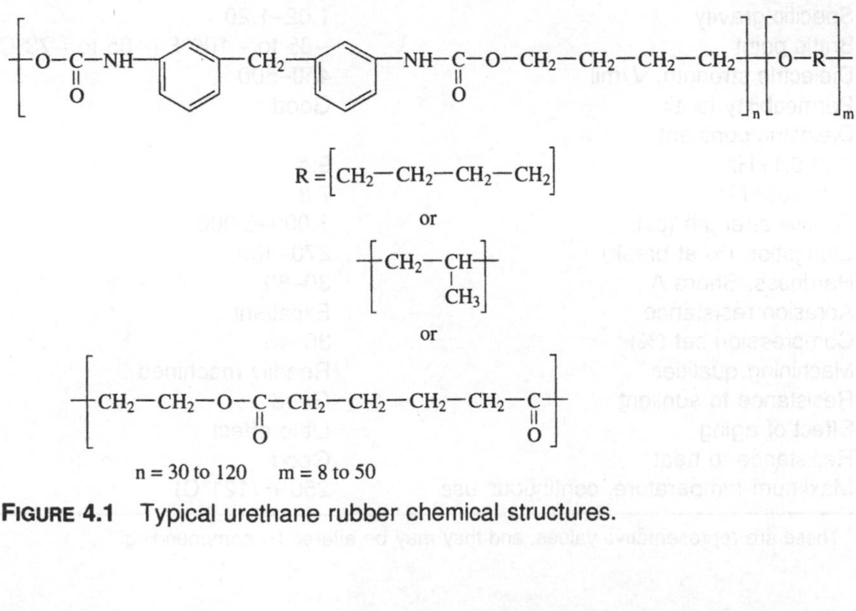

FIGURE 4.1 Typical urethane rubber chemical structures.

resistance (low heat buildup, or hysteresis), thermal stability, and hydrolytic stability.

TPUs deteriorate slowly but noticeably between 265 and 340°F/130 and 170°C by both chemical degradation and morphological changes. Melting of the hard phase causes morphological changes and is reversible, while oxidative degradation is slow and irreversible. As the temperature increases, both processes become progressively more rapid. TPUs with polyether soft blocks have greater resistance to thermal and oxidative attack than TPU based on polyester blocks.

TPUs are noted for their outstanding abrasion resistance and low coefficient of friction on other surfaces. They have specific gravities comparable to that of carbon black-filled rubber, and they do not enjoy the density advantage over thermoset rubber compounds.

There are other factors that can have an adverse effect on aging. Such factors as elevated temperature, dynamic forces, contact with chemical agents, or any combination of these factors will lead to a loss in physical and mechanical properties.

Typical mechanical properties are given in Table 4.32.

Urethane rubbers exhibit excellent recovery from deformation. Parts made of urethane rubbers may build up heat when subjected to high-frequency deformation, but their high strength and load-bearning capacity may permit the use of sections thin enough to disssipate the heat as fast as it is generated.

Urethane rubbers are produced from a number of polyurethane polymers. The properties exhibited are dependent upon the specific polymer and the

TABLE 4.32 Physical and Mechanical Properties of Urethane Rubbers[a]

Specific gravity	1.02–1.20
Brittle point	−85 to −100°F (−65 to −73°C)
Dielectric strength, V/mil	450–500
Permeability to air	Good
Dielectric constant	
at 0.1 kHz	9.4
at 100 kHz	7.8
Tensile strength (psi)	3,000–8,000
Elongation (% at break)	270–400
Hardness, Shore A	30–80
Abrasion resistance	Excellent
Compression set (%)	30–45
Machining qualities	Readily machined
Resistance to sunlight	Good
Effect of aging	Little effect
Resistance to heat	Good
Maximum temperature, continuous use	250°F (121°C)

[a] These are representative values, and they may be altered by compounding.

compounding. Urethane (AU) rubber is a unique material that combines many of the advantages of rigid plastics, metals, and ceramics, yet still has the extensibility and elasticity of rubber. It can be formulated to provide a variety of products with a wide range of physical properties.

Compositions with a Shore A hardness of 95 (harder than a typewriter platen) are elastic enough to withstand stretching to more than four times their own lengths.

At room temperature, a number of raw polyurethane polymers are liquid, simplifying the production of many large and intricately shaped molded products. When cured, these elastomeric parts are hard enough to be machined on standard metalworking equipment. Cured urethane does not require fillers or reinforcing agents.

A. Physical and Mechanical Properties

Urethane rubbers are known for their toughness and durability. Because of high resistance to abrasion, urethane rubbers are used where severe wear is a problem. Urethane rubber in actual service has outworn ordinary rubbers and plastics by a factor of as high as 8 to 1. One such application is that of wear pads used to prevent the marring of multiton rolls of sheet metal.

Most applications use material with a hardness of from 80 Shore A to 75 Shore D. The D scale is used to measure hardnesses greater than 95 Shore A. Most elastomers have hardness between 30 and 80 on the A scale, while structural plastics begin at 55 on the D scale. Therefore, it can be seen that the urethane rubbers bridge this gap.

Depending upon the formulation, resilience values as low as 15% or as high as 80% can be achieved. Urethane rubbers maintain their resilience with changing temperature better than most other rubbers through the range of 50–212°F (10–100°C). This stability is valuable in certain shock mounting applications.

Specific polymers and/or formulations of these rubbers can produce appreciably better impact resistance than structural plastics. Standard compounds, including the hardest types, exhibit good low-temperature impact resistance and low brittle points.

Urethane has a low unlubricated coefficient of friction that decreases sharply as hardness increases. This property, combined with its abrasion resistance and load-carrying ability, is an important reason why urethane elastomer is used in bearings and bushings. If necessary, special compounding can lower the coefficient of friction even further.

Urethane rubbers are readily machined using standard metalworking equipment. Operations such as drilling, routing, sawing, and turning are typical.

Some urethane polymers are naturally translucent. This offers several advantages in certain applications, such as potting or encapsulation. The flexibility and strength of these elastomers help protect electrical assemblies or

TABLE 4.33 Compatibility of Urethane Rubbers with Selected Corrodents[a]

Chemical	Maximum temp.		Chemical	Maximum temp.	
	°F	°C		°F	°C
Acetamide	x	x	Citric acid, concd		
Acetic acid 10%	x	x	Copper sulfate	90	32
Acetic acid 50%	x	x	Corn oil		
Acetic acid 80%	x	x	Cottonseed oil	90	32
Acetic acid, glacial	x	x	Cresol	x	x
Acetone	x	x	Dextrose	x	x
Ammonia gas			Dibutyl phthalate	x	x
Ammonium chloride 10%	90	32	Ethers, general	x	x
Ammonium chloride 28%	90	32	Ethyl acetate	x	x
Ammonium chloride 50%	90	32	Ethyl alcohol	x	x
Ammonium chloride, sat.	90	32	Ethylene chloride		
Ammonium hydroxide 10%	90	32	Ethylene glycol	90	32
Ammonium hydroxide 25%	90	32	Fomaldehyde, dil	x	x
Ammonium hydroxide, sat.	80	27	Fomaldehyde 37%	x	x
Ammonium sulfate 10–40%			Formaldehyde 50%	x	x
Amyl alcohol	x	x	Glycerine	90	32
Aniline	x	x	Hydrochloric acid, dil	x	x
Benzene	x	x	Hydrochloric acid 20%	x	x
Benzoic acid	x	x	Hydrochloric acid 35%	x	x
Bromine water, dil	x	x	Hydrochloric acid 38%	x	x
Bromine water, sat.	x	x	Hydrochloric acid 50%	x	x
Butane	100	38	Isopropyl alcohol	x	x
Butyl acetate	x	x	Jet fuel JP 4	x	x
Butyl alcohol	x	x	Jet fuel JP 5	x	x
Calcium chloride, dil	x	x	Kerosene	x	x
Calcium chloride, sat.	x	x	Lard oil	90	32
Calcium hydroxide 10%	x	x	Linseed oil	90	32
Calcium hydroxide 20%	90	32	Magnesium hydroxide	90	32
Calcium hydroxide 30%	90	32	Mercury	90	32
Calcium hydroxide, sat.	90	32	Methyl alcohol	90	32
Cane sugar liquors			Methyl ethyl ketone	x	x
Carbon bisulfide			Methyl isobutyl ketone	x	x
Carbon tetrachloride	x	x	Mineral oil	90	32
Carbonic acid	90	32	Muriatic acid	x	x
Castor oil	90	32	Nitric acid	x	x
Cellosolve	x	x	Oils and fats	80	27
Chlorine water, sat.	x	x	Phenol	x	x
Chlorobenzene	x	x	Potassium bromide 30%	90	32
Chloroform	x	x	Potassium chloride 30%	90	32
Chromic acid 10%	x	x	Potassium hydroxide to	90	32
Chromic acid 30%	x	x	50%		
Chromic acid 40%	x	x	Potassium sulfate 10%	90	32
Chromic acid 50%	x	x	Silver nitrate	90	32
Citric acid 5%	x	x	Soaps	x	x
Citric acid 10%	x	x	Sodium carbonate	x	x
Citric acid 15%			Sodium chloride	80	27

TABLE 4.33 Continued

Chemical	Maximum temp. °F	Maximum temp. °C	Chemical	Maximum temp. °F	Maximum temp. °C
Sodium hydroxide to 50%	90	32	Trisodium phosphate	90	32
Sodium hypochlorite, all concn	x	x	Turpentine	x	x
			Water, salt	x	x
Sulfuric acid, all concn	x	x	Water, sea	x	x
Toluene	x	x	Whiskey	x	x
Trichloroethylene	x	x	Xylene	x	x

[a] The chemicals listed are in the pure state or in a saturated solution unless otherwise indicated. Compatibility is shown to the maximum allowable temperature for which data are available. Incompatibility is shown by an x. A blank space indicates that data are unavailable.
Source: Schweitzer PA. Corrosion Resistance Tables, 4th ed., Vols. 1–3. New York: Marcel Dekker, 1995.

sheathed glass elements from damage, while their translucence permits easy inspection. If desired, urethane rubbers can be pigmented, which is quite commonplace. Products made of urethane rubbers have little or no resistance to burning. Some slight improvement can be made in this area by special compounding, but the material will still ignite in an actual fire situation. If fire safety is a factor in the selection of an elastomer, this property must be taken into account and the potential hazards must be evaluated.

Standard formulations do not support fungus growth and are generally resistant to fungus attack. However, certain of the special formulations are susceptible to fungus attack. To compensate for this, commercial fungicides must be added to these formulations.

Urethane products are stable under high-vacuum conditions. Outgassing tests produce very little weight loss. This property, coupled with their excellent low-temperature performance, makes urethane rubbers suitable for aerospace applications.

Table 4.32 lists the physical and mechanical properties of urethane rubbers.

Urethane rubbers possess many desirable mechanical properties that are superior to those of other elastomeric materials. Their load-bearing capacity is greater than that of other rubbers of comparable hardness. Because of their strength and toughness they can be used in thin sections. They also resist cracking under repeated flexing.

These properties are retained over a wide temperature range. At very low temperatures, the material remains flexible and has good resistance to thermal shock. The standard compositions can be used at temperatures as low as −80°F (−62°C) without becoming brittle, although stiffness gradually increases as the

TABLE 4.34 Compatibility of Polyether Urethane Rubbers with Selected Corrodents[a]

Chemical	Maximum temp.		Chemical	Maximum temp.	
	°F	°C		°F	°C
Acetamide	x	x	Citric acid 15%	90	32
Acetic acid 10%	x	x	Cottonseed Oil	130	38
Acetic acid 50%	x	x	Cresylic acid	x	x
Acetic acid 80%	x	x	Cyclohexane	80	27
Acetic acid, glacial	x	x	Dioxane	x	x
Acetone	x	x	Diphenyl	x	x
Acetyl chloride	x	x	Ethyl acetate	x	x
Ammonium chloride 10%	130	54	Ethyl benzene	x	x
Ammonium chloride 28%	130	54	Ethyl chloride	x	x
Ammonium chloride 50%	130	54	Ethylene oxide	x	x
Ammonium chloride, sat.	130	54	Ferric chloride 75%	100	38
Ammonium hydroxide 10%	110	43	Fluorosilicic acid	100	38
Ammonium hydroxide 25%	110	43	Formic acid	x	x
Ammonium hydroxide, sat.	100	38	Freon F-11	x	x
Ammonium sulfate 10–40%	120	49	Freon F-12	130	38
Amyl alcohol	x	x	Glycerine	130	38
Aniline	x	x	Hydrochloric acid, dil	x	x
Benzene	x	x	Hydrochloric acid 20%	x	x
Benzoic acid	x	x	Hydrochloric acid 35%	x	x
Benzyl alcohol	x	x	Hydrochloric acid 38%	x	x
Borax	90	32	Hydrochloric acid 50%	x	x
Butane	80	27	Hydrofluoric acid, dil	x	x
Butyl acetate	x	x	Hydrofluoric acid 30%	x	x
Butyl alcohol	x	x	Hydrofluoric acid 40%	x	x
Calcium chloride, dil	130	54	Hydrofluoric acid 50%	x	x
Calcium chloride, sat.	130	54	Hydrofluoric acid 70%	x	x
Calcium hydroxide 10%	130	54	Hydrofluoric acid 100%	x	x
Calcium hydroxide 20%	130	54	Isobutyl alcohol	x	x
Calcium hydroxide 30%	130	54	Isopropyl acetate	x	x
Calcium hydroxide, sat.	130	54	Lactic acid, all concn	x	x
Calcium hypochlorite 5%	130	54	Methyl ethyl ketone	x	x
Carbon bisulfide	x	x	Methyl isobutyl ketone	x	x
Carbon tetrachloride	x	x	Monochlorobenzene	x	x
Carbonic acid	80	27	Muriatic acid	x	x
Castor oil	130	54	Nitrobenzene	x	x
Caustic potash	130	54	Olive oil	120	49
Chloroacetic acid	x	x	Phenol	x	x
Chlorobenzene	x	x	Potassium acetate	x	x
Chloroform	x	x	Potassum chloride 30%	130	54
Chromic acid 10%	x	x	Potassium hydroxide	130	54
Chromic acid 30%	x	x	Potassium sulfate 10%	130	54
Chromic acid 40%	x	x	Propane	x	x
Chromic acid 50%	x	x	Propyl acetate	x	x
Citric acid 5%	90	32	Propyl alcohol	x	x
Citric acid 10%	90	32	Sodium chloride	130	54

TABLE 4.34 Continued

Chemical	Maximum temp. °F	°C	Chemical	Maximum temp. °F	°C
Sodium hydroxide 30%	140	60	Trichloroethylene	x	x
Sodium hypochlorite	x	x	Water, demineralized	130	54
Sodium peroxide	x	x	Water, distilled	130	54
Sodium phosphate, acid	130	54	Water, salt	130	54
Sodium phosphate, alk.	130	54	Water, sea	130	54
Sodium phosphate, neut.	130	54	Xylene	x	x
Toluene	x	x			

[a] The chemicals listed are in the pure state or in a saturated solution unless otherwise indicated. Compatibility is shown to the maximum allowable temperature for which data are available. Incompatibility is shown by an x. A blank space indicates that data are unavailable. *Source:* Schweitzer PA. Corrosion Resistance Tables, 4th ed., Vols. 1–3. New York: Marcel Dekker, 1995.

temperature goes below 0°F (−18°C). Special compositions can be formulated that will permit some flexibility to be retained as low as −125°F (−87°C).

On the high-temperature side, formulations can be provided that will permit continuous operation at temperatures of 200–250°F (93–121°C). Even at these temperatures, there is only a minor decrease in physical and mechanical properties. However, caution is advised in arbitrarily applying urethane rubbers at these elevated temperatures. Each potential application should be evaluated beforehand by consultation with the supplier and by testing.

Under static, room-temperature conditions, time has little effect upon urethane rubbers. Shelf aging is not a problem, based on samples stored for at least 15 years and then put into service. Neither is prolonged use in normal service.

B. Resistance to Sun, Weather, and Ozone

Urethane rubbers exhibit excellent resistance to ozone attack and have good resistance to weathering. However, extended exposure to ultraviolet light will reduce their physical properties and will cause the rubber to darken. This can be prevented by the use of pigments or ultraviolet screening agents.

C. Chemical Resistance

Urethane rubbers are resistant to most mineral and vegetable oils, greases and fuels, and to aliphatic, and chlorinated hydrocarbons. This makes these materials particularly suited for service in contact with lubricating oils and automotive fuels.

Tests conducted at the Hanford Laboratories of General Electric Co. indicated that adiprene urethane rubber, as manufactured by DuPont, offered the greatest resistance to the effects of gamma-ray irradiation of the many elastomers and plastics tested. Satisfactory service was given even when the material was exposed to the relatively large gamma-ray dosage of 1×10^9 roentgens. When irradiated with gamma rays, these rubbers were more resistant to stress cracking than other elastomers and retained a large proportion of their original flexibility and toughness.

Aromatic hydrocarbons, polar solvents, esters, ethers, and ketones will attack urethane. Alcohols will soften and swell urethane rubbers.

Urethane rubbers have limited service in weak acid solutions and cannot be used in concentrated acids. Neither are they resistant to steam or caustic, but they are resistant to the swelling and deteriorating effects of being immersed in water.

Refer to Table 4.33 for the compatibility of urethane rubber with selected corrodents and to Table 4.34 for the compatibility of polyether urethane rubber with selected corrodents.

D. Applications

The versatility of urethane has led to a wide variety of applications. Products made from urethane rubber are available in three basic forms, solid, cellular, and films and coatings.

Included under the solid category are those products that are cast or molded, comprising such items as large rolls, impellers, abrasion-resistant parts for textile machines, O-rings, electrical encapsulations, gears, tooling pads, and small intricate parts.

Cellular products are such items as shock mountings, impact mountings, shoe soles, contact wheels for belt grinders, and gaskets.

It is possible to apply uniform coatings or films of urethane rubber to a variety of substrate materials including metal, glass, wood, fabrics, and paper. Examples of products to which a urethane film or coating is often applied are tarpaulins, bowling pins, pipe linings, tank linings, and exterior coatings for protection against atmospheric corrosion. These films also provide abrasion resistance.

Filtration units, clarifiers, holding tanks, and treatment sumps constructed of reinforced concrete are widely used in the treatment of municipal, industrial, and thermal generating station wastewater. In many cases, particularly in anaerobic, industrial, and thermal generating systems, urethane linings are used to protect the concrete from severe chemical attack and prevent seepage into the concrete of chemicals that can attack the reinforcing steel. These linings provide protection from abrasion and erorsion and act as a waterproofing system to combat leakage of the equipment resulting from concrete movement and shrinkage.

The use of urethane rubbers in manufactured products has been established as a result of their many unique properties of high tensile and tear strengths, resiliency, impact resistance, and load-bearing capacity. Other products made from urethane rubbers include bearings, gear couplings, mallets and hammers, solid tires, conveyer belts, and many other miscellaneous items.

4.17 POLYAMIDES

Polyamides are produced under a variety of trade names, the most popular of which are Nylons (Nylon 6, etc.) manufactured by DuPont. Although the polyamides find their greatest use as textile fibers, they can also be formulated into thermoplastic molding compounds with many attractive properties. Their relatively high price tends to restrict their use.

There are many varieties of polyamides in production, but the four major types are Nylon 6, Nylon 6/6, Nylon 11, and Nylon 12. Of these, Nylons 11 and 12 find application as elastomeric materials.

The highest performance class of thermoplastic elastomers are block copolymeric elastomeric polyamides (PEBAs). Amide linkages connect the hard and soft segments of these TPEs, and the soft segment may have a polyester or a polyether structure. The structures of these commercial PEBAs are shown in Figure 4.2. The amide linkages connecting the hard and soft blocks are more resilient to chemical attack than either an ester of urethane bond. Consequently the PEBAs have higher temperature resistance than the urethane elastomers, and their cost is greater.

The structure of the hard and soft blocks also contributes to their performance characteristics. The soft segment may consist of polyester, polyether, or polyetherester chains. Polyether chains provide better low-temperature properties and resistance to hydrolysis, while polyester chains provide better fluid resistance and resistence to oxidation at elevated temperatures.

A. Physical and Mechanical Properties

Polyamide materials have an unusual combination of high tensile strength, ductility, and toughness. They can withstand extremely high impact, even at extremely low temperatures. According to ASTM test D-746, the cold brittleness temperature is $-94°F$ ($-70°C$). Nylons 11 and 12 have an extremely high elastic memory that permits them to withstand repeated stretching and flexing over long periods of time.

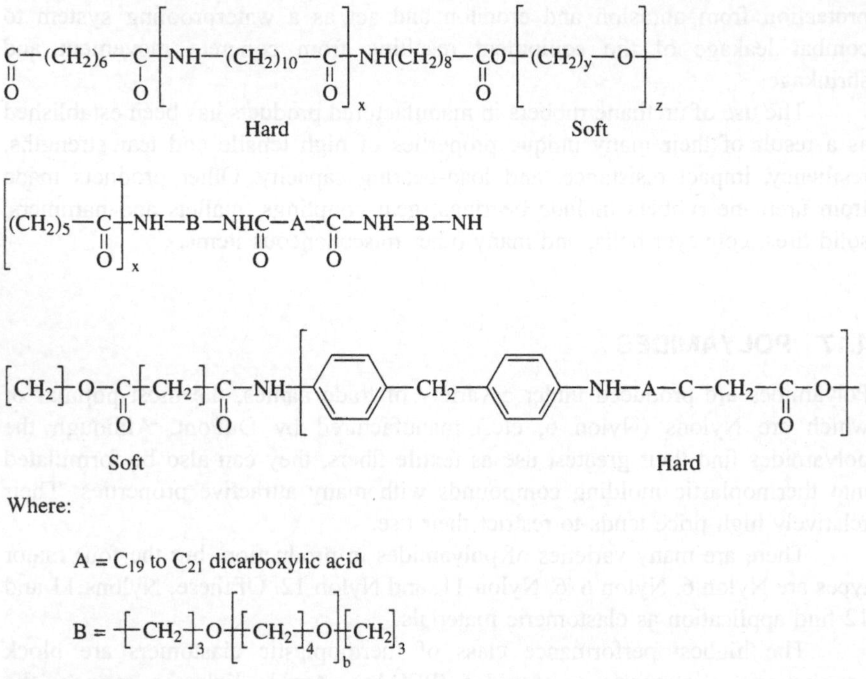

$$\underset{\underset{O}{\|}}{C}-(CH_2)_6-\underset{\underset{O}{\|}}{C}\left[NH-(CH_2)_{10}-\underset{\underset{O}{\|}}{C}\right]_x NH(CH_2)_8-\underset{\underset{O}{\|}}{C}O\left[(CH_2)_y-O\right]_z$$

Hard   Soft

$$\left[(CH_2)_5-\underset{\underset{O}{\|}}{C}\right]_x NH-B-NHC-\underset{\underset{O}{\|}}{A}-\underset{\underset{O}{\|}}{C}-NH-B-NH$$

$$\left[CH_2\right]_\overline{}O-\underset{\underset{O}{\|}}{C}\left[CH_2\right]_\overline{}\underset{\underset{O}{\|}}{C}-NH-\hexagon-CH_2-\hexagon-NH-\underset{\underset{O}{\|}}{A}-\underset{\underset{O}{\|}}{C}-CH_2-\underset{\underset{O}{\|}}{C}-O$$

Soft   Hard

Where:

A = C$_{19}$ to C$_{21}$ dicarboxylic acid

$$B = \left[CH_2\right]_3 O\left[\left[CH_2\right]_4 O\right]_b \left[CH_2\right]_3$$

FIGURE 4.2 Structures of three elastomeric polyamides.

They also exhibit good abrasion resistance. Their absorption of moisture is very low, which means that parts produced from these materials will have good dimensional stability regardless of the humidity of the environment. Moisture also has little effect on the mechanical properties, particularly the modulus of elasticity.

The electrical insulation properties of the polyamides grade 11 and grade 12 are good. They have stable volume resistivity and offer excellent resistance to arc tracking. These materials are also self-extinguishing.

The physical and mechanical properties of polyamides are given in Table 4.35.

PEBAs have a hardness ranging from a high of 65 Shore D to a low of 60 Shore A. These elastomeric PEBAs have useful tensile properties at ambient temperatures and excellent retention of these properties at higher temperatures. For example, a 90 Shore A PEBA retains more than 90% of its tensile strength and modulus at 212°F/100°C. Annealing of the elastomer above the melting point of the hard phase can result in significant increases in tensile strength, modulus, and ultimate elongation. These elastomers are second only to urethane

TABLE 4.35 Physical and Mechanical Properties of Polyamide Elastomers[a]

	Grade 11	Grade 12
Specific gravity	1.03–1.05	1.01–1.02
Specific heat (Btu/lb-°F)	0.3–0.5	0.3–0.5
Thermal conductivity		
$\quad$ (10^{-4} cal/cm-s-°C	5.2	5.2
$\quad$ Btu/hr-ft^2-°F-in.)	1.5	1.5
Coefficient of linear thermal expansion		
$\quad$ (10^{-5}/°C)	10.0	10.0
$\quad$ (10^{-5}/°F)	5.1	5.3
Continuous service temperature		
$\quad$ (°F)	180–300	180–250
$\quad$ (°C)	82–149	82–121
Brittle point		
$\quad$ (°F)	−94	_[b]
$\quad$ (°C)	−70	_[b]
Volume resistivity (ohm-cm)	10^{13}	10^{13}
Dielectric strength (3 mm thick)	17	18
Dielectric constant at 10^3-10^6 Hz	3.2–3.7	3.8
Dissipation factor	0.05	5×10^{12}
Water absorption (%)	0.3	0.25
Tensile strength (psi)	8,000	8,000–9,500
Elongation (% at break)	300	300
Hardness, Rockwell	R108	R106–R109
Abrasion resistance	Good	Good
Impact resistance (kg-cm/cm)	9.72	10.88–29.92
Resistance to compression set	Good	Good
Tear resistance	Good	Good
Machining qualities	Can be machined	
Resistance to sunlight	Good	Good
Effect of aging	Nil	Nil
Resistance to heat	Good	Good

[a]These are representative values, and they may be altered by compounding.
[b]Data not available.

rubbers in abrasion resistance and show excellent fatigue resistance and tear strength.

B. Resistance to Sun, Weather, and Ozone

Polyamides are resistant to sun, weather, and ozone. Many metals are coated with polyamide to provide protection from harsh weather.

TABLE 4.36 Compatibility of Nylon 11 with Selected Corrodents[a]

Chemical	Maximum temp.		Chemical	Maximum temp.	
	°F	°C		°F	°C
Acetamide	140	60	Citric acid, concd	200	93
Acetic acid 10%	x	x	Copper nitrate	x	x
Acetic acid 50%	x	x	Corn oil		
Acetic acid 80%	x	x	Cottonseed oil	200	93
Acetic acid, glacial	x	x	Cresol	x	x
Acetone			Cyclohexanone	200	93
Ammonia gas			Dibutyl phthalate	80	27
Ammonium chloride 10%	200	93	Ethers, general		
Ammonium chloride 28%	200	93	Ethyl acetate		
Ammonium chloride 50%	200	93	Ethyl alcohol	200	93
Ammonium chloride, sat.			Ethylene chloride	200	93
Ammonium hydroxide 10%	200	93	Ethylene glycol	200	93
Ammonium hydroxide 25%	200	93	Ethylene oxide	100	38
Ammonium hydroxide, sat.	200	93	Ferric chloride	x	x
Ammonium sulfate 10–40%	80	27	Formic acid	x	x
Amyl alcohol	100	38	Glycerine	200	93
Aniline	x	x	Hydrochloric acid, dil	80	27
Benzene	200	93	Hydrochloric acid 20%	x	x
Benzoic acid 10%	200	93	Hydrochloric acid 35%	x	x
Bromine, gas	x	x	Hydrochloric acid 38%	x	x
Bromine, liquid	x	x	Hydrochloric acid 50%	x	x
Butane	140	60	Hydrofluoric acid, dil	x	x
Butyl acetate			Hydrofluoric acid 30%	x	x
Butyl alcohol	200	93	Hydrofluoric acid 40%	x	x
Calcium chloride, dil	200	93	Hydrofluoric acid 50%	x	x
Calcium chloride, sat.	200	93	Hydrofluoric acid 70%	x	x
Calcium hydroxide 10%	140	60	Hydrofluoric acid 100%	x	x
Calcium hydroxide, 20%	140	60	Hydrogen sulfide	x	x
Calcium hydroxide 30%	140	60	Iodine	x	x
Calcium hydroxide, sat.	140	60	Lactic acid, all concn	200	93
Calcium sulfate	x	x	Methyl ethyl ketone	160	71
Carbon bisulfide	80	27	Methyl isobutyl ketone	100	38
Carbon tetrachloride	140	60	Monochlorobenzene	x	x
Carbonic acid	80	27	Naphtha	120	49
Caustic potash 50%	140	60	Nitric acid	x	x
Chlorine gas	x	x	Oxalic acid	140	60
Chlorine water, sat.	x	x	Phenol	x	x
Chlorobenzene			Phosphoric acid to 10%	80	27
Chloroform	x	x	above 10%	x	x
Chromic acid 10%	x	x	Potassium hydroxide to	200	93
Chromic acid 30%	x	x	50%		
Chromic acid 40%	x	x	Potassium sulfate 10%	80	27
Chromic acid 50%	x	x	Propane	80	27
Citric acid 5%	200	93	Silicone oil	80	27
Citric acid 10%	200	93	Sodium carbonate	220	104
Citric acid 15%	200	93	Sodium chloride	200	93

TABLE 4.36 Continued

Chemical	Maximum temp. °F	°C	Chemical	Maximum temp. °F	°C
Sodium hydroxide, all concn	200	93	Water, acid mine	80	27
Sodium hypochlorite, all concn	x	x	Water, distilled	80	27
			Water, salt	140	60
Sulfuric acid, all concn	x	x	Water, sea	140	60
Toluene	140	60	Whiskey	80	27
Trichloroethylene	160	71	Zinc chloride	x	x

[a] The chemicals listed are in the pure state or in a saturated solution unless otherwise indicated. Compatibility is shown to the maximum allowable temperature for which data are available. Incompatibility is shown by an x. A blank space indicates that data are unavailable.
Source: Schweitzer PA. Corrosion Resistance Tables, 4th ed., Vols. 1–3. New York: Marcel Dekker, 1995.

C. Chemical Resistance

Polyamides exhibit excellent resistance to a broad range of chemicals and harsh environments. They have good resistance to most inorganic alkalies, particularly ammonium hydroxide and ammonia even at elevated temperatures and sodium potassium hydroxide at ambient temperatures. They also display good resistance to almost all inorganic salts and almost all hydrocarbons and petroleum-based fuels.

They are also resistant to organic acids (citric, lactic, oleic, oxalic, stearic, tartaric, and uric) and most aldehydes and ketones at normal temperatures. They display limited resistance to hydrochloric, sulfonic, and phosphoric acids at ambient temperatures.

Refer to Table 4.36 for the compatibility of Nylon 11 with selected corrodents.

D. Applications

Polyamides find many diverse applications resulting from their many advantageous properties. A wide range of flexibility permits material to be produced that is soft enough for high-quality bicycle seats and other materials whose strength and rigidity are comparable to those of many metals. Superflexible grades are also available that are used for shoe soles, gaskets, diaphragms, and seals. Because of the high elastic memory of polyamides, these parts can withstand repeated stretching and flexing over long periods of time.

Since polyamides can meet specification SAE J844 for air-brake hose, coiled tubing is produced for this purpose for use on trucks. Coiled air-brake hose produced from this material has been used on trucks that have traveled over 2 million miles without a single reported failure. High-pressure hose and fuel lines are also produced from this material.

Corrosion-resistant and wear-resistant coverings for aircraft control cables, automotive cables, and electrical wire are also produced.

4.18   POLYESTER (PE) ELASTOMER

This elastomer combines the characteristics of thermoplastics and elastomers. It is structurally strong, resilient, and resistant to impact and flexural fatigue. Its physical and mechanical properties vary depending upon the hardness of the elastomer. Hardnesses range from 40 to 72 on the Shore D scale. The standard hardnesses to which PE elastomers are formulated are 40, 55, 63, and 72 Shore D. Combained are such features as resilience, high resistance to deformation under moderate strain conditions, outstanding flex-fatigue resistance, good abrasion resistance, retention of flexibility at low temperatures, and good retention of properties at elevated temperatures. Polyester elastomers can successfully replace other thermoset rubbers at lower cost in many applications by taking advantage of their higher strength and by using thinner cross sections.

A.   Physical and Mechanical Properties

When modulus is an important design consideration, thinner sections of PE elastomers can be used since its tensile stress at low strain shows higher modulus values than other elastomeric materials. Polyester elastomers yield at approximately 25% strain and beyond the yield point elongate to 300–500% with some permanent set. To take full advantage of the elastomeric properties of PE, parts should be designated to function with deformations that do not exceed the yield point (up to approximately 20% strain).

The Bashore resilience of PE exceeds 60% for the 40D hardness grade. Even as the hardness approaches that of plastics (63D), these materials still have a resilience of 40%.

The load-bearing properties of the PE elastomers under compression are good. As the hardness increases, so does the compression modulus. The compression modulus of PE elastomers is 50–100% greater than that of other rubbers of comparable hardness.

Table 4.37 lists the compression set of PE elastomers at constant load (1,350 psi) and at varying temperatures. These conditions were chosen because they very closely simulate the conditions under which parts fabricated from PE elastomers operate. The compression set of these rubbers measured for 22 hr at

TABLE 4.37 Physical and Mechanical Properties of Polyester Elastomers (PE)[a]

Specific gravity	1.17–1.25
Brittle point	−94°F (−70°C)
Resilience (%)	42–62
Coefficient of linear expansion (in./in.-°C)	$2 \times 10^{-5} - 21 \times 10^{-5}$
Dielectric strength (V/mil)	645–900
Dielectric constant at 1 kHz	4.16–6
Permeability to air	Low
Tear strength	
(lb/in.)	631–1,055
(kg/m)	110–185
Water absorption (%/24 hr)	0.6–0.03
Tensile strength (psi)	3,700–5,700
Elongation (% at break)	350–450
Hardness, Shore A	40–72
Abrasion resistance	Good
Maximum temperature, continuous use	302°F (150°C)
Impact resistance	Good
Compression set (%)	
at 73°F (23°C)	1–11
at 158°F (70°C)	2–27
at 212°F (100°C)	4–44
Resistance to sunlight	Good
Effect of aging	Nil
Resistance to heat	Excellent
Flexural modulus (psi)	7,000–75,000

[a] These are representative values, and they may be altered by compounding.

158°F (70°C) with a constant deflection of 25% is 60% for materials having a hardness of 40D and 56% for materials of 55D hardness. Because of the high load-bearing capability of these elastomers, the majority of compression applications employ deflections well below 25%. At 25% deflection, most formulations of PE elastomers are deformed beyond their yield point or at least beyond the limits of good design practice. Under these conditions, PE elastomers are generally comparable to urethane elastomers. Annealing of PE parts will reduce the compression set values to 40% or less with 25% deflection.

Very little hysteresis loss is exhibited by PE elastomers when they are stressed below the yield point. Applications at low strain levels can usually be expected to exhibit complete recovery and continue to do so in cyclic applications with little heat buildup. Because of the high resilience and low heat buildup of these elastomers, their resistance to cut growth is outstanding. This means that a part made of PE, engineered to operate at low strain levels, can usually be expected to exhibit complete recovery from deformation and to continue to do so under repeated cycling for extremely long periods of time without heat buildup or distortion.

The mechanical properties of these elastomers are maintained up to 302°F (150°C), better than many rubbers, particularly the harder polymers. Above 248°F (120°C), their tensile strengths far exceed those of other rubbers. Their hot strength and good resistance to hot fluids make PE elastomers suitable for many applications involving fluid containment.

All of the PE formulations have brittle points below −94°F (−70°C) and exhibit high impact strength and resistance to stiffening at temperatures down to −40°F (−40°C). As would be expected, the softer members exhibit better low-temperature flexibility.

The physical and mechanical properties of polyester elastomers are given in Table 4.38.

B. Resistance to Sun, Weather, and Ozone

Polyester rubbers possess excellent resistance to ozone. When formulated with appropriate additives, their resistance to sunlight aging is very good. Resistance to general weathering is good.

C. Chemical Resistance

In general, the fluid resistance of polyester rubbers increases with increasing hardness. Since these rubbers contain no plasticizers, they are not susceptible to the solvent extraction or heat volatilization of such additives. Many fluids and chemicals will extract plasticizers from elastomers, causing a significant increase in stiffness (modulus) and volume shrinkage.

Overall, PE elastomers are resistant to the same classes of chemicals and fluids as polyurethanes are. However, PE has better high-temperature properties than polyurethanes and can be used satisfactorily at higher temperatures in the same fluids.

Polyester elastomers have excellent resistance to nonpolar materials such as oils and hydraulic fluids, even at elevated temperatures. At room temperature, elastomers are resistant to most polar fluids, such as acids, base, amines, and glycols. Resistance is very poor at temperatures of 158°F (70°C) or above. These rubbers should not be used in applications requiring continuous exposure to polar fluids at elevated temperatures.

Polyester elastomers also have good resistance to hot, moist atmospheres. Their hydrolytic stability can be further improved by compounding.

TABLE 4.38 Compatibility of Polyester Elastomers with Selected Corrodents[a]

	Maximum temp.			Maximum temp.	
Chemical	°F	°C	Chemical	°F	°C
Acetic acid 10%	80	27	Ethyl alcohol	80	27
Acetic acid 50%	80	27	Ethylene chloride	x	x
Acetic acid 80%	80	27	Ethylene glycol	80	27
Acetic acid, glacial	100	38	Formaldehyde, dil	80	27
Acetic acid, vapor	90	32	Formaldehyde 37%	80	27
Ammonium chloride 10%	90	32	Formic acid 10 to 85%	80	27
Ammonium chloride 28%	90	32	Glycerine	80	27
Ammonium chloride 50%	90	32	Hydrochloric acid, dil.	80	27
Ammonium chloride, sat.	90	32	Hydrochloric acid 20%	x	x
Ammonium sulfate 10–40%	80	27	Hydrochloric acid 35%	x	x
Amyl acetate	80	27	Hydrochloric acid 38%	x	x
Aniline	x	x	Hydrochloric acid 50%	x	x
Beer	80	27	Hydrofluoric acid, dil.	x	x
Benzene	80	27	Hydrofluoric acid 30%	x	x
Borax	80	27	Hydrofluoric acid 40%	x	x
Boric acid	80	27	Hydrofluoric acid 50%	x	x
Bromine, liquid	x	x	Hydrofluoric acid 70%	x	x
Butane	80	27	Hydrofluoric acid 100%	x	x
Butyl acetate	80	27	Hydrogen	80	27
Calcium chloride, dil	80	27	Hydrogen sulfide, dry	80	27
Calcium chloride, sat.	80	27	Lactic acid	80	27
Calcium hydroxide 10%	80	27	Methyl ethyl ketone	80	27
Calcium hydroxide 20%	80	27	Methylene chloride	x	x
Calcium hydroxide 30%	80	27	Mineral oil	80	27
Calcium hydroxide, sat.	80	27	Muriatic acid	x	x
Calcium hypochlorite 5%	80	27	Nitric acid	x	x
Carbon bisulfide	80	27	Nitrobenzene	x	x
Carbon tetrachloride	x	x	Ozone	80	27
Castor oil	80	27	Phenol	x	x
Chlorine gas, dry	x	x	Potassium dichromate 30%	80	27
Chlorine gas, wet	x	x	Potassium hydroxide	80	27
Chlorobenzene	x	x	Pyridine	80	27
Chloroform	x	x	Soap solutions	80	27
Chlorosulfonic acid	x	x	Sodium chloride	80	27
Citric acid 5%	80	27	Sodium hydroxide, all concn	80	27
Citric acid 10%	80	27	Stannous chloride 15%	80	27
Citric acid 15%	80	27	Stearic acid	80	27
Citric acid, concd	80	27	Sulfuric acid, all concn	x	x
Copper sulfate	80	27	Toluene	80	27
Cottonseed oil	80	27	Water, demineralized	160	71
Cupric chloride to 50%	80	27	Water, distilled	160	71
Cyclohexane	80	27	Water, salt	160	71
Dibutyl phthalate	80	27	Water, sea	160	71
Dioctyl phthalate	80	27	Xylene	80	27
Ethyl acetate	80	27	Zinc chloride	80	27

[a] The chemicals listed are in the pure state or in a saturated solution unless otherwise indicated. Compatibility is shown to the maximum allowable temperature for which data are available. Incompatibility is shown by an x. A blank space indicates that data are unavailable.

Source: Schweitzer PA. Corrosion Resistance Tables, 4th ed., Vols. 1–3. New York: Marcel Dekker, 1995.

D. Applications

Applications for PE elastomers are varied. Large quantities of PE materials are used for liners for tanks, ponds, swimming pools, and drums. Because of their low permeability to air, they are also used for inflatables. Their chemical resistance to oils and hydraulic fluids coupled with their high heat resistance make PE elastomers very suitable for automotive hose applications.

Since the PE elastomers do not contain any plasticizers, hoses and tubing produced from them do not stiffen with age. Other PE products include seals, gaskets, speciality belting, noise-damping devices, low-pressure tires, industrial solid tires, wire and cable jacketing, pump parts, electrical connectors, flexible shafts, sports equipment, piping clamps and cushions, gears, flexible couplings, and fasteners.

4.19 THERMOPLASTIC ELASTOMERS (TPE) OLEFINIC TYPE (TEO)

Thermoplastic elastomers contain sequences of hard and soft repeating units in the polymer chain. Elastic recovery occurs when the hard segments act to pull back the more soft and rubbery segments. Cross-linking is not required. The six generic classes of TPEs are, in order of increasing cost and performance, styrene block copolymers, polyolefin blends, elastomeric alloys, thermoplastic urethanes, thermoplastic copolyesters, and thermoplastic polyamides.

The family of thermoplastic olefins (TEO) are simple blends of a rubbery polymer (such as EPDM or NBR) with a thermoplastic such as PP or PVC. Each polymer will have its own phase, and the rubber phase will have little or no cross-linking (that is, vulcanization). The polymer present in the larger amount will usually be the continuous phase, and the thermoplastic is favored because of its lower viscosity. The discontinuous phase should have a smaller particle size for the best performance of the TEO. EPDM rubber and PP are the constituents of the most common TEOs. Blends of NBR and PVC are also significant but less common in Europe and North America than in Japan.

TEOs are similar to thermoset rubbers since they can be compounded with a variety of the same additives and fillers to meet specific applications. These additives include carbon black, plasticizers, antioxidants, and fillers, all of which tend to concentrate in the soft rubber phase of the TEO.

A. Physical and Mechanical Properties

The most important performance characteristics of the TPEs are their good toughness, elasticity, and load-bearing properties. They have a service temperature range of −40 to 227°F (−40 to 136°C) and can tolerate intermittent exposure at 302°F (150°C) for short periods of time.

Many desirable elastic properties are exhibited by these elastomers in the 90 Shore A to 50 Shore D hardness range. Their recovery from deformation, both in tension and compression, is comparable to that of hard rubber compounds and superior to that of flexible thermoplastics, which do not recover at all. They also exhibit good resistance to creep under moderate stress. Under tensile stress, the TPEs initially react like a soft plastic, requiring stresses of 1,000–2,000 psi to produce strains of 25–50%. At this point, the slope of the stress–strain curve gradually changes to a rubbery plateau, where small increases in stress produce large increases in strain or elongation, Even at high strain levels, they continue to exhibit elastic recovery. A well-defined yield point is not present.

In general, the TPEs exhibit good strength and toughness, particularly tear resistance and abrasion resistance. They retain functional strength to 302°F (150°C) and flexibility for low-temperature service to at least −40°F (−40°C).

When exposed to high temperature for an extended period of time, they slowly increase in hardness, modulus, and tensile strength while losing some elongation. These elastomers can be colored.

Physical and mechanical properties are given in Table 4.39.

B. Resistance to Sun, Weather, and Ozone

TPEs possess good resistance to sun and ozone and have excellent weatherability. Their water resistance is excellent, showing essentially no property changes after prolonged exposure to water at elevated temperatures.

C. Chemical Resistance

As a result of the low level of cross-linking, a TEO is highly vulnerable to fluids with a similar solubility parameter (or polarity). The EPDM/PP TEOs have very poor resistance to hydrocarbon fluids such as the alkanes, alkenes, or alkyl substituted benzenes, especially at elevated temperatures.

The absence of unsaturation in the polymer backbone of both EPDM and PP makes these TEO blends very resistant to oxidation.

The nonpolar nature of EPDM/PP TEOs make them highly resistant to water, aqueous solutions, and other polar fluids such as alcohols and glycols, but they swell excessively with loss of properties when exposed to halocarbons and hydrocarbons such as oils and fuels. Blends with NBR and PVC are more resistant to aggressive fluids with the exception of the halocarbons.

D. Applications

These elastomeric compounds are found in a variety of applications, including reinforced hose, seals, gaskets and profile extrusions, flexible and supported tubing, automotive trim, functional parts and under-the-hood components, mechanical goods, and wire and cable jacketing.

TABLE 4.39 Physical and Mechanical Properties of Thermoplastic
Elastomers, Olefinic Type[a]

Specific gravity	0.88–1.0
Brittle point	−67 to −76°F
	(−55 to −60°C)
Tensile strength (psi)	
at 73°F (23°C)	1,700–2,150
at 212°F (100°C)	1,040–1,170
at 277°F(136°C)	730–770
at 302°F (150°C)	640–750
Elongation (% at break)	
at 73°F (23°C)	210–300
at 212°F (100°C)	290–670
at 277°F (136°C)	380–870
at 302°F (150°C)	300–750
Hardness, Shore	92A–54D
Abrasion resistance	Good
Maximum temperature, continuous use	277°F (136°C)
Compression set, method A (%)	
after 22 hr at 73°F (23°C)	8–18
after 22 hr at 212°F (100°C)	47–48
Tear resistance	Good
Resistance to sunlight	Good
Resistance to heat	Good
Effect of aging	Small
Electrical properties	Excellent

[a] These are representative values, and they may be altered by compounding.

4.20 SILICONE (SI) AND FLUOROSILICONE (FSI) RUBBERS

Silocone rubbers, also known as polysiloxanes, are a series of compounds whose
polymer structure consists of silicon and oxygen atoms rather than the carbon
structures of most other elastomers. The silicones are derivatives of silica, SiO_2 or
$O=Si=O$. When the atoms are combined so that the double linkages are broken
and methyl groups enter the linkages, silicone rubber is produced:

$$\left[\begin{array}{cc} CH_3 & CH_3 \\ | & | \\ O & O \\ Si & Si \\ | & | \\ CH_3 & CH_3 \end{array} \right]_n$$

Silicon is in the same chemical group as carbon but is a more stable element, and therefore more stable compounds are produced. The basic structure can be modified with vinyl or fluoride groups, which improve such properties as tear resistance, oil resistance, and chemical resistance. This results in a family of silicones that covers a wide range of physical and environmental requirements.

A. PHYSICAL AND MECHANICAL PROPERTIES

Silicones are some of the most heat-resistant elastomers available and the most flexible at low temperatures. Their effective operating temperature range is from −60 to 450°F (−51 to 232°C). They exhibit excellent properties even at the lowest temperature.

Fluorosilicones have an effective operating temperature range of −100 to 375°F (−73 to 190°C).

Silicone rubbers possess outstanding electrical properties, superior to those of most elastomers. The decomposition product of carbon-based elastomers is conductive carbon black, which can sublime and thus leave nothing for insulation, whereas the decomposition product of the silicone rubbers is an insulating silicone dioxide. This property is taken advantage of in the insulation of electric motors. The polysiloxanes have poor abrasion resistance, tensile strength, and tear resistance, but they exhibit good compression set resistance and rebound properties in both cold and hot environments. Their resistance to flame is good.

The physical and mechanical properties of silicone rubbers are given in Table 4.40.

Fluorosilicones (FSIs) have essentially the same physical and mechanical properties as silicones but with some improvement in adhesion to metals and impermeability. Table 4.41 lists the physical and mechanical properties of fluorosilicones.

B. Resistance to Sun, Weather, and Ozone

Silicone and fluorosilicone rubbers display excellent resistance to sun, weathering, and ozone. Their properties are virtually unaffected by long-term exposure.

C. Chemical Resistance

Silicone rubbers can be used in contact with dilute acids and alkalies, alcohols, animal and vegetable oils, and lubricating oils. They are also resistant to aliphatic hydrocarbons, but aromatic solvents such as benzene, toluene, gasoline, and chlorinated solvents will cause excessive swelling. Although they have excellent resistance to water and weathering, they are not resistant to high-pressure, high-temperature steam.

The fluorosilicone rubbers have better chemical resistance than the silicone rubbers. They have excellent resistance to aliphatic hydrocarbons and good

TABLE 4.40   Physical and Mechanical Properties of Silicone (SI) Rubbers[a]

Specific gravity	1.05–1.94
Brittle point	−75°F (−60°C)
Water absorption (%/24 h)	0.02–0.6
Dielectric strength (V/mil)	350–590
Dissipation (power) factor	
at 60 Hz	0.0007
at 1 MHz	8.5×10^{-3}–2.6×10^{-3}
Dielectric constant	
at 60 Hz	2.91
at 1 MHz	2.8–3.94
Volume resistivity (ohm-cm)	1×10^{14}–1×10^{16}
Tensile strength (psi)	1,200–6,000
Elongation (% at break)	800
Hardness, Shore A	20–90
Abrasion resistance	Poor
Maximum temperature, continuous use	450°F (232°C)
Tear resistance	Fair to good
Compression set (%)	10–15
Impact resistance, notch 1/8-in. specimen (ft-lb/in.)	0.25–0.30
Resistance to sunlight	Excellent
Effect of aging	Nil
Resistance to heat	Excellent

[a] These are representative values, and they may be altered by compounding.

TABLE 4.41   Physical and Mechanical Properties of Fluorosilicones[a]

Specific gravity	1.4
Hardness, Shore A	40–75
Tensile strength (psi)	1,000–5,400
Elongation (% at break)	100–500
Compression set (%)	15
Tear resistance	Poor to fair
Maximum temperature, continuous use	375°F (190°C)
Abrasion resistance	Poor
Resistance to sunlight	Excellent
Effect of aging	Nil
Resistance to heat	Excellent

[a] These are representative values, and they may be altered by compounding.

resistance to aromatic hydrocarbons, oil and gasoline, animal and vegetable oils, dilute acids and alkalies, and alcohols and fair resistance to concentrated alkalies.

Refer to Table 4.42 for the compatibility of silicone rubbers with selected corrodents.

D. Applications

Because of their unique thermal stability and/or their insulating values, silicone rubbers find many uses in the electrical industries, primarily in appliance, heaters, furnaces, aerospace devices, and automotive parts.

Their excellent weathering qualities and wide temperature range have also resulted in their employment as caulking compounds.

When silicone or fluorosilicone rubbers are infused with a high-density conductive filler, an electric path is created. These conductive elastomers are used as part of an EMI/RFI/EMP shielding process in forms such as O-rings and gaskets to provide

1. Shielding for containment to prevent the escape of EMI internally generated by the device.
2. Shielding for exclusion to prevent the intrusion of EMI/RF/EMP created by outside sources into the protected device.
3. Exclusion or containment plus pressure or vacuum sealing to provide both EMI/EMP attenuation and pressure containment and/or weatherproofing.
4. Grounding and contacting to provide a dependable low-impedance connection to conduct electric energy to ground, often used where mechanical mating is imperfect or impractical.

The following equipment is either capable of generating EMI or susceptible to EMI:

Aircraft and aerospace electronics
Analog instrumentation
Automotive electronics
Business machines
Communication systems
Digital instrumentation and process control systems
Home appliances
Radio-frequency instrumentation and radar
Medical electronics
Military and marine electronics
Security systems (military and commercial)

Table 4.43 lists some typical properties of these infused elastomers.

TABLE 4.42 Compatibility of Silicone Rubbers with Selected Corrodents[a]

Chemical	Maximum temp. °F	Maximum temp. °C	Chemical	Maximum temp. °F	Maximum temp. °C
Acetamide	80	27	Ethyl chloride	x	x
Acetic acid 10%	90	32	Ethylene chloride	x	x
Acetic acid 20%	90	32	Ethylene diamine	400	204
Acetic acid 50%	90	32	Ethylene glycol	400	204
Acetic acid 80%	90	32	Ferric chloride	400	204
Acetic acid, glacial	90	32	Fluosilicic acid	x	x
Acetic acid, vapors	90	32	Formaldehyde, all concn	200	93
Acetone	110	43	Fuel oil	x	x
Acetone, 50% water	110	43	Gasoline	x	x
Acetophenone	x	x	Glucose (corn syrup)	400	204
Acrylic acid 75%	80	27	Glycerine	410	210
Acrylonitrile	x	x	Green liquor	400	204
Aluminum acetate	x	x	Hexane	x	x
Aluminium phosphate	400	204	Hydrobromic acid	x	x
Aluminum sulfate	410	210	Hydrochloric acid, dil	90	32
Ammonia gas	x	x	Hydrochloric acid 20%	90	32
Ammonium chloride 10%	80	27	Hydrochloric acid 35%	x	x
Ammonium chloride 29%	80	27	Hydrofluoric acid	x	x
Ammonium chloride, sat.	80	27	Hydrogen peroxide, all	200	93
Ammonium hydroxide 10%	210	99	concn		
Ammonium hydroxide, sat.	400	204	Lactic acid, all concn	80	27
Ammonium nitrate	80	27	Lead acetate	x	x
Amyl acetate	x	x	Lime sulfur	400	204
Amyl alcohol	x	x	Linseed oil	x	x
Aniline	80	27	Magnesium chloride	400	204
Aqua regia 3 : 1	x	x	Magnesium sulfate	400	204
Barium sulfide	400	204	Mercury	80	27
Benzene	x	x	Methyl alcohol	410	210
Benzyl chloride	x	x	Methyl cellosolve	x	x
Boric acid	390	189	Methyl chloride	x	x
Butyl alcohol	80	27	Methyl ethyl ketone	x	x
Calcium acetate	x	x	Methylene chloride	x	x
Calcium bisulfite	400	204	Mineral oil	300	149
Calcium chloride, all concn	300	149	Naptha	x	x
Calcium hydroxide to 30%	210	99	Nickel acetate	x	x
Calcium hydroxide, sat.	400	204	Nickel chloride	400	204
Carbon bisulfide	x	x	Nickel sulfate	400	204
Carbon monoxide	400	204	Nitric acid 5%	80	27
Carbonic acid	400	204	Nitric acid 10%	80	27
Chlorobenzene	x	x	Nitric acid 20%	x	x
Chlorosulfonic acid	x	x	Nitric acid, anhydrous	x	x
Dioxane	x	x	Nitrous acid–sulfuric acid	x	x
Ethane	x	x	50 : 50		
Ethers, general	x	x	Nitrobenzene	x	x
Ethyl acetate	170	77	Nitrogen	400	204
Ethyl alcohol	400	204	Nitromethane	x	x

TABLE 4.42 Continued

Chemical	Maximum temp. °F	°C	Chemical	Maximum temp. °F	°C
Oils, vegetable	400	204	Sodium acetate	x	x
Oleic acid	x	x	Sodium bisulfite	410	210
Oleum	x	x	Sodium borate	400	204
Oxalic acid to 50%	80	27	Sodium carbonate	300	149
Ozone	400	204	Sodium chloride 10%	400	204
Palmitic acid	x	x	Sodium hydroxide, all concn	90	32
Paraffin	x	x	Sodium peroxide	x	x
Peanut oil	400	204	Sodium sulfate	400	204
Perchloric acid	x	x	Sodium thiosulfate	400	204
Phenol	x	x	Stannic chloride	80	27
Phosphoric acid	x	x	Styrene	x	x
Picric acid	x	x	Sulfite liquors	x	x
Potassium chloride 30%	400	204	Sulfuric acid	x	x
Potassium cyanide 30%	410	210	Sulfurous acid	x	x
Potassium dichromate	410	210	Tartaric acid	400	204
Potassium hydroxide	210	99	Tetrahydrofuran	x	x
to 50%			Toluene	x	x
Potassium hydroxide 90%	80	27	Tributyl phosphate	x	x
Potassium nitrate to 80%	400	204	Turpentine	x	x
Potassium sulfate 10%	400	204	Vinegar	400	204
Potassium sulfate, pure	400	204	Water, acid mine	210	99
Propane	x	x	Water, demineralized	210	99
Propyl acetate	x	x	Water, distilled	210	99
Propyl alcohol	400	204	Water, salt	210	99
Propyl nitrate	x	x	Water, sea	210	99
Pyridine	x	x	Xylene	x	x
Silver nitrate	410	210	Zinc chloride	400	204

[a] The chemicals listed are in the pure state or in a saturated solution unless otherwise indicated. Compatibility is shown to the maximum allowable temperature for which data are available. Incompatibility is shown by an x. A blank space indicates that data are unavailable. *Source:* Schweitzer PA. Corrosion Resistance Tables, 4th ed., Vols. 1–3. New York: Marcel Dekker, 1995.

4.21 VINYLIDENE FLUORIDE (HFP, PVDF)

Polyvinylidene fluoride (PVDF) is a homopolymer of 1,1-difluoroethene with alternating CH_2 and CF_2 groups along the polymer chain. These groups impart a unique polarity that influences its solubility and electrical properties. The polymer has the characteristic stability of fluoropolymers when exposed to aggressive thermal, chemical, and ultraviolet conditions.

TABLE 4.43　Typical Properties of Conductive Elastomers[a]

Property	SI	FSI
Volume resistivity (ohm-cm)	0.002–0.01	0.002–0.01
Shielding effectiveness (dB)		
at 200 kHz (H field)	30–75	60–75
at 100 kHz (E field)	70–120	95–120
at 500 kHz (E field)	60–120	90–120
at 2 GHz (plane wave)	40–120	80–115
at 10 GHz (plane wave)	30–120	75–120
Heat aging (ohm-cm)	0.006–0.012	0.006–0.015
Electrical stability after break (ohm-cm)	0.003–0.015	0.003–0.015
Vibration resistance (ohm-cm)		
during	0.004–0.008	0.004–0.008
after	0.002–0.010	0.002–0.005
EMP survivability (kA/in. perimeter)	0.9	0.9
Specific gravity	1.9–4.5	2.1–4.1
Hardness, Shore A	50–85	70–85
Tensile strength (psi)	175–600	200–500
Elongation (% at break)	20–300	70–300
Tear strength (psi)	20–75	40–50
Compression set (%)	22–40	24–29
Operating temperature		
(min °F/°C)	−85/−65	−85/−65
(max °F/°C)	392/200	392/200

[a]These are typical values and they may be altered by compounding. SI = silicone; FSI = fluorosilicone.

In general, PVDF is one of the easiest fluoropolymers to process and can be easily recycled without affecting its physical and chemical properties. As with other elastomeric materials, compounding can be used to improve certain specific properties. Cross-linking of the polymer chain and control of the molecular weight are also done to improve particular properties.

PVDF possesses mechanical strength and toughness, high abrasion resistance, high thermal stability, high dielectric strength, high purity, resistance to most chemicals and solvents, resistance to ultraviolet and nuclear radiation, resistance to weathering, and resistance to fungi. It can be used in applications intended for repeated contact with food per Title 21, Code of Federal Regulations, Chapter 1, Part 177.2520. PVDF is also permitted for use in processing or storage areas on contact with meat or poultry food products prepared under federal inspection according to the U.S. Department of Agriculture (USDA). Use is also permitted under "3-A Sanitary Standards for Multiple-Use Plastic Materials Used as Product Contact Surfaces for Dairy Equipment Serial No. 2000." This material has the ASTM designation of MFP.

A. Physical and Mechanical Properties

PVDF elastomers have high tensile and impact strengths. The ambient temperature tensile strength at yield of 4,000–7,000 psi (28–48 Mpa) and the unnotched impact strength, 15–80 ft-lb/in. (800–4270 kJ/m), indicate that all grades of this polymer are strong and tough. These properties are retained over a wide temperature range.

Excellent resistance to creep and fatigue are also exhibited by PVDF, while thin sections such as films, filament, and tubing are flexible and transparent. PVDF wire insulation has excellent resistance to cut-through. Where load bearing is important, these polymers are rigid and resistant to creep under mechanical load. Resistance to deformation under load is extremely good over the temperature range of −112 to 302°F (−80 to 150°C).

PVDF finds application as thin-wall primary insulation and as a jacket for control wiring as a result of its high dielectric strength and good mechanical properties. Because of its high dissipation factor, PVDF has limited use at high frequencies. However, this property becomes an advantage in components utilizing dielectric heating techniques.

In order for PVDF to burn, it is necessary to have a 44% oxygen environment. The Underwriters Laboratories give PVDF a vertical burn rating of 94-V-0. PVDF is fire resistant and self-extinguishing.

PVDF will not support the growth of fungi. However, if additives that will support the growth of fungi are used in compounding, then fungicides should also be added to overcome this problem.

Because of its extremely low weight loss when exposed to high vacuum, PVDF can be used in high-vacuum applications. At 212°F (100°C) and a pressure of 5×10^{-6} torr, the measured rate of weight loss is 13×10^{-11} g/cm^2-sec.

PVDF can also be pigmented.

The physical and mechanical properties of PVDF are given in Table 4.44.

PVDF is suitable for applications where load-bearing characteristics are important, particularly at elevated temperatures. Long-term deflections at 140°F (60°C) with a load of 172 psi vary from 0.8 to 119%, while at 194°F (90°C) with a load of 530 psi, the deflection ranges from 1.2 to 3.6%.

PVDF has a wide useful temperature range with a brittle point at −80°F (−62°C) and a maximum continuous operating temperature of 302°F (150°C). Over this entire range, its properties remain virtually unaffected.

Special formulations are available for use in plenum cable applications. These formulations produce products having the impact strengths of 12–18

TABLE 4.44 Physical and Mechanical Properties of Vinylidene
Fluoride (PVDF)[a]

Specific gravity	1.76–1.78
Refractive index	1.42
Specific heat (Btu/lb-°F)	0.30–0.34
Brittle point	−80°F (−62°C)
Coefficient of linear expansion	
per °F	7.8×10^{-5}
per °C	1.4×10^{-4}
Thermal conductivity	
(BTU-in./hr-ft^2°F	0.70–0.87
(cal/cm-sec-°C	3×10^{-4}
Dielectric strength (kV/mm)	63–67
Dielectric constant at 77°F (23°C)	
at 100 Hz (ohm-cm)	9.9
at 1 kHz (ohm-cm)	9.3
at 100 kHz (ohm-cm)	8.5
Tensile strength (psi)	4,000–7,000
Elongation (% at break)	25–650
Hardness, Shore A	77–83
Abrasion resistance, Armstrong (ASTM	0.3
D 1242) 30-lb load volume loss (cm^3)	
Maximum temperature, continuous use	302°F (150°C)
Impact resistance, Izod notched, (ft-lb/in.)	3–18
Compression set	Good
Machining qualities	Excellent
Resistance to sunlight	Excellent
Effect of aging	Nil
Resistance to heat	Good

[a]These are representative values, and they may be altered by
compounding.

ft-lb/in. and elongations of 400–650%. They also have improved stress crack
resistance.

B. Resistance to Sun, Weather, and Ozone

PVDF is highly resistant to the effects of sun, weather, and ozone. Its mechanical
properties are retained, while the percent elongation to break decreases to a lower
level and then remains constant.

C. Chemical Resistance

In general, PVDF is completely resistant to chlorinated solvents, aliphatic
solvents, weak bases and salts, strong acids, halogens, strong oxidants, and
aromatic solvents. Strong bases will attack the material.

The broader molecular weight of PVDF gives it greater resistance to stress cracking than many other materials, but it is subject to stress cracking in the presence of sodium hydroxide.

PVDF also exhibits excellent resistance to nuclear radiation. The original tensile strength is essentially unchanged after exposure to 1000 Mrads of gamma irradiation from a cobalt-60 source at 122°F (50°C) and in high vacuum (10^{-6} torr). Because of cross-linking, the impact strength and elongation are slightly reduced. This resistance makes PVDF useful in plutonium reclamation operations.

Refer to Table 4.45 for the compatibility of PVDF with selected coorodents.

D. Applications

PVDF finds many applications where its properties of corrosion resistance, wide allowable operating temperature range, mechanical strength and toughness, high abrasion resistance, high dielectric strength, and resistance to weathering, ultraviolet light, radiation, and fungi are useful.

In the electrical and electronics fields, PVDF is used for multiwire jackcting, plenum cables, heat-shrinkable tubing, anode lead wire, computer wiring, and cable and cable ties. Because of its acceptance in the handling of foods and pharmaceuticals, transfer hoses are lined with PVDF. Its corrosion resistance is also a factor in these applications.

In fluid-handling systems, PVDF finds applications as gasketing material, valve diaphragms, and membranes for microporous filters and ultrafiltration.

As a result of its resistance to fungi and its exceptional corrosion resistance, it is also used as the insulation material for underground anode bed installations.

4.22 FLUOROELASTOMERS (FKM)

Fluoroelastomers are fluorine-containing hydrocarbon polymers with a saturated structure obtained by polymerizing fluorinated monomers such as vinylidene fluoride, hexafluoropropene, and tetrafluoroethylene. The result is a high-performance synethetic rubber with exceptional resistance to oils and chemicals at elevated temperatures. Initially, this material was used to produce O-rings for use in severe conditions. Although this remains a major area of application, these compounds have found wide use in other applications because of their chemical resistance at high temperatures and other desirable properties.

As with other rubbers, fluoroelastomers are capable of being compounded with various additives to enhance specific properties for particular applications. Fuoroelastomers are suitable for all rubber processing applications, including compression molding, injection molding, injection/compression molding, transfer molding, extrusion, calendering, spreading, and dipping.

TABLE 4.45 Compatibility of PVDF with Selected Corrodents[a]

Chemical	Maximum temp.		Chemical	Maximum temp.	
	°F	°C		°F	°C
Acetaldehyde	150	66	Barium sulfate	280	138
Acetamide	90	32	Barium sulfide	280	138
Acetic acid 10%	300	149	Benzaldehyde	120	49
Acetic acid 50%	300	149	Benzene	150	66
Acetic acid 80%	190	88	Benzenesulfonic acid 10%	100	38
Acetic acid, glacial	190	88	Benzoic acid	250	121
Acetone	x	x	Benzyl alcohol	280	138
Acetyl chloride	120	49	Benzyl chloride	280	138
Acrylic acid	150	66	Borax	280	138
Acrylonitrile	130	54	Boric acid	280	138
Adipic acid	280	138	Bromine gas, dry	210	99
Allyl alcohol	200	93	Bromine gas, moist	210	99
Allyl chloride	200	93	Bromine, liquid	140	60
Alum	180	82	Butadiene	280	138
Aluminum acetate	250	121	Butyl acetate	140	60
Aluminum chloride, aqueous	300	149	Butyl alcohol	280	138
			n-Butylamine	x	x
Aluminum chloride, dry	270	132	Butyric acid	230	110
Aluminum fluoride	300	149	Calcium bisulfide	280	138
Aluminum hydroxide	260	127	Calcium bisulfite	280	138
Aluminum nitrate	300	149	Calcium carbonate	280	138
Aluminum oxychloride	290	143	Calcium chlorate	280	138
Aluminum sulfate	300	149	Calcium chloride	280	138
Ammonia gas	270	132	Calcium hydroxide 10%	270	132
Ammonium bifluoride	250	121	Calcium hydroxide, sat.	280	138
Ammonium carbonate	280	138	Calcium hypochlorite	280	138
Ammonium chloride 10%	280	138	Calcium nitrate	280	138
Ammonium chloride 50%	280	138	Calcium oxide	250	121
Ammonium chloride, sat.	280	138	Calcium sulfate	280	138
Ammonium fluoride 10%	280	138	Caprylic acid	220	104
Ammonium fluoride 25%	280	138	Carbon bisulfide	80	27
Ammonium hydroxide 25%	280	138	Carbon dioxide, dry	280	138
Ammonium hydroxide, sat.	280	138	Carbon dioxide, wet	280	138
Ammonium nitrate	280	138	Carbon disulfide	80	27
Ammonium persulfate	280	138	Carbon monoxide	280	138
Ammonium phosphate	280	138	Carbon tetrachloride	280	138
Ammonium sulfate 10–40%	280	138	Carbonic acid	280	138
Ammonium sulfide	280	138	Cellosolve	280	138
Ammonium sulfite	280	138	Chloroacetic acid, 50% water	210	99
Amyl acetate	190	88			
Amyl alcohol	280	138	Chloroacetic acid	200	93
Amyl chloride	280	138	Chlorine gas, dry	210	99
Aniline	200	93	Chlorine gas, wet, 10%	210	99
Antimony trichloride	150	66	Chlorine, liquid	210	99
Aqua regia 3:1	130	54	Chlorobenzene	220	104
Barium carbonate	280	138	Chloroform	250	121
Barium chloride	280	138	Chlorosulfonic acid	110	43
Barium hydroxide	280	138	Chromic acid 10%	220	104

TABLE 4.45 Continued

Chemical	Maximum temp. °F	°C	Chemical	Maximum temp. °F	°C
Chromic acid 50%	250	121	Manganese chloride	280	138
Chromyl chloride	110	43	Methyl chloride	x	x
Citric acid 15%	250	121	Methyl ethyl ketone	x	x
Citric acid, concd	250	121	Methyl isobutyl ketone	110	43
Copper acetate	250	121	Muriatic acid	280	138
Copper carbonate	250	121	Nitric acid 5%	200	93
Copper chloride	280	138	Nitric acid 20%	180	82
Copper cyanide	280	138	Nitric acid 70%	120	49
Copper sulfate	280	138	Nitric acid, anhydrous	150	66
Cresol	210	99	Nitrous acid, concd	210	99
Cupric chloride 5%	270	132	Oleum		x
Cupric chloride 50%	270	132	Perchloric acid 10%	210	99
Cyclohexane	250	121	Perchloric acid 70%	120	49
Cyclohexanol	210	99	Phenol	200	93
Dibutyl phthalate	80	27	Phosphoric acid 50–80%	220	104
Dichloroacetic acid	120	49	Picric acid	80	27
Dichloroethane	280	138	Potassium bromide 30%	280	138
(ethylene dichloride)			Salicylic acid	220	104
Ethylene glycol	280	138	Silver bromide 10%	250	121
Ferric chloride	280	138	Sodium carbonate	280	138
Ferric chloride 50% in water	280	138	Sodium chloride	280	138
Ferric nitrate 10–50%	280	138	Sodium hydroxide 10%	230	110
Ferrous chloride	280	138	Sodium hydroxide 50%	220	104
Ferrous nitrate	280	138	Sodium hydroxide, concd	150	66
Fluorine gas, dry	80	27	Sodium hypochlorite 20%	280	138
Fluorine gas, moist	80	27	Sodium hypochlorite, concd	280	138
Hydrobromic acid, dil	260	127	Sodium sulfide to 50%	280	138
Hydrobromic acid 20%	280	138	Stannic chloride	280	138
Hydrobromic acid 50%	280	138	Stannous chloride	280	138
Hydrochloric acid 20%	280	138	Sulfuric acid 10%	250	121
Hydrochloric acid 38%	280	138	Sulfuric acid 50%	220	104
Hydrocyanic acid 10%	280	138	Sulfuric acid 70%	220	104
Hydrofluoric acid 30%	260	127	Sulfuric acid 90%	210	99
Hydrofluoric acid 70%	200	93	Sulfuric acid 98%	140	60
Hydrofluoric acid 100%	200	93	Sulfuric acid 100%	x	x
Hypochlorous acid	280	138	Sulfuric acid, fuming	x	x
Iodine solution 10%	250	121	Sulfurous acid	220	104
Ketones, general	110	43	Thionyl chloride	x	x
Lactic acid 25%	130	54	Toluene	x	x
Lactic acid, concd	110	43	Trichloroacetic acid	130	54
Magnesium chloride	280	138	White liquor	80	27
Malic acid	250	121	Zinc chloride	260	127

[a]The chemicals listed are in the pure state or in a saturated solution unless otherwise indicated. Compatibility is shown to the maximum allowable temperature for which data are available. Incompatibility is shown by an x. A blank space indicates that data are unavailable.
Source: Schweitzer PA. Corrosion Resistance Tables, 4th ed., Vols. 1–3. New York: Marcel Dekker, 1995.

These compounds possess the rapid recovery from deformation, or resilience, of a true elastomer and exhibit mechanical properties of the same order of magnitude as those of conventional synthetic rubbers.

Fluoroelastomers are manufactured under various trade names by different manufacturers. Three typical materials are listed below.

Trade name	Manufacturer
Viton	Dupont
Technoflon	Ausimont
Fluorel	3M

These elastomers have the ASTM designation FKM.

A. Physical and Mechanical Properties

The general physical and mechanical properties of fluoroelastomers are similar to those of other synthetic rubbers. General-purpose compounds have a hardness of 70–75 Shore A. Formulations are produced that have hardnesses ranging from 45 to 95 Shore A. At elevated temperatures (250–500°F; 121–260°C), hardness may decrease by 5–15 points depending upon the polymer and the formulation. These variations must be taken into account when specifying the hardness of products to be used at elevated temperatures.

Fluoroelastomer compounds have good tensile strength, ranging from 1,800 to 2,900 psi. In general, the tensile strength of any elastomer tends to decrease at elevated temperatures; however, this loss in tensile strength is much less with the fluoroelastomers. Percent elongation at break is an indication of operating life. A high percentage is essential when high resistance to bending stress is required for the application. These elastomers have a range of 100–400%. The ability of fluoroelastomers to recover their original dimension after compression and their exceptional thermal resistance make it possible to fabricate cured items with very low set compression values even under the most severe operating conditions. These values become even more meaningful at elevated temperatures when it is realized that most rubbers have a maximum service temperature of less than 250°F (121°C). Table 4.46 lists the values at 300°F (149°C) and 392°F (200°C).

The resilience of fluoroelastomers makes them suitable for application as vibration isolators at elevated temperatures and as vibration dampers (energy absorbers) at room temperature. In the latter case, because of cost, they would normally be used only in extremely corrosive atmospheres. These rubbers can be applied as coatings to fabrics or adhered to a variety of metals to provide fluid resistant to the substrate.

The temperature resistance of fluoroelastomers is exceptionally good over a wide temperature range. At high temperatures, their mechanical properties are

TABLE 4.46 Physical and Mechanical Properties of Fluoroelastomers (FKM)[a]

Specific gravity	1.8
Specific heat	0.395
Brittle point	−25 to −75°F
	(−32 to −59°C)
Coefficient of linear expansion	$88 \times 10^{-6}/°F$,
	$16 \times 10^{-5}/°C$
Thermal conductivity	
(BTU-in./hr-ft^2°F at 100°F)	1.58
(kg-cal/cm-cm^2-°C-hr at 38°C)	1.96
Electrical properties	
Dielectric constant at 1,000 Hz	
at 75°F (24°C)	10.5
at 300°F (149°C)	7.1
at 390°F (199°C)	9.1
Dissipation factor at 1,000 Hz	
at 75°F (24°C)	0.034
at 300°F (149°C)	0.273
at 390°F (199°C)	0.39–1.19
Permeability (cm^3/cm^2-cm-sec-atm)	
at 75°F (24°C)	
to air	0.0099×10^{-7}
to helium	0.892×10^{-7}
to nitrogen	0.0054×10^{-7}
at 86°F (30°C)	
to carbon dioxide	0.59×10^{-7}
to oxygen	0.11×10^{-7}
Tensile strength (psi)	1,800–2,900
Elongation (% at break)	400
Hardness, Shore A	45–95
Abrasion resistance	Good
Maximum temperature, continuous use	400°F (205°C)
Compression set (%)	
at 70°F (21°C)	21
at 300°F (149°C)	32
at 392°F (200°C)	98
Tear resistance	Good
Resistance to sunlight	Excellent
Effect of aging	Nil
Resistance to heat	Excellent

[a] These are representative values, and they may be altered by compounding.

retained better than those of any other elastomer. Compounds remain usefully elastic indefinitely when exposed to aging up to 400°F (204°C). Continuous service limits are generally considered to be

        3,000 hr at 450°F (232°C)
        1,000 hr at 500°F (260°C)
          240 hr at 550°F (288°C)
            48 hr at 600°F (315°C)

On the low-temperature side these rubbers are generally serviceable in dynamic applications down to −10°F (−23°C). Flexibility at low temperature is a function of the material thickness. The thinner the cross section, the less stiff the material is at every temperature. The brittle point at a thickness of 0.075 in. (1.9 mm) is in the neighborhood of −50°F (−45°C). This temperature can have a range of −25 to −75°F (−32 to −59°C) depending upon the thickness and hardness of the material. Fluoroelastomers are relatively impermeable to air and gases, ranking about midway between the best and poorest elastomers in this respect. Their permeability can be modified considerably by the way they are compounded. In all cases, permeability increases rapidly with increasing temperature. Table 4.46 provides some data on the permeability of the fluoroelastomers.

    Being halogen-containing polymers, these elastomers are more resistant to burning than exclusively hydrocarbon rubbers. Normally compounded material will burn when directly exposed to flame but will stop burning when the flame is removed. Natural rubber and synthetic hydrocarbon rubbers under the same conditions will continue to burn when the flame is removed. However, it must be remembered that under an actual fire condition, fluoroelastomers will burn. During combustion, fluorinated products such as hydrofluoric acid can be given off. Special compounding can improve the flame resistance. One such formulation has been developed for the space program that will not ignite under conditions of the NASA test, which specifies 100% oxygen at 6.2 psi (absolute).

    Fluoroelastomers will increase in stiffness and hardness when exposed to gamma radiation from a cobalt-60 source. For dynamic applications, radiation exposure should not exceed 1×10^7 roentgens. Higher dosages are permissible for static applications. There is no evidence of radiation-induced stress cracking. There are other elastomers that exhibit superior radiation resistance. However, high temperatures are frequently encountered along with exposure to radiation, and in many cases these elevated temperatures will rule out the more radiation-resistant elastomers.

    Fluoroelastomers are particularly recommended when resistance to ozone, high temperatures, or highly corrosive fluids is required in addition to radiation resistance.

    The dielectric properties of fluoroelastomers permit them to be used as insulating materials at low tension and frequency in high-temperature applications and in the presence of higher concentrations of ozone and highly aggressive chemicals. The value of the individual properties can be greatly influenced by formulation but are generally in the following ranges:

Direct current resistivity	2×10^{13} ohm-cm
Dielectric constant	10–15
Dissipation factor	0.01–0.05
Dielectric strength	500 V/mil (2000 V/mm)

Additional electrical properties will be found in Table 4.46.

Fluoroelastomers have been approved by the U.S. Food and Drug Administration for use in repeated contact with food products. More details are available in the *Federal Register*, Vol. 33 No. 5, Tuesday January 9, 1968, Part 121—Food Additives, Subpart F—Food Additives Resulting from Contact with Containers or Equipment and Food Additives Otherwise Affecting Food—Rubber Articles Intended for Repeated Use.

The biological resistance of fluoroelastomers is excellent. A typical compound tested against specification MIL-E-5272C showed no fungus growth after 30 days. This specification covers four common fungus groups.

Additional physical and mechanical properties are given in Table 4.46.

B. Resistance to Sun, Weather, and Ozone

Because of their chemically saturated structure, fluoroelastomers exhibit excellent weathering resistance to sunlight and especially to ozone. After 13 years of exposure in Florida in direct sunlight, samples showed little or no change in properties or appearance. Similar results were experienced with samples exposed to various tropical conditions in Panama for a period of 10 years. Products made of this elastomer are unaffected by ozone concentrations as high as 100 ppm. No cracking occurred in a bent loop test after 1 year of exposure to 100 ppm of ozone in air at 100°F (38°C) or in a sample held at 356°F (180°C) for several hundred hours. This property is particularly important considering that standard tests, for example, in the automotive industry, require resistance only to 0.5 ppm ozone.

C. Chemical Resistance

Fluoroelastomers provide excellent resistance to oils, fuels, lubricants, most mineral acids, many aliphatic and aromatic hydrocarbons (carbon tetrachloride, benzene, toluene, xylene) that act as solvents for other rubbers, gasoline, naphtha, chlorinated solvents, and pesticides. Special formulation, can be produced to obtain resistance to hot mineral acids, steam, and hot water.

These elastomers are not suitable for use with low molecular weight esters and ethers, ketones, certain amines, or hot anhydrous hydrofluoric or chlorosulfonic acids. Their solubility in low molecular weight ketones is an advantage in producing solution coatings of fluoroelastomers.

Refer to Table 4.47 for the compatibility of fluoroelastomers with selected corrodents.

TABLE 4.47 Compatibility of Fluoroelastomers with Selected Corrodents[a]

Chemical	Maximum temp. °F	°C	Chemical	Maximum temp. °F	°C
Acetaldehyde	x	x	Barium hydroxide	400	204
Acetamide	210	99	Barium sulfate	400	204
Acetic acid 10%	190	88	Barium sulfide	400	204
Acetic acid 50%	180	82	Benzaldehyde	x	x
Acetic acid 80%	180	82	Benzene	400	204
Acetic acid, glacial	x	x	Benzenesulfonic acid 10%	190	88
Acetic anhydride	x	x	Benzoic acid	400	204
Acetone	x	x	Benzyl alcohol	400	204
Acetyl chloride	400	204	Benzyl chloride	400	204
Acrylic acid	x	x	Borax	190	88
Acrylonitrile	x	x	Boric acid	400	204
Adipic acid	190	88	Bromine gas, dry, 25%	180	82
Allyl alcohol	190	88	Bromine gas, moist, 25%	180	82
Allyl chloride	100	38	Bromine, liquid	350	177
Alum	190	88	Butadiene	400	204
Aluminum acetate	180	82	Butyl acetate	x	x
Aluminum chloride, aqueous	400	204	Butyl alcohol	400	204
			n-Butylamine	x	x
Aluminum fluoride	400	204	Butyric acid	120	49
Aluminum hydroxide	190	88	Calcium bisulfide	400	204
Aluminum nitrate	400	204	Calcium bisulfite	400	204
Aluminum oxychloride	x	x	Calcium carbonate	190	88
Aluminum sulfate	390	199	Calcium chlorate	190	88
Ammonia gas	x	x	Calcium chloride	300	149
Ammonium bifluoride	140	60	Calcium hydroxide 10%	300	149
Ammonium carbonate	190	88	Calcium hydroxide, sat.	400	204
Ammonium chloride 10%	400	204	Calcium hypochlorite	400	204
Ammonium chloride 50%	300	149	Calcium nitrate	400	204
Ammonium chloride, sat.	300	149	Calcium sulfate	200	93
Ammonium fluoride 10%	140	60	Carbon bisulfide	400	204
Ammonium fluoride 25%	140	60	Carbon dioxide, dry	80	27
Ammonium hydroxide 25%	190	88	Carbon dioxide, wet	x	x
Ammonium hydroxide, sat.	190	88	Carbon disulfide	400	204
Ammonium nitrate	x	x	Carbon monoxide	400	204
Ammonium persulfate	140	60	Carbon tetrachloride	350	177
Ammonium phosphate	180	82	Carbonic acid	400	204
Ammonium sulfate 10–40%	180	82	Cellosolve	x	x
Ammonium sulfide	x	x	Chloroacetic acid, 50% water	x	x
Amyl acetate	x	x			
Amyl alcohol	200	93	Chloroacetic acid	x	x
Amyl chloride	190	88	Chlorine gas, dry	190	88
Aniline	230	110	Chlorine gas, wet	190	88
Antimony trichloride	190	88	Chlorine, liquid	190	88
Aqua regia 3 : 1	190	88	Chlorobenzene	400	204
Barium carbonate	250	121	Chloroform	400	204
Barium chloride	400	204	Chlorosulfonic acid	x	x

TABLE 4.47   Continued

Chemical	Maximum temp. °F	Maximum temp. °C	Chemical	Maximum temp. °F	Maximum temp. °C
Chromic acid 10%	350	177	Manganese chloride	180	82
Chromic acid 50%	350	177	Methyl chloride	190	88
Citric acid 15%	300	149	Methyl ethyl ketone	x	x
Citric acid, concd	400	204	Methyl isobutyl ketone	x	x
Copper acetate	x	x	Muriatic acid	350	177
Copper carbonate	190	88	Nitric acid 5%	400	204
Copper chloride	400	204	Nitric acid 20%	400	204
Copper cyanide	400	204	Nitric acid 70%	190	88
Copper sulfate	400	204	Nitric acid, anhydrous	190	88
Cresol	x	x	Nitrous acid, concd	90	32
Cupric chloride 5%	180	82	Oleum	190	88
Cupric chloride 50%	180	82	Perchloric acid 10%	400	204
Cyclohexane	400	204	Perchloric acid 70%	400	204
Cyclohexanol	400	204	Phenol	210	99
Dibutyl phthalate	80	27	Phosphoric acid 50–80%	300	149
Dichloroethane (ethylene dichloride)	190	88	Picric acid	400	204
Ethylene glycol	400	204	Potassium bromide 30%	190	88
Ferric chloride	400	204	Salicylic acid	300	149
Ferric chloride 50% in water	400	204	Sodium carbonate	190	88
Ferric nitrate 10–50%	400	204	Sodium chloride	400	204
Ferrous chloride	180	82	Sodium hydroxide 10%	x	x
Ferrous nitrate	210	99	Sodium hydroxide 50%	x	x
Fluorine gas, dry	x	x	Sodium hydroxide, concd	x	x
Fluorine gas, moist	x	x	Sodium hypochlorite 20%	400	204
Hydrobromic acid, dil	400	204	Sodium hypochlorite, concd	400	204
Hydrobromic acid 20%	400	204	Sodium sulfide to 50%	190	88
Hydrobromic acid 50%	400	204	Stannic chloride	400	204
Hydrochloric acid 20%	350	177	Stannous chloride	400	204
Hydrochloric acid 38%	350	177	Sulfuric acid 10%	350	177
Hydrocyanic acid 10%	400	204	Sulfuric acid 50%	350	177
Hydrofluoric acid 30%	210	99	Sulfuric acid 70%	350	177
Hydrofluoric acid 70%	350	177	Sulfuric acid 90%	350	177
Hydrofluoric acid 100%	x	x	Sulfuric acid 98%	350	177
Hypochlorous acid	400	204	Sulfuric acid 100%	180	88
Iodine solution 10%	190	88	Sulfuric acid, fuming	200	93
Ketones, general	x	x	Sulfurous acid	400	204
Lactic acid 25%	300	149	Thionyl chloride	x	x
Lactic acid, concd	400	204	Toluene	400	204
Magnesium chloride	390	199	Trichloroacetic acid	190	88
Malic acid	390	199	White liquor	190	88
			Zinc chloride	400	204

[a] The chemicals listed are in the pure state or in a saturated solution unless otherwise indicated. Compatibility is shown to the maximum allowable temperature for which data are available. Incompatibility is shown by an x. A blank space indicates that data are unavailable.
Source: Schweitzer PA. Corrosion Resistance Tables, 4th ed., Vols. 1–3. New York: Marcel Dekker, 1995.

D. Applications

The main applications for fluoroelastomers are in those products requiring
resistance to high operating temperatures together with high chemical resistance
to aggressive fluids and to those characterized by severe operating conditions that
no other elastomer can withstand. By proper formulation, cured items can be
produced that will meet the rigid specifications of the industrial, aerospace, and
military communities.

Recent changes in the automotive industry that have required reduction in
environmental pollution, reduced costs, energy saving, and improved reliability
have resulted in higher operating temperatures, which in turn require a higher
performance elastomer. The main innovations resulting from these requirements
are

Turbocharging
More compact, more efficient and faster engines
Catalytic exhausts
Cx reduction
Soundproofing

In addition, the use of lead-free fuels, alternative fuels, sour gasoline
lubricants, and antifreeze fluids has caused automotive fluids to be more corrosive
to elastomers. At the present time, fluoroelastomers are being applied as shaft
seals, valve stem seals, O-ring (water-cooled cylinders and injection pumps),
engine head gaskets, filter casing gaskets, diaphragms for fuel pumps, water
pump gaskets, turbocharge lubricating circuit bellows, carburetor accelerating
pump diaphragms, carburetor needle-valve tips, fuel hoses, and seals for exhaust
gas pollution control equipment.

In the field of aerospace applications, the reliability of materials under
extreme exposure conditions is of prime importance. The high- and low-
temperature properties of the fluoroelastomers have permitted them to give
reliable performance in a number of aircraft and missile components, specifically
manifold gaskets, coated fabrics, firewall seals, heat-shrinkable tubing and fittings
for wire and cable, mastic adhesive sealants, protective coatings, and numerous
types of O-ring seals.

The ability of fluoroelastomers to seal under extreme vacuum conditions in
the range of 10^{-9} mmHg is an additional feature that makes these materials useful
for components used in space.

The exploitation of oilfields in difficult areas, such as deserts or offshore
sites, has increased the problems of high temperatures and pressures, high
viscosities, and high acidity. These extreme operating conditions require elastomers that have high chemical resistance, thermal stability, and overall reliability
to reduce maintenance. The same problems exist in the chemical industry.

Fluoroelastomers provide a solution to these problems and are used for O-rings, V-rings, U-rings, gaskets, valve seats, diaphragms for metering pumps, hoses, expansion joints, safety clothing and gloves, linings for valves, and maintenance coatings.

Another important application of these elastomers is in the production of coatings and linings. Their chemical stability solves the problems of chemical corrosion by making it possible to use them for such purposes as a protective lining for power station stacks operated with high-sulfur fuels, a coating on rolls for the textile industry to permit scouring of fabrics, and tank linings for the chemical industry.

4.23 ETHYLENE–TETRAFLUOROETHYLENE (ETFE) ELASTOMER

This elastomer is sold under the trade name Tefzel by DuPont. Ethylene–tetrafluoroethylene (ETFE) is a modified partially fluorinated copolymer of ethylene and polytetrafluoroethylene (PTFE). Since it contains more than 75% TFE by weight, it has better resistance to abrasion and cut-through than TFE while retaining most of the corrosion resistance properties.

TABLE 4.48 Physical and Mechanical Properties ETFE[a]

Specific gravity	1.7
Hardness range, Rockwell	R-50 to D-75
Tensile strength (psi)	6,500
Elongation (% at break)	100–400
Tear resistance	Good
Maximum temperature, continuous use	300°F (149°C)
Brittle point	−150°F (−101°C)
Water absorption (%/24 hr)	0.029
Abrasion resistance	Good
Volume resistivity (ohm-cm)	$> 10^{16}$
Dielectric strength (kV/mm)	16 (3 mm)
Dielectric constant (10^{-3} to 10^6 Hz range)	2.6
Dissipation (power) factor	8×10^{-4}
Resistance to sunlight	Excellent
Resistance to heat	Good
Machining qualities	Good

[a] These are representative values, and they may be altered by compounding.

TABLE 4.49 Compatibility of ETFE with Selected Corrodents[a]

Chemical	Maximum temp. °F	Maximum temp. °C	Chemical	Maximum temp. °F	Maximum temp. °C
Acetaldehyde	200	93	Barium sulfide	300	149
Acetamide	250	121	Benzaldehyde	210	99
Acetic acid 10%	250	121	Benzene	210	99
Acetic acid 50%	250	121	Benzenesulfonic acid 10%	210	99
Acetic acid 80%	230	110	Benzoic acid	270	132
Acetic acid, glacial	230	110	Benzyl alcohol	300	149
Acetic anhydride	300	149	Benzyl chloride	300	149
Acetone	150	66	Borax	300	149
Acetyl chloride	150	66	Boric acid	300	149
Acrylonitrile	150	66	Bromine gas, dry	150	66
Adipic acid	280	138	Bromine water 10%	230	110
Allyl alcohol	210	99	Butadiene	250	121
Allyl chloride	190	88	Butyl acetate	230	110
Alum	300	149	Butyl alcohol	300	149
Aluminum chloride, aqueous	300	149	n-Butylamine	120	49
			Butyric acid	250	121
Aluminum chloride, dry	300	149	Calcium bisulfide	300	149
Aluminum fluoride	300	149	Calcium carbonate	300	149
Aluminum hydroxide	300	149	Calcium chlorate	300	149
Aluminum nitrate	300	149	Calcium chloride	300	149
Aluminum oxychloride	300	149	Calcium hydroxide 10%	300	149
Aluminum sulfate	300	149	Calcium hydroxide, sat.	300	149
Ammonium bifluoride	300	149	Calcium hypochlorite	300	149
Ammonium carbonate	300	149	Calcium nitrate	300	149
Ammonium chloride 10%	300	149	Calcium oxide	260	127
Ammonium chloride 50%	290	143	Calcium sulfate	300	149
Ammonium chloride, sat.	300	149	Caprylic acid	210	99
Ammonium fluoride 10%	300	149	Carbon bisulfide	150	66
Ammonium fluoride 25%	300	149	Carbon dioxide, dry	300	149
Ammonium hydroxide 25%	300	149	Carbon dioxide, wet	300	149
Ammonium hydroxide, sat.	300	149	Carbon disulfide	150	66
Ammonium nitrate	230	110	Carbon monoxide	300	149
Ammonium persulfate	300	149	Carbon tetrachloride	270	132
Ammonium phosphate	300	149	Carbonic acid	300	149
Ammonium sulfate 10–40%	300	149	Cellosolve	300	149
Ammonium sulfide	300	149	Chloroacetic acid, 50% water	230	110
Amyl acetate	250	121			
Amyl alcohol	300	149	Chloroacetic acid	230	110
Amyl chloride	300	149	Chlorine gas, dry	210	99
Aniline	230	110	Chlorine gas, wet	250	121
Antimony trichloride	210	99	Chlorine, water	100	38
Aqua regia 3:1	210	99	Chlorobenzene	210	99
Barium carbonate	300	149	Chloroform	230	110
Barium chloride	300	149	Chlorosulfonic acid	80	27
Barium hydroxide	300	149	Chromic acid 10%	150	66
Barium sulfate	300	149	Chromic acid 50%	150	66

TABLE 4.49 Continued

Chemical	Maximum temp. °F	Maximum temp. °C	Chemical	Maximum temp. °F	Maximum temp. °C
Chromyl chloride	210	99	Muriatic acid	300	149
Citric acid 15%	120	49	Nitric acid 5%	150	66
Copper chloride	300	149	Nitric acid 20%	150	66
Copper cyanide	300	149	Nitric acid 70%	80	27
Copper sulfate	300	149	Nitric acid, anhydrous	x	x
Cresol	270	132	Nitrous acid, concd	210	99
Cupric chloride 5%	300	149	Oleum	150	66
Cyclohexane	300	149	Perchloric acid 10%	230	110
Cyclohexanol	250	121	Perchloric acid 70%	150	66
Dibutyl phthalate	150	66	Phenol	210	99
Dichloroacetic acid	150	66	Phosphoric acid 50–80%	270	132
Ethylene glycol	300	149	Picric acid	130	54
Ferric chloride 50% in water	300	149	Potassium bromide 30%	300	149
Ferric nitrate 10–50%	300	149	Salicylic acid	250	121
Ferrous chloride	300	149	Sodium carbonate	300	149
Ferrous nitrate	300	149	Sodium chloride	300	149
Fluorine gas, dry	100	38	Sodium hydroxide 10%	230	110
Fluorine gas, moist	100	38	Sodium hydroxide 50%	230	110
Hydrobromic acid, dil	300	149	Sodium hypochlorite 20%	300	149
Hydrobromic acid 20%	300	149	Sodium hypochlorite, concd	300	149
Hydrobromic acid 50%	300	149	Sodium sulfide to 50%	300	149
Hydrochloric acid 20%	300	149	Stannic chloride	300	149
Hydrochloric acid 38%	300	149	Stannous chloride	300	149
Hydrocyanic acid 10%	300	149	Sulfuric acid 10%	300	149
Hydrofluoric acid 30%	270	132	Sulfuric acid 50%	300	149
Hydrofluoric acid 70%	250	121	Sulfuric acid 70%	300	149
Hydrofluoric acid 100%	230	110	Sulfuric acid 90%	300	149
Hypochlorous acid	300	149	Sulfuric acid 98%	300	149
Lactic acid 25%	250	121	Sulfuric acid 100%	300	149
Lactic acid, concd	250	121	Sulfuric acid, fuming	120	49
Magnesium chloride	300	149	Sulfurous acid	210	99
Malic acid	270	132	Thionyl chloride	210	99
Manganese chloride	120	49	Toluene	250	121
Methyl chloride	300	149	Trichloroacetic acid	210	99
Methyl ethyl ketone	230	110	Zinc chloride	300	149
Methyl isobutyl ketone	300	149			

[a] The chemicals listed are in the pure state or in a saturated solution unless otherwise indicated. Compatibility is shown to the maximum allowable temperature for which data are available. Incompatibility is shown by an x. A blank space indicates that data are unavailable.
Source: Schweitzer PA. Corrosion Resistance Tables, 4th ed., Vols. 1–3. New York: Marcel Dekker, 1995.

A. Physical and Mechanical Properties

Ethylene–tetrafluoroethylene has excellent mechanical strength, stiffness, and abrasion resistance, with a service temperature range of −370 to 300°F (−223 to 149°C). It also exhibits good tear resistance and good electrical properties. However, its outstanding property is its resistance to a wide range of corrodents.

The physical and mechanical properties of ETFE are given in Table 4.48.

B. Resistance to Sun, Weather, and Ozone

Ethylene–tetrafluoroethylene has outstanding resistance to sunlight, ozone, and weather. This feature, coupled with its wide range of corrosion resistance, makes the material particularly suitable for outdoor applications subject to atmospheric corrosion.

C. Chemical Resistance

Ethylene–tetrafluoroethylene is inert to strong mineral acids, inorganic bases, halogens, and metal salt solutions. Even carboxylic acids, anhydrides, aromatic and aliphatic hydrocarbons, alcohols, aldehydes, ketones, ethers, esters, chlorocarbons, and classic polymer solvents have little effect on ETFE.

Very strong oxidizing acids near their boiling points, such as nitric acid at high concentration, will affect ETFE in varying degrees, as will organic bases such as amines and sulfonic acids.

Refer to Table 4.49 for the compatibility of ETFE with selected corrodents.

D. Applications

The principal applications for ETFE are found in such products as gaskets, packings, and seals (O-rings, lip and X-rings) in areas where corrosion is a problem. The material is also used for sleeve, split curled, and thrust bearings, and for bearing pads for pipe and equipment support where expansion and contraction or movement may occur.

4.24 ETHYLENE–CHLOROTRIFLUOROETHYLENE (ECTFE) ELASTOMER

Ethylene–chlorotrifluoroethylene (ECTFE) elastomer is a 1 : 1 alternating copolymer of ethylene and chlorotrifluoroethylene. This chemical structure gives the polymer a unique combination of properties. It possesses excellent chemical resistance, good electrical properties, and broad-use temperature range [from cryogenic to 340°F (171°C)] and meets the requirements of the UL-94V-0 vertical flame test in thicknesses as low as 7 mils. ECTFE is a tough material with excellent impact strength over its entire operating temperature range. Of all the fluoropolymers, ECTFE ranks among the best for abrasion resistance.

Most techniques used in processing polyethylene can be used to process ECTFE. It can be extruded, injection molded, rotomolded, and applied by ordinary fluidized bed for electrostatic coatings techniques.

A. Physical and Mechanical Properties

Ethylene–chlorotrifluoroethylene possesses advantageous electrical properties. It has high resistivity and low loss. The dissipation factor varies somewhat with frequency, particularly above 1 kHz. The ac loss properties of ECTFE are superior to those of vinylidene fluoride and come close to those of PTFE. The dielectric constant is stable across a broad temperature and frequency range. (Refer to Table 4.50)

According to ASTM D-149, the dielectric strength of ECTFE has a value of 2000 V/mil in 1-mil thickness and 500 V/mil in $\frac{1}{8}$-in. thickness, which are similar to those obtained for polyethylene or PTFE.

The resistance to permeation by oxygen, carbon dioxide, chlorine gas, or hydrochloric acid is superior to that of PTFE or (fluorinated ethylene–propylene) FEP, being 10–100 times better. Water absorption is less than 0.1%.

Other important physical properties include a low coefficient of friction, excellent machinability, and the ability to be pigmented. In thicknesses as low as 7 mils, ECTFE has a UL-94-V-0 rating. The oxygen index (ASTM D2863) is 60 on a $\frac{1}{16}$-in. thick sample and 48 on a 0.0005-in. filament yarn. ECTFE is a strong, highly impact-resistant material that retains useful properties over a wide range of temperatures. Outstanding in this respect are properties related to impact at low temperatures. ECTFE can be applied at elevated temperatures in the range of 300–340°F (149–171°C). (Refer to Table 4.50.)

In addition to its excellent impact properties, ECTFE also possesses good tensile, flexural, and wear-related properties.

The resistance of ECTFE to degradation by heat is excellent. It can resist temperatures of 300–340°F (149–171°C) for extended periods of time without degradation. It is one of the most radiation-resistant polymers. Laboratory testing has determined that the following useful life can be expected at the temperatures indicated:

| Temperature | | Useful life, |
°F	°C	years
329	165	10
338	170	4.5
347	175	2
356	180	1.25

TABLE 4.50 Physical and Mechanical Properties of ECTFE Elastomer[a]

Specific gravity	1.68
Refractive index, nD	1.44
Specific heat (Btu/lb-°F)	0.28
Brittle point	−105°F (−76°C)
Insulation resistance (ohms)	$> 10^{15}$
Thermal conductivity at 203°F (93°C)	
(Btu-in./h-ft^2-°F)	1.09
Coefficient of linear expansion (°F^{-1} or °C^{-1})	
−22 to 122°F	4.4×10^{-5}
122 to 185°F	5.6×10^{-5}
185 to 257°F	7.5×10^{-5}
257 to 356°F	9.2×10^{-5}
−30 to 50°C	8×10^{-5}
50 to 85°C	10×10^{-5}
85 to 125°C	13.3×10^{-5}
125 to 180°C	16.5×10^{-5}
Dielectric strength (V/mil)	
0.0001 in. thick	2,000
$\frac{1}{8}$ in. thick	490
Dielectric constant	
at 60 Hz	2.6
at 10^3 Hz (1 kHz)	2.5
at 10^6 Hz (1 MHz)	2.5
Dissipation factor at 60 Hz	< 0.0009
at 10^3 Hz (1 kHz)	0.005
at 10^6 Hz (1 MHz)	0.003
Arc resistance (sec)	135
Moisture absorption (%)	< 0.1
Tensile strength (psi)	6,000–7,000
Elongation at break (% at room temperature)	200–300
Hardness, Shore D	75
Impact resistance (ft-lb/in. notch)	
at 73°F (23°C)	No break
at −40°F (−40°C)	2–3
Abrasion resistance, Armstrong (ASTM D1242)	
30 lb load, volume loss (cm^3)	0.3
Coefficient of friction	
static	0.15
dynamic (50 cm/sec)	0.65
Maximum temperature, continuous use	340°F (171°C)
Machining qualities	Excellent
Resistance to sunlight	Excellent
Effect of aging	Good
Resistance to heat	Good

[a] These are representative values, and they may be altered by compounding.

B. Resistance to Sun, Weather, and Ozone

Ethylene–chlorotrifluoroethylene is extremely resistant to sun, weather, and ozone attack. Its physical properties undergo very little change after long exposures.

C. Chemical Resistance

The chemical resistance of ECTFE is outstanding. It is resistant to most of the common corrosive chemicals encountered in industry. Included in this list of chemicals are strong mineral and oxidizing acids, alkalies, metal etchants, liquid oxygen, and practically all organic solvents except hot amines (aniline, dimethylamine, etc.). No known solvent dissolves or stress cracks ECTFE at temperatures up to 250°F (120°C).

Some halogenated solvents can cause ECTFE to become slightly plasticized when it comes into contact with them. Under normal circumstances, this does not impair the usefulness of the polymer. When the part is removed from contact with the solvent and allowed to dry, its mechanical properties return to their original values, indicating that no chemical attack has taken place.

As with other fluoropolymers, ECTFE will be attacked by metallic sodium and potassium.

The useful properties of ECTFE are maintained on exposure to cobalt-60 radiation of 200 Mrads.

Refer to Table 4.51 for the compatibility of ECTFE with selected corrodents.

D. Applications

This elastomer finds many applications in the electrical industry as wire and cable insulation and jacketing; plenum cable insulation, oil well wire and cable insulation; logging wire jacketing and jacketing for cathodic protection; aircraft, mass transit, and automotive wire; connectors; coil forms; resistor sleeves; wire tie wraps; tapes, tubing; and flexible printed circuitry and flat cable.

Applications are also found in other industries as diaphragms, flexible tubing, closures, seals, gaskets, convoluted tubing, and hose, particularly in the chemical, cryogenic, and aerospace industries.

Materials of ECTFE are also used for lining vessels, pumps, and other equipment.

4.25 PERFLUOROELASTOMERS (FPM)

Perfluoroelastomers provide the elastomeric properties of fluoroelastomers and the chemical resistance of PTFE. These compounds are true rubbers. Compared

TABLE 4.51 Compatibility of ECTFE with Selected Corrodents[a]

Chemical	Maximum temp.		Chemical	Maximum temp.	
	°F	°C		°F	°C
Acetic acid 10%	250	121	Benzene	150	66
Acetic acid 50%	250	121	Benzenesulfonic acid 10%	150	66
Acetic acid 80%	150	66	Benzoic acid	250	121
Acetic acid, glacial	200	93	Benzyl alcohol	300	149
Acetic anhydride	100	38	Benzyl chloride	300	149
Acetone	150	66	Borax	300	149
Acetyl chloride	150	66	Boric acid	300	149
Acrylonitrile	150	66	Bromine gas, dry	x	x
Adipic acid	150	66	Bromine, liquid	150	66
Allyl chloride	300	149	Butadiene	250	121
Alum	300	149	Butyl acetate	150	66
Aluminum chloride,	300	149	Butyl alcohol	300	149
aqueous			Butyric acid	250	121
Aluminum chloride, dry			Calcium bisulfide	300	149
Aluminum fluoride	300	149	Calcium bisulfite	300	149
Aluminum hydroxide	300	149	Calcium carbonate	300	149
Aluminum nitrate	300	149	Calcium chlorate	300	149
Aluminum oxychloride	150	66	Calcium chloride	300	149
Aluminum sulfate	300	149	Calcium hydroxide 10%	300	149
Ammonia gas	300	149	Calcium hydroxide, sat.	300	149
Ammonium bifluoride	300	149	Calcium hypochlorite	300	149
Ammonium carbonate	300	149	Calcium nitrate	300	149
Ammonium chloride 10%	290	143	Calcium oxide	300	149
Ammonium chloride 50%	300	149	Calcium sulfate	300	149
Ammonium chloride, sat.	300	149	Caprylic acid	220	104
Ammonium fluoride 10%	300	149	Carbon bisulfide	80	27
Ammonium fluoride 25%	300	149	Carbon dioxide, dry	300	149
Ammonium hydroxide 25%	300	149	Carbon dioxide, wet	300	149
Ammonium hydroxide, sat.	300	149	Carbon disulfide	80	27
Ammonium nitrate	300	149	Carbon monoxide	150	66
Ammonium persulfate	150	66	Carbon tetrachloride	300	149
Ammonium phosphate	300	149	Carbonic acid	300	149
Ammonium sulfate 10–40%	300	149	Cellosolve	300	149
Ammonium sulfide	300	149	Chloroacetic acid,	250	121
Amyl acetate	160	71	50% water		
Amyl alcohol	300	149	Chloroacetic acid	250	121
Amyl chloride	300	149	Chlorine gas, dry	150	66
Aniline	90	32	Chlorine gas, wet	250	121
Antimony trichloride	100	38	Chlorine, liquid	250	121
Aqua regia 3 : 1	250	121	Chlorobenzene	150	66
Barium carbonate	300	149	Chloroform	250	121
Barium chloride	300	149	Chlorosulfonic acid	80	27
Barium hydroxide	300	149	Chromic acid 10%	250	121
Barium sulfate	300	149	Chromic acid 50%	250	121
Barium sulfide	300	149	Citric acid 15%	300	149
Benzaldehyde	150	66	Citric acid, concd	300	149

TABLE 4.51 Continued

Chemical	°F	°C	Chemical	°F	°C
Copper carbonate	150	66	Nitric acid 5%	300	149
Copper chloride	300	149	Nitric acid 20%	250	121
Copper cyanide	300	149	Nitric acid 70%	150	66
Copper sulfate	300	149	Nitric acid, anhydrous	150	66
Cresol	300	149	Nitrous acid, concd	250	121
Cupric chloride 5%	300	149	Oleum	x	x
Cupric chloride 50%	300	149	Perchloric acid 10%	150	66
Cyclohexane	300	149	Perchloric acid 70%	150	66
Cyclohexanol	300	149	Phenol	150	66
Ethylene glycol	300	149	Phosphoric acid 50–80%	250	121
Ferric chloride	300	149	Picric acid	80	27
Ferric chloride 50% in water	300	149	Potassium bromide 30%	300	149
Ferric nitrate 10–50%	300	149	Salicylic acid	250	121
Ferrous chloride	300	149	Sodium carbonate	300	149
Ferrous nitrate	300	149	Sodium chloride	300	149
Fluorine gas, dry	x	x	Sodium hydroxide 10%	300	149
Fluorine gas, moist	80	27	Sodium hydroxide 50%	250	121
Hydrobromic acid, dil	300	149	Sodium hydroxide, concd	150	66
Hydrobromic acid 20%	300	149	Sodium hypochlorite 20%	300	149
Hydrobromic acid 50%	300	149	Sodium hypochlorite, concd	300	149
Hydrochloric acid 20%	300	149	Sodium sulfide to 50%	300	149
Hydrochloric acid 38%	300	149	Stannic chloride	300	149
Hydrocyanic acid 10%	300	149	Stannous chloride	300	149
Hydrofluoric acid 30%	250	121	Sulfuric acid 10%	250	121
Hydrofluoric acid 70%	240	116	Sulfuric acid 50%	250	121
Hydrofluoric acid 100%	240	116	Sulfuric acid 70%	250	121
Hypochlorous acid	300	149	Sulfuric acid 90%	150	66
Iodine solution 10%	250	121	Sulfuric acid 98%	150	66
Lactic acid 25%	150	66	Sulfuric acid 100%	80	27
Lactic acid, concd	150	66	Sulfuric acid, fuming	300	149
Magnesium chloride	300	149	Sulfurous acid	250	121
Malic acid	250	121	Thionyl chloride	150	66
Methyl chloride	300	149	Toluene	150	66
Methyl ethyl ketone	150	66	Trichloroacetic acid	150	66
Methyl isobutyl ketone	150	66	White liquor	250	121
Muriatic acid	300	149	Zinc chloride	300	149

[a] The chemicals listed are in the pure state or in a saturated solution unless otherwise indicated. Compatibility is shown to the maximum allowable temperature for which data are available. Incompatibility is shown by an x. A blank space indicates that data are unavailable.
Source: Schweitzer PA. Corrosion Resistance Tables, 4th ed., Vols. 1–3. New York: Marcel Dekker, 1995.

with other elastomeric compounds, they are more resistant to swelling and embrittlement and retain their elastomeric properties over the long term. In difficult environments, there are no other elastomers that can outperform the perfluoroelastomers. These synthetic rubbers provide the sealing force of a true elastomer and the chemical inertness and thermal stability of polytetrafluoroethylene.

As with other elastomers, perfluoroelastomers are compounded to modify certain of their properties. Such materials as carbon black, perfluorinated oil, and various fillers are used for this purpose.

The ASTM designations for these elastomers is FPM.

A. Physical and Mechanical Properties

One of the outstanding physical properties of the perfluoroelastomers is their thermal stability. They retain their elasticity and recovery properties up to 600°F (316°C) in long-term service and up to 650°F (343°C) in intermittent service. This is the highest temperature rating of any elastomer.

In general, the physical properties of the perfluoroelastomers are similar to those of the fluoroelastomers. As with most compounding, the enhancement of one property usually has an opposite effect on another property. For example, as the coefficient of friction increases, the hardness decreases. Because of these factors the physical and mechanical properties are given in Table 4.52 in ranges where they are compound-dependent. Special compounds are available for applications requiring thermal cycling, increased resistance to strong oxidizing environments, other pressures, different hardnesses, or other specific physical properties. Selection of specific compounds for special applications should be done in cooperation with the manufacturer.

These elastomers provide excellent performance in high-vacuum environments. They exhibit negligible outgassing over a wide temperature range. This is an important property in any application where freedom from contamination of process streams is critical. Typical applications include semiconductor manufacturing operations, aerospace applications, and analytical instruments.

Perfluoroelastomers have good mechanical properties (see Table 4.52). Because these elastomers are based on expensive monomers and require complex processing, they cost more than other elastomeric materials. As a result these materials are limited to use in extremely hostile environments and/or applications where high heat will quickly attack other elastomers. Perfluoro rubbers retain their elastic properties in long-term service at elevated temperatures.

One such perfluoroelastomer is sold by DuPont under the trade name Kalrez. As with other perfluoroelastomers, Kalrez is available in different compounds to meet specific needs. Table 4.53 lists the physical properties of standard compounds.

TABLE 4.52 Physical and Mechanical Properties of Perfluoroelastomers[a]

Specific gravity	1.9–2.0
Specific heat at 122–302°F (50–150°C) (cal/g)	0.226–0.250
Brittle point	−9 to −58°F (−23 to −50°C)
Coefficient of friction (to steel)	0.25–0.60
Tear strength (psi)	1.75–27
Coefficient of linear expansion	
(in./°F)	1.3×10^{-4}
(in./°C)	2.3×10^{-4}
Thermal conductivity (BTU-in./hr-°F-ft²)	
at 122°F (50°C)	1.3
at 212°F (100°C)	1.27
at 392°F (200°C)	1.19
at 572°F (300°C)	1.10
Dielectric constant (kV/mm)	17.7
Dielectric constant at 1,000 Hz	4.9
Dissipation factor at 1,000 Hz	5×10^{-3}
Permeability ($\times 10^{-9}$ cm³-cm/S-cm²-cmHg P)	
to nitrogen, at room temperature	0.05
to oxygen, at room temperature	0.09
at helium, at room temperature	2.5
to hydrogen, at 199°F (93°C)	113
Tensile strength (psi)	1,850–3,800
Elongation (% at break)	20–190
Hardness, Shore A	65–95
Maximum temperature, continuous use	600°F (316°C)
Abrasion resistance, NBS	121
Compression set (%)	
at room temperature	15–40
at 212°F (100°C)	32–54
at 400°F (204°C)	63–82
at 500°F (260°C)	63–79
Resistance to sunlight	Excellent
Effect of aging	Nil
Resistance to heat	Excellent

[a] These are representative values, and they may be altered by compounding.

1. Compound 4079

This is a carbon black filled compound having low compression set with excellent chemical resistance, good mechanical properties, and outstanding hot air aging properties. It exhibits low swell in organic and inorganic acids and aldehydes and has good response to temperature cycling effects. This compound is not

TABLE 4.53 Physical Properties of Standard Kalrez Compounds

Compound	4079	1050LF	1050
Hardness, Shore A ± 5	75	82	80
100% Modulus (psi)	1,050	1,800	1,400
Tensile Strength at break (psi)	2,450	2,700	2,300
Elongation at break (%)	150	125	135
Compression set (%) 70 hrs at 400°F/204°C	25	35	45
Brittle point (°F)	−58	−42	−35
(°C)	−50	−41	−37

recommended for use in hot water/steam applications or in contact with certain hot aliphatic amines, ethylene oxide, and propylene oxide.

It has a maximum continuous operating temperature of 600°F/316°C with short-term exposures at higher temperatures.

Applications include O-rings, diaphragms, seals, and other parts used in the process and aircraft industries.

2. Compound 1050LF

This compound has good water/steam resistance, excellent amine resistance, and compression set properties. It is the suggested compound for use with ethylene oxide and propylene oxide. It is not recommended for use with organic or inorganic acids at high temperatures. The maximum recommended continuous operating temperature is 550°F/288°C.

This compound finds applications as O-rings, seals, and other parts used in the chemical process industry.

3. Compound 1050

This compound is carbon black filled. It exhibits good mechanical properties, excellent all around chemical resistance, and is slightly harder than compound 4079. This compound should not be used in pure water/steam applications at elevated temperatures.

The maximum allowable continuous operating temperature is 500°F/260°C with short-term exposures at 550°F/288°C in nonoxidizing environments.

Applications incluce O-rings, seals, and other parts in the chemical process industry.

4. Specialty Compounds

In addition to the three standard compounds, there are six specialty compounds available. These have been modified to meet specific needs, but it is recommended that the manufacturer be consulted before ordering. If the standard compounds do not meet your needs, contact the manufacturer for a recommendation on the use of a specialty compound.

B. Resistance to Sun, Weather, and Ozone

Perfluoroelastomers provide excellent resistance to sun, weather, and ozone. Long-term exposure under these conditions has no effect on them.

C. Chemical Resistance

Perfluoroelastomers have outstanding chemical resistance. They are virtually immune to chemical attack at ambient and elevated temperatures. Typical corrodents to which perfluoroelastomers are resistant include the following:

Chlorine, wet and dry
Fuels (ASTM Reference Fuel C, JP-5 jet fuel, aviation gas, kerosene)
Heat transfer fluids
Hot mercury
Hydraulic fluids
Inorganic and organic acids (hydrochloric, nitric, sulfuric, trichloroacetic)
    and bases (hot caustic soda)
Inorganic salt solutions
Metal halides (titanium tetrachloride, diethylaluminium chloride)
Oil well sour gas (methane, hydrogen sulfide, carbon dioxide, steam)
Polar solvents (ketones, esters, ethers)
Steam
Strong organic solvents (benzene, dimethyl formamide, perchloroethylene,
    tetrahydrofuran)
Strong oxidizing agents (dinitrogen tetroxide, fuming nitric acid)

These perfluoroelastomers should not be exposed to molten or gaseous alkali metals such as sodium because a highly exothermic reaction may occur. Service life can be greatly reduced in fluids containing high concentrations of some diamines, nitric acid, and basic phenol when the temperature exceeds 212°F (100°C). Uranium hexafluoride and fully halogenated Freons (F-11 and F-12) cause considerable swelling.

The corrosion resistance given above is for the base polymer. Since the polymer is quite often compounded with fillers and curatives, these additives may interact with the environment, even though the polymer is resistant. Therefore, a knowledge of the additives present is essential in determining the material's suitability for a particular application. A corrosion testing program is the best method whereby this evaluation can be undertaken.

Kalrez compounds have virtually universal chemical resistance. They withstand attack by more than 1,600 chemicals, solvents, and plasmas. Table 4.54 lists the compatibilities of Kalrez compounds with selected corrodents. The listing is based on standard compounds 4079 and 1050LF for the majority of the fluids. When no compound is specified, any Kalrez compound may be used. Kalrez will be resistent to the corrodents listed up to a maximum temperature of

TABLE 4.54 Compatibility of Kalrez Compounds with Selected Corrodents[a]

Abietic acid	Ammonium salicylate	Bromic acid
Acetic acid, glacial (4079)	Ammonium sulfate	Bromine, anhydrous
Acetic acid, 30% (4079)	Ammonium sulfide	Butadiene
Acetone	Ammonium sulfite	Butane
Acetonitrile	Ammonium thiosulfate	Butyl acetate
Acetophenetidine	Amyl acetate	Butyl chloride
Acetophenone	Amyl alcohol	Butyl ether
Acetyl bromide	Amyl chloride	Butyl lactate
Acetyl chloride	Amyl mercaptan	Butyl mercaptan
Acrylic acid	Amyl naphthalene	Butyl chloride
Acrylonitrile (1050LF)	Amyl nitrate	Cadmium chloride
Alkyl acetone	Amyl nitrite	Cadmium nitrate
Alkyl alcohol	Amyl phenol	Cadmium sulfate
Alkyl benzene	Aniline hydrdochloride	Cadmium sulfide
Alkyl chloride	Aniline sulfate	Calcium acetate
Alkyl sulfide	Aniline sulfite	Calcium arsenate
Aluminum acetate	Animal fats	Calcium bicarbonate
Aluminum bromide	Animal oils	Calcium bisulfide
Aluminum chlorate	Anthraquinone	Calcium hydrosulfide
Aluminum chloride	Antimony sulfate	Calcium carbide
Aluminum ethylate	Antimony tribromide	Calcium carbonate
Aluminum fluoride	Antimony trichloride	Calcium chlorate
Aluminum fluorosilicate	Antimony trioxide	Calcium chloride
Aluminum formate	Aqua regia	Calcium chromate
Aluminum hydroxide	Arsenic oxide	Calcium fluoride
Aluminum nitrate	Arsenic trichloride	Calcium hydroxide
Aluminum oxalate	Arsenic trioxide	Calcium hypochlorite
Aluminum phosphate	Arsenic trisulfide	Calcium nitrate
Aluminum sulfate	Ascorbic acid	Calcium oxide
Alum	Barium carbonate	Calcium peroxide
Ammonia, anhydrous (1050LF)	Barium chlorate	Calcium phosphate
Ammonium acetate	Barium chloride, aqueous	Calcium stearate
Ammonium arsenate	Barium cyanide	Calcium sulfate
Ammonium benzoate	Barium hydroxide	Calcium sulfide
Ammonium bicarbonate	Barium iodide	Calcium sulfite
Ammonium bisulfite	Barium nitrate	Cane sugar liquors
Ammonium bromide	Barium oxide	Capric acid
Ammonium carbonate	Barium peroxide	Carbamate
Ammonium chloride	Barium salts	Carbon bisulfide
Ammonium citrate	Beet sugar liquors	Carbon dioxide
Ammonium dichromate	Benzaldehyde	Carbon disulfide
Ammonium fluoride (1050LF)	Benzene	Carbon fluorides
Ammonium fluorosilicate	Benzenesulfonic acid	Carbon monoxide
Ammonium formate	Benzocatechol	Caustic lime
Ammonium hydride,	Benzoyl chloride	Caustic potash
concd (1050LF)	Benzyl chloride	Chloric acid
Ammonium iodide	Bismuth carbonate	Chlorine, dry
Ammonium lactate	Bismuth nitrate	Chloroacetic acid
Ammonium nitrate	Bismuth oxychloride	Chloroacetone
Ammonium nitrite	Boric acid	Chloroform
Ammonium oxalate	Brine	Chlorosilanes

TABLE 4.54 Continued

Chlorosulfonic acid
Chromic acid
Chromic chloride
Chromic fluorides
Chromic hydroxide
Chromic nitrates
Chromic oxides
Chromic phosphate
Chromic sulfate
Coconut oil
Cod liver oil
Coke oven gas
Copper carbonate
Copper chloride
Copper cyanide
Copper nitrate
Copper oxide
Copper sulfate
Corn oil
Coconut oil
Cresylic acid
Crude oil
Cutting oils
Cyclohexane
Cyclohexanol
Cyclohexanone
Cyclohexene
Denatured alcohol
Detergent solution
Dextrain
Dextrose
Diacetone
Diallyl ether
Diallyl phthalate
Dichloroacetic acid
Dichloroaniline
o-Dichlorobenzene
Dichloroethane
Dichloroethylene
Dichloromethane
Dichlorophenol
Diesel oil
Diethanolamine (1050LF)
Diethylamine (1050LF)
Diethyl sulfate
Diethylbenzene
Diethylene glycol
Difluoroethane
Difluoromonochloroethane
Diisobutyl ketone

Diisobutylcarbinol
Diisobutylene
Diisopropyl ether
Diisopropyl ketone
Epichlorohydrin
Ethane
Ethanol
Ethanolamine (1050LF)
Ethers
Ethyl acetate
Ethyl acetoacetate
Ethyl acrylate
Ethyl alcohol
Ethylamine (1050LF)
Ethyl benzene
Ethyl benzoate
Ethyl bromide
Ethyl butylate (4079)
Ethyl cellulose
Ethyl chloride
Ethyl ether
Ethyl formate (4079)
Ethyl nitrite
Eththyl oxalate
Ethylene
Ethylene chloride
Ethylene dibromide
Ethylene dichloride
Ethylene glycol
Ethylene oxide (2035)
Fatty acids
Ferric acetate
Ferric hydroxide
Ferric sulfate
Ferrous carbonate
Ferrous chloride
Fuel oils
Fuming sulfuric acid
Gallic acid
Gasoline
Gelatin
Glauber's salt
Gluconic acid
Glucose
Glue
Glycerine (glycerol)
Helium
Heptane
Hexane
Hydroiodic acid (4079)

Hydrobromic acid
Hydrobromic acid 40%
Hydrochloric acid 37%, cold
Hydrochloric acid, concd
Hydrochloric acid 37%, hot
Hydrocyanic acid
Hydrofluoric acid, anhydrous
Hydrofluoric acid, concd, cold concd, hot (4079)
Hydrogen peroxide 90%
Iodic acid
Iodine
Iodoform
Isopropyl chloride
Isopropyl ether
Isopropylamine (1050LF)
Kerosene
Lacquer solvents
Lacquers
Lactic acid, cold
Lactic acid, hot
Lauric acid
Lead, molten
Lead acetate
Lead arsenate
Lead bromide
Lead carbonate
Lead chloride
Lead chromate
Lead dioxide
Lead nitrate
Lead oxide
Lime bleach
Lime sulfur
Linoleic acid
Linseed oil
Lithium carbonate
Lithium chloride
Lithium hydroxide
Lithium nitrate
Lithium nitrite
Lithium salicylate
Lithopone
Lye
Magnesium chloride
Magnesium hydroxide
Magnesium sulfate

(Continued)

TABLE 4.54 Continued

Magnesium sulfite	Nitric acid 50–100% (4079)	Piperazine (1050LF)
Maleic acid	Nitroaniline	Piperidine
Manganese acetate	Nitrobenzene	Plating solution, chrome
Maleic anhydride	Nitrocellulose	
Manganese carbonate	Nitrochlorobenzene	Potassium acid sulfate
Manganese dioxide	Nitrogen	Potassum alum
Manganese chloride	Nitrogen oxides	Potassum aluminum
Manganese phosphate	Nitrogen peroxide	sulfate
Manganese sulfate,	Nitroglycerine	Potassium bicarbonate
aqueous	Nitroglycerol	Potassium bichromate
Mercuric cyanide	Nitromethane	Potassium bifluoride
Mercuric iodide	Nitrophenol	Potassium bisulfate
Mercuric nitrate	Nitrotoluene	Potassium bisulfite
Mercuric sulfate	Nitrous acid	Potassium bitartrate
Mercuric sulfite	Nonane	Potassium bromide
Mercurous nitrate	n-Octane	Potassium carbonate
Mercury	Octyl acetate	Potassium chlorate
Mercury chloride	Octyl alcohol	Potassium chloride
Methane	Octyl chloride	Potassium chromates
Methyl acetate	Olefins	Potassium citrate
Methyl acrylate	Oleic acid	Potassium cyanate
Methyl alcohol (methanol)	Oleum	Silicone tetrachloride,
Methyl benzoate	Olive oil	dry
Methyl butyl ketone	Oxalic acid	Silicone tetrachloride,
Methyl chloride	Oxygen, cold (4079)	wet
Methyl ether	Paint thinner	Silver bromide
Methyl ethyl ketone	Palmitic acid	Silver chloride
Methyl isobutyl ketone	Paraffins	Silver cyanide
Methyl salicylate	Peanut oil	Silver nitrate
Methylene bromide	Pectin (liquor)	Silver sulfate
Methylene chloride	Penicillin	Soap solutions
Methylene iodide	Peracetic acid	Soda ash
Mineral oil	Perchloric acid (4079)	Sodium acetate
Mixed acids	Perchloroethylene	Sodium benzoate
Morpholine	Petroleum	Sodium bicarbonate
Motor oils	Petroleum ether	Sodium bichromate
Mustard gas	Petroleum crude (1050LF)	Sodium bifluoride
Myristic acid	Phenyl acetate	Sodium bisulfate
Naphtha	Phosgene	Sodium bisulfide
Naphthalene	Phosphoric acid 20%	Sodium bisulfite
Natural gas	Phosphoric acid 45%	Sodium borate
Neon	Phosphorous, molten	Sodium bromate
Nickel acetate aqueous	Phosphorous oxychloride	Sodium bromide
Nickel chloride aqueous	Phosphorous trichloride	Sodium carbonate
Nickel cyanide	Phthalic acid	Sodium chlorate
Nickel nitrate	Phthalic anhydride	Sodium chloride
Nickel salts	Pickling solution	Sodium chlorite
Nicotine	Picric acid	Sodium chloroacetate
Nicotone sulfate	Pine oil	Sodium chromate
Niter cake	Pine tar	Sodium citrate
Nitric acid 0–50% (4079)	Pinene	Sodium cyanate

TABLE 4.54 Continued

Sodium cyanide	Steam below 300°F/149°C	Trichloroethylene
Sodium ferricyanide	(2035)	Triethanolamine (1050LF)
Sodium ferrocyanide	Stearic acid	Triethyl phosphate
Sodium fluoride	Stoddard solvent	Triethylamine (1050LF)
Sodium fluorosilicate	Strontium acetate	Trimethylamine (1050LF)
Sodium hydride	Strontium carbonate	Trimethylbenzene
Sodium hydrosulfide	Strontium chloride	Trinitrotoluene
Sodium hydroxide	Strontium hydroxide	Trisodium phosphate
Sodium hypochlorite	Strontium nitrate, aqueous	Tung oil
Sodium lactate	Styrene (3018)	Turbine oil
Sodium nitrate	Succinic acid	Turpentine
Sodium oleate	Sucrose solution	Uranium sulfate
Sodium oxalate	Sulfamic acid	Uric acid
Sodium perborate	Sulfite liquors	Valeric acid
Sodium percarbonate	Sulfur chloride	Vanilla extract (2035)
Sodium perchlorate	Sulfur dioxide, dry	Vegetable oils
Sodium peroxide	Sulfur dioxide, liquified	Vinyl benzene
Sodium persulfate	Sulfur dioxide, wet	Vinyl benzoate
Sodium phenolate	Sulfur trioxide	Vinyl chloride
Sodium phosphate	Sulfuric acid, concd	Vinyl fluoride
Sodium salicylate	Sulfuric acid, dil	Vinylidene chloride
Sodium salts	Sulfurous acid	Water, cold
Sodium silicate	Sulfuryl chloride	Water, hot (1050LF)
Sodium stannate	Sulfonated oil	White oil
Sodium sulfate	Tallow	White pine oil
Sodium sulfide	Tannic acid	Wood oil
Sodium sulfite	Tar, bituminous	Xenon
Sour crude oil (1050LF)	Tartaric acid	Xylene
Sour natural gas (1050LF)	Tetraethyl lead	Zinc acetate
Soybean oil	Tetrahydrofuran	Zinc chloride
Stannic ammonium chloride	Thionyl chloride	Zinc chromate
Stannic chloride, aqueous	Thiourea	Zinc nitrate
Stannic tetrachloride	Toluene	Zinc oxide
Stannous bromide	Transformer oil	Zinc sulfate
Stannous chloride, aqueous	Transmission fluid type A	Zinc sulfide
Stannous fluoride	Triacetin	Zirconium nitrate
Stannous sulfate	Trichloroacetic acid (4079)	
Steam above 300°F/149°C	Trichlorobenzene	
(1050LF)	Trichloroethane	

[a] Kalrez is resistant to the above chemicals up to a maximum tempeature of 212°F/100°C. In some chemicals, a higher operating temperature may be allowable. The listing is based on standard compounds 4079 and 1050LF for the majority of the fluids. When no compound is specified, any Kalrex compound may be used.

212°F/100°C. In contact with some chemicals, a higher operating temperature may be allowable.

D. Applications

Perfluoroelastomer parts are a practical solution wherever the sealing performance of rubber is desirable but not feasible because of severe chemical or thermal conditions. In the petrochemical industry, FPM is widely used for O-ring seals on equipment. O-rings of FPM are employed in mechanical seals, pump housings, compressor casings, valves, rotameters, and other instruments. Custom-molded parts are also used as valve seats, packings, diaphragms, gaskets, and miscellaneous sealing elements including U-cups and V-rings.

Other industries where FPM contributes importantly are aerospace (versus jet fuels, hydrazine, N_2O_4 and other oxidizers, Freon-21 fluorocarbon, etc.); nuclear power (versus radiation, high temperature); oil, gas, and geothermal drilling (versus sour gas, acidic fluids, amine-containing hydraulic fluids, extreme temperatures and pressures); and analytical and process instruments (versus high vacuum, liquid and gas chromotography exposures, high-purity reagents, high-temperature conditions).

The semiconductor industry makes use of FPM O-rings to seal the aggressive chemical reagents and specialty gases required for producing silicon chips. Also the combination of thermal stability and low outgassing characeristics are desirable in furnaces for growing crystals and in high-vacuum applications.

The chemical transportation industry is also a heavy user of FPM components in safety relief and unloading valves to prevent leakage from tank trucks and trailers, rail cars, ships, and barges carrying hazardous and corrosive chemicals.

Other industries that also use FPM extensively include pharmaceuticals, agricultural chemicals, oil and gas recovery, and analytical and process control instrumentation.

Because of their cost, perfluoroelastomers are used primarily as seals where their corrosion- and/or heat-resistance properties can be utilized and other elastomeric materials will not do the job or where high maintenance costs results if other elastomeric materials are used.

4.26 EPICHLORHYDRIN RUBBER

The epichlorhydrin polymer group includes the homopolymer (CO), the copolymer with ethylene oxide (ECO), and terpolymers (GECO).

A. Physical and Mechanical Properties

In general, epichlorhydrin rubber possesses a blend of the good properties of neoprene and nitrile rubber. Epichlorhydrin rubbers have excellent thermal stability with maximum operating temperatures for CO rubber of 300°F/150°C and for ECO rubber of 275°F/135°C. The various polymers exhibit widely different performance levels with regard to resilience and low-temperature flexibility. The ECO version has dynamic properties similar to those found in natural rubber. They also have good aging properties. Their high chlorine content imparts good to fair flame resistance. Some typical physical and mechanical properties for the CO and ECO polymers are as follows:

Specific gravity	1.27 to 1.36
Resilience	Fair
Compression	Fair
Abrasion resistance	Fair to good
Tensile strength (psi)	2,500
Elongation at break (%)	400
Hardness, Shore A	40–90

B. Resistance to Sun, Weather, and Ozone

These rubbers have excellent resistance to sun, weathering, and ozone.

C. Chemical Resistance

Epichlorhydrin rubbers are resistant to hydrocarbon fuels, to swelling in oils, and resistant to acids, bases, water, and aromatic hydrocarbons. However, they are susceptible to devulcanization in the presence of oxidized fuels such as sour gasoline.

D. Applications

Epichlorhydrin rubber applications include seals and tubes in air-conditioning and fuel systems.

4.27 ETHYLENE-VINYLACETATE COPOLYMER (EVM)

EVM can only be cured by peroxide. The elastic properties of EVM are due to the absence of crystallinity in the copolymer.

A. Physical and Mechanical Properties

EVM has excellent thermal stability, even surpassing EPDM. Although it has poor set resistance below room temperature, it has outstanding set resistance at elevated temperatures. The best resilience is also obtained at elevated temperatures.

As with other elastomers, the properties of EVM are dependent upon compounding. With proper compounding, EVM can achieve levels of volume resistivity and dielectric constant making it suitable for low-voltage insulation and cable sheathing.

EVM has excellent adhesion to metals and can be difficult to mold.

B. Resistance to Sun, Weather, and Ozone

EVM has excellent resistance to sun, weathering, and ozone.

C. Chemical Resistance

The resistance of EVM to hydrocarbon fluids is dependent upon the vinyl acetate content. Higher levels of vinyl acetate increase the polarity of the polymer and therefore increase the hydrocarbon resistance.

D. Applications

EVM finds application as wire insulation in the automotive industry. Ethylene vinyl acetate is also used in the production of thermoplastic elastomers as the soft phase.

4.28 CHLORINATED POLYETHYLENE (CM)

Unmodified polyethylene cannot be converted into an elastomer because of its molecular crystallinity. Chlorination of the polyethylene polymer disrupts the crystallinity and allows the polymer to become elastic upon vulcanization. CM can be vulcanized by peroxides or certain nitrogen-containing organic compounds that cross-link the chlorine atoms.

A. Physical and Mechanical Properties

CM is flame retardant and heat resistant. Since it does not contain sulfur, the vulcanized elastomer has good colorability in lead-stabilized, moisture-resistant formulations.

The following are typical physical and mechanical properties:

Specific gravity	1.26–1.31
Tensile strength at break (psi)	1,400–3,000

| Elongation at break (%) | 300–900 |
| Hardness, Shore A | 60–76 |

B. Resistance to Sun, Weather, and Ozone

CM has excellent resistance to sun, weatheriing, and ozone.

C. Chemical Resistance

Chlorinated polyethylene has reasonably good resistance to fuels and oils and excellent resistance to atmospheric pollutants.

D. Applications

CM finds application as vinyl siding, window profiles, hoses, protective boots, flexible dust shields, and various wire and cable applications.

5

Thermoplastic Piping Systems

5.1 INTRODUCTION AND DESIGN CONSIDERATIONS

A combination of corrosion resistance and low cost has been the reason for the wide application of plastic piping systems. The U.S. pipe market grew from 4.9 billion feet in 1977 to 6.6 billion feet in 1988 and is expected to reach 9.8 billion feet in the year 2000.

Pipe made of thermoplastic resins can be bent or shaped by the application of heat whereas pipe made of thermosetting resins cannot be so processed.

Plastic piping systems do not have specific compositions. In order to produce the piping system, it is necessary to add various ingredients to the base resin to aid in the manufacturing operation and/or to alter mechanical or corrosion resistant properties. In so doing, one property is improved at the expense of another. Each piping system carries the name of the base resin from which it is manufactured. Because of the compounding and the various formulations, mechanical and corrosion resistant properties can vary from one manufacturer to another. The tabulated values given are averages and are to be used as guides. Specific recommendations for an application should be obtained from the manufacturer of the piping system.

Additional details on plastic piping systems can be found in Schweitzer PA. Corrosion Resistant Piping Systems. New York: Marcel Dekker, 1994.

389

A. Working Pressures of Thermoplastic Piping Systems

The governing factor in the determination of the allowable working pressure of thermoplastic piping systems is the hoop or circumferential stress. The two most useful forms for expressing this stress are

$$S = \frac{P[D_o - t]}{2t}$$

or

$$P = \frac{2St}{D_o - t}$$

where

S = stress (psi)
P = internal pressure (psi)
D_o = outside pipe diameter (in.)
t = minimum wall thickness (in.)

Standard methods have been developed to determine the long-term hydrostatic strength of plastic pipe. From these tests, S values are calculated and a hydrostatic design basis (HDB) is established. This HDB is then multiplied by a service factor to obtain the hydrostatic design stress.

The service factor is based on two groups of conditions. The first is based on manufacturing and testing variables, such as normal variations in material, manufacture, dimensions, good handling techniques, and evaluation procedures. Normal variations within this group are usually within 1 to 10%. The second group takes into account the application or use, including such items as installation, environment (both inside and outside of the pipe), temperature, hazard involved, life expectancy desired, and the degree of reliability selected.

A piping system handling water at 73°F (23°C) would have a service factor of 0.4. This means that the pressure rating of the pipe would be 1.25 times the operating pressure.

When a plastic piping system is to operate at elevated temperatures or is to handle a corrodent other than water, a service factor other than 0.4 must be used. Temperature effects have been studied and temperature corrrection factors developed by the manufacturers. These factors are used with the allowable operating pressures at 73°F (23°C) and are discussed under the heading for each individual system. The determination of allowable operating pressure for various corrodent systems is best left to the piping manufacturer.

After the service design factor has been determined, the maximum allowable operating pressure can be calculated from the following equation:

$$P_W = \frac{2S_W t}{D_o - t}$$

where

P_W = pressure rating (psi)
S_W = hydrostatic design basis × service factor (psi)
 t = wall thickness (in.)
D_o = outside diameter of pipe (in.)

Thermoplastic piping systems are available in schedule 40 or schedule 80 designations and have wall thicknesses corresponding to that of steel pipe with the same designations. Table 5.1 shows the dimensions of schedule 40 and schedule 80 thermoplatic pipe.

TABLE 5.1 Thermoplastic Pipe Dimensions

| Nominal pipe size (in.) | Schedule 40 (in.) | | | Schedule 80 (in.) | | |
	Outside diameter	Inside diameter	Wall thickness	Outside diameter	Inside diameter	Wall thickness
$\frac{1}{4}$	0.540	0.364	0.088	0.540	0.302	0.119
$\frac{3}{8}$	0.675	0.493	0.091	0.675	0.423	0.126
$\frac{1}{2}$	0.840	0.622	0.109	0.840	0.546	0.147
$\frac{3}{4}$	1.050	0.824	0.113	1.050	0.742	0.154
1	1.315	1.049	0.133	1.315	0.957	0.179
$1\frac{1}{4}$	1.660	1.380	0.140	1.660	1.278	0.191
$1\frac{1}{2}$	1.900	1.610	0.145	1.900	1.500	0.200
2	2.375	2.067	0.154	2.375	1.939	0.218
$2\frac{1}{2}$	2.875	2.469	0.203	2.875	2.323	0.276
3	3.500	3.068	0.216	3.500	2.900	0.300
4	4.500	4.026	0.237	4.500	3.826	0.337
6	6.625	6.065	0.280	6.625	5.761	0.432
8	8.625	7.981	0.322	8.625	7.625	0.500
10	10.750	10.020	0.365	10.750	9.564	0.593
12	12.750	11.938	0.406	12.750	11.376	0.687
14	14.000	13.126	0.437			
16	16.000	15.000	0.500			

Because of the wide range of differences in the physical properties of the various thermoplastics, there is no correlation between the schedule number and the allowable working pressure. For example, depending upon the specific thermoplastic, the operating pressure of a 1-inch schedule 80 pipe at 73°F (23°C) can vary from 130 to 630 psi. In addition, within the same plastic piping system the allowable operation pressure can vary from 850 psi for $\frac{1}{2}$-inch-diameter schedule 80 pipe to 230 psi for 12-inch-diameter schedule 80 pipe.

Because of these problems, the industry is attempting to correct this condition by the use of Standard Dimension Ratio (SDR) ratings. These ratios are so arranged that a piping system designed to these dimensions will maintain a uniform pressure rating at a specified temperature regardless of pipe diameters. The allowable operating pressure varies among different piping materials for the same SDR rating.

Allowable working pressures and temperature correction factors for each particular system are discussed under the heading of that system.

B. Joining of Thermoplastic Pipe

A variety of methods are available for joining thermoplastic piping. The specific method selected will depend upon the function of the pipe, the corrodent being handled, and the material characteristics of the pipe. The available methods may be divided into two general categories, permanent joint techniques and nonpermanent joint techniques. Included in the permanent joint techniques are

Solvent cementing
Butt fusion
Socket fusion

Nonpermanent methods include

Threading
Flanging
Bell-ring-gasket joint
Compression insert joint
Grooved-end mechanical joint

As would be expected, there are advantages and disadvantages to the use of each system, as well as specific types of installations for which each method is best suited.

1. Solvent Cementing

This is the most popular method used for the joining of PVC, CPVC, ABS, and other styrene materials. No special tools are required, and the solvent-cemented joint is as strong as the pipe itself. Sufficient drying time must be allowed before the joint can be tested or pressurized.

One word of caution: Specific solvent cements are recommended for specific piping systems and specific piping sizes. Manufacturers' recommendations for specific primers and cements should always be followed. So-called universal cements should be avoided.

The advantages of this method of joining are

1. Pipe is less expensive
2. Optimal connection for type of material
3. Pull-out resistant connection
4. Large selection of fittings and valves available
5. Pressure resistant up to burst pressure
6. Good chemical resistance (joint does not become a "weak" link in the system)
7. No thread cutting
8. Easy installation

The disadvantages of this technique are

1. Instructions must be followed carefully
2. Connection cannot be disassembled
3. Leaky joints difficult to repair
4. Revision to the piping system more difficult since all joints are permanent

2. Butt Fusion

This method results in a joint having excellent strength and permits the piping system to be put into service as soon as the last joint has cooled. Materials joined by this technique are those made from polyolefins, such as polypropylene, polyethylene, polybutylenes, and polyvinylidene fluoride. In this method the pipe ends to be butted together are heated and partially melted, butted together, and fused.

The advantages of this technique are

1. Optimal connection method for the polyolefins
2. Pull-out resistance
3. Pressure resistance beyond the burst pressure of the pipe
4. Excellent chemical resistance (a "weak" link is not formed at the joint)

The primary disadvantages are

1. Cost of the fusion equipment
2. Bulkiness of the equipment required

3. Connection cannot be disassembled
4. Instructions for fusing must be followed carefully

3. Socket Fusion

This technique is also used for polyolefins such as polypropylene, polyethylene, polybutylene, and polyvinylidene fluoride. Two different methods are available to produce these thermally fused joints: electric-resistance fusion and socket heat fusion.

The first of these methods utilizes heat from an electrified copper coil to soften the outside surface of the pipe end and the inside surface of the fitting socket. Some manufacturers furnish the fittings with the coils imbedded. Where this is not the case, the coils are attached manually. This technique is used primarily for polypropylene piping systems in acid waste drainage installations.

The primary advantage of this method is that a piping system can be dry fitted and assembled before permanent joints are made.

The disadvantages to this system are

1. Imperfect heat distribution possible, resulting in low joint strength and possible corrosion resistance problems
2. System cannot be disassembled
3. Joining instructions must be followed closely
4. Cumbersome equipment required

Socket heat fusion is the preferred technique to use for systems handling corrodents. This method involves the use of an electrically heated tool, which softens the outside surface of the pipe and the inside surface of the fitting. The joint produced is as strong as the pipe or fitting. This technique is used on all of the polyolefins.

The advantages of this method are

1. Flexibility of installation (heat tool may be held by hand or in a bench vise while making a joint)
2. Joint is as strong as the pipe or fitting
3. Equipment needed is least expensive and least cumbersome of any of the fusion techniques
4. Optimal connection for polyolefin materials
5. Pull-out-resistant connection
6. Excellent chemical resistance

The primary disadvantage stems from one of the advantages. The high level of flexibility afforded by this technique requires that the installer have a high degree of skill and dexterity. Also, the connection cannot be disassembled.

4. Threaded Joints

Threading of thermoplastic piping can be utilized provided schedule 80 pipe or pipe of thicker wall is used and the pipe diameter does not exceed 4 inches. With a diameter of less than 4 inches, the pipe out of roundness can be more easily controlled.

Even with smaller diameters, leakage at the joints poses a problem. Consequently, threaded joints are recommended only for use in applications where leakage will not pose a hazard or in low-pressure systems.

The only advantage to threaded joints is the fact that the system can easily be dismantled for periodic cleaning and/or modification.

The disadvantages to a threaded system are numerous:

1. Allowable operating pressure of the pipe reduced by 50%
2. Threaded joints in the polyolefins pose high threat of leakage if operating pressure exceeds 20 psig due to the low modulus of elasticity of these materials
3. Heavier, more expensive pipe must be used (schedule 80 vs. schedule 40, which is used for solvent-cementing or thermal fusion)
4. Allowing for expansion and contraction due to thermal changes difficult

5. Flanging

Flanges are available for most thermoplastic piping systems. They are affixed to the pipe by any of the previous jointing techniques discussed. A flanged joint is used primarily for one or more of the following conditions:

1. Connection to pumps, equipment, or metallic piping systems
2. Temporary piping system
3. Process lines that require periodic dismantling
4. To reduce field labor time and/or expertise, since these joints can be made up in the shop and sent to the field for installation
5. When weather conditions or lack of utilities prevent other methods from being used

When making a flanged joint, it is important to use the correct bolting torque. Table 5.2 shows the recommended bolting torques based on well-lubricated bolts, which will give a bolt stress of 10,000–15,000 psi. It is also necessary to provide a gasket of appropriate corrosion resistant properties and compressibility between mating flanges.

The disadvantages of flanged joints are

1. High material and labor costs
2. Bulkiness

TABLE 5.2 Recommended Bolt Torques for Plastic Flanges

Flange size (in.)	Bolt diameter (in.)	Torque (ft-lb)
$\frac{1}{2}$	$\frac{1}{2}$	10–15
$\frac{3}{4}$	$\frac{1}{2}$	10–15
1	$\frac{1}{2}$	10–15
$1\frac{1}{4}$	$\frac{1}{2}$	10–15
$1\frac{1}{2}$	$\frac{1}{2}$	10–15
2	$\frac{5}{8}$	20–30
$2\frac{1}{2}$	$\frac{5}{8}$	20–30
3	$\frac{5}{8}$	20–30
4	$\frac{5}{8}$	20–30
6	$\frac{3}{4}$	35–40
8	$\frac{3}{4}$	35–40
10	$\frac{7}{8}$	55–75
12	1	80–110

Until recently, flanging was the only method whereby thermoplastic pipe could be connected to metallic pipe. Now a transition union is available. The union consists of a metallic male threaded section, which joins a female threaded connection on the metallic pipe. The other end of the union consists of a plastic tailpiece with either a socket or female threaded end. A platic nut screws onto the tailpiece, which secures the metal end connector tightly against an O-ring face seal on the tailpiece.

6.   Bell-Ring-Gasket Joints

The most common use of this joint is on underground PVC pressure-rated piping systems handling water. This joint can also be used to connect PVC pipe to metallic pipe.

An elastomeric ring is retained in a groove in the female joint section. The ring becomes compressed as the pipe is inserted into the joint. It is common practice to anchor these joints in concrete at every directional change of the piping to prevent the pipe from backing out of the joint or heaving. This joint would not be used on underground pipe used to convey corrosive materials since it is very difficult to achieve a bubble-tight joint seal.

The advantages of this joint are

1.   Simplicity (quick to make)
2.   Ease of making with reduced labor cost

The disadvantages are

1. Difficult to make leak free
2. Danger of pull-out

7. Grooved-End Mechanical Joint

A groove is cut or rolled around the perimeter near the ends of each pipe to be joined together. A metal coupling consisting of two identical sleeves, each of which has a ridge corresponding to the groove in the pipe ends, is fitted over the pipe. The two halves are connected by means of a bolt and hinge. Retained between the pipe and sleeves is an elastomeric ring, which forms the seal.

This type of joint is especially useful with PVC or CPVC pipe, since the modulus of elasticity of these materials and the material strength are sufficient to retain the integrity of the grooves.

The grooved-end mechanical joint is used primarily for field assembly, where the frequent dismantling of the lines is required, because of the ease with which the joint may be disassembled and reassembled. It is also used for temporary piping installations.

The primary disadvantage is the difficulty in finding an elastomer that will be compatible with the corrodent being handled.

8. Compression Insert Joints

This type of joint is not generally used when handling corrodents. Its primary application is in water service such as irrigation systems, swimming pools, and agricultural service, and it is used with flexible PVC or PE tubing in sizes under 2 inches. The joint consists of a barbed fitting inserted into the tubing and held in place by means of a stainless steel clamp attached around the tubing to maintain pressure on the fitting barb.

C. Effect of Thermal Change on Thermoplastic Pipe

As the temperature changes above and below the installation temperature, plastic piping undergoes dimensional changes. The amount of expansion or contraction that occurs is dependent upon the piping material and its coefficient of linear expansion, the temperature differential, and the length of pipe run between direction changes. Thermoplastic piping materials have a considerably larger coefficient of thermal expansion than do metallic piping materials. Consequently, temperature changes that would not affect metallic materials can be detrimental to plastic materials if not compensated for. The change in length resulting from a thermal change can be calculated using the following formula:

$$\Delta L = \frac{Y[T_1 - T_2]}{10} \times \frac{L}{100}$$

where

ΔL = dimensional change due to thermal expansion or contraction (in.)

Y = expansion coefficient (in./$10°F$/100 ft)

$T_1 - T_2$ = difference between installation temperature and the maximum or minimum system temperature, whichever provides the greatest differential (°F)

L = length of pipe run between changes in direction (ft)

If the movement that results from thermal changes is restricted by equipment or supports, the resulting stresses and forces can cause damage to the equipment or to the pipe itself. The following formulas provide an estimate of the magnitude of the resultant forces:

$$S = EC\,(T_1 - T_2)$$

where

S = stress (psi)

E = modulus of elasticity (psi)

C = coefficient of thermal expansion (in./in./°F $\times 10^{-5}$)

$(T_1 - T_2)$ = difference between the installation temperature and the maximum or minimum system temperature

and

$$F = S \times A$$

where

F = force (lb)

S = stress (psi)

$$A = \text{cross-sectional area (in.}^2) = \left[\left(\frac{\text{Pipe OD}}{2}\right)^2 - \left(\frac{\text{Pipe ID}}{2}\right)^2\right]3.14$$

The stresses and forces resulting from thermal expansion or contraction can be reduced or eliminated by providing flexibility in the piping system. This can be accomplished by incorporating sufficient changes in direction to provide the flexibility. Most piping installations are automatically installed with sufficient changes in direction to compensate for these stresses.

The following formula can be used to determine the required flexibility in leg R (see Figure 5.1). Since both pipe legs will expand and contract, the shortest

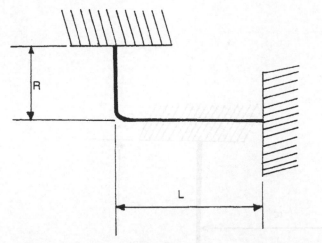

FIGURE 5.1   Rigidly installed pipe with a change of direction.

leg must be selected for the flexibility test when analyzing inherent flexibility in naturally occurring offsets.

$$R = 2.041\sqrt{D\,\Delta L}$$

where

   R = length of opposite leg to be flexed (ft)
   D = actual outside diameter of pipe (in.)
   ΔL = dimensional change in adjacent leg due to thermal expansion or
         contraction (in.)

Flexural offsets must be incorporated into systems where straight runs of pipe are long or where the ends are restricted from movement. An expansion loop, such as shown in Figure 5.2, can be calculated using the following formula:

$$R^{1} = 2.041\sqrt{\frac{D\,\Delta L}{2}}$$

or

$$R^{1} = 1.443\sqrt{D\,\Delta L}$$

Keep in mind that rigid supports or restraints should not be placed within the length of an expansion loop, offset, or bend.

   Thermal expansion stresses and forces can also be compensated for by the installation of expansion joints. Initially these joints were produced from natural rubber, but at the present time they are available in a wide range of elastomeric materials such as neoprene, Buna-N, Butyl, EPDM, and TFE. When using these

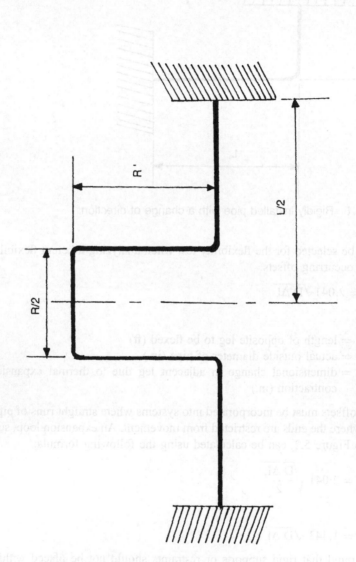

FIGURE 5.2 Rigidly installed pipe with an expansion loop.

joints, care must be taken that the material of construction of the tube section is compatible with the fluid being handled. Figure 5.3 shows the typical construction of an expansion joint. These joints provide axial compression, axial elongation, angular movement, torsional movement, and lateral movement. Operating pressures and temperatures are dependent upon the specific elastomer.

D. Burial of Thermoplastic Pipe

The primary criterion for the safe design of a buried thermoplastic pipe system is that of maximum allowable deflection. On rare occasions buckling of the pipe and/or excessive loading may control the design. In the latter case the safe stress under continuous compression may conservatively be assumed to equal the hydrodynamic design stress for the end use conditions. Control of the deflection of the pipe, along with proper installation procedures, then becomes the key

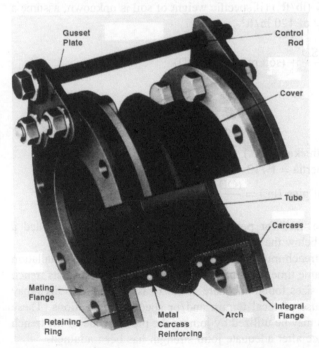

FIGURE 5.3 Typical construction of elastomeric expansion joint. (Courtesy of the Metraflex Co., Chicago, IL.)

factor in an underground pipe installation. The percentage of deflection can be calculated as follows:

$$\% \text{ deflection} = \frac{\text{Calculated deflection} \times 100}{\text{Outside diameter of pipe}}$$

The calculated deflection is found from the following equation:

$$x = \frac{K_x DW}{[0.149 \times PS + 0.061 \times E_1]}$$

where

OD = outside diameter of pipe (in.)
K_x = bedding factor (see Table 5.6)
D = deflection lag factor (see Table 5.5)
W = weight per lineal inch (lb/in.)
PS = pipe stiffness (psi)
E_1 = soil modulus (see Table 5.3 and 5.4)
E = flexural modulus (psi)
H = height of cover (ft)
SD = soil density (lb/ft^3) (If specific weight of soil is unknown, assume a soil density of 120 lb/ft^3.).

$$W = \frac{H \times OD \times SD}{144} + (\text{soil pressure} \times OD)$$

$$PS = \frac{E \times I}{0.149\, R^3}$$

where

t = average wall thickness (in.)
I = moment of inertia = $t^3/12$
R = mean radii of pipe (in.) = $\dfrac{(OD - t)}{2}$

Pipe that is to be used for potable water service should be installed a minimum of 12 inches below the expected frost line.

The width of the trench must be sufficient to allow convenient installation of the pipe, but at the same time be as narrow as possible. Tables 5.7 lists trench widths for various pipe sizes found adequate in most cases. However, sizes may have to be varied because of local terrain and/or specific applications. These minimum trench widths may be utilized by joining the pipe outside of the trench and lowering it into place after adequate joint strength has been attained.

The trench bottom should be relatively smooth and free of rocks. If hardpan, boulders, or rocks are encountered, it is advisable to provide a minimum

TABLE 5.3 Values of E_1 Soil Modulus for Various Soils

Soil type and pipe[a] bedding material	E_1 for degree of compaction of bedding (lb/in.^{2})			
	Dumped	Slight <85% Proctor <40% rel. den.	Moderate 85–95% Proctor 40–70% rel. den.	High >95% Proctor >70% rel. den.
Fine-grained soils (LL > 50) with medium to high plasticity. CH,MH,CH-MH	No data available—consult a soil engineer or use $E_1 = 0$			
Fine-grained soils (LL < 50) with medium to no plasticity. CL,ML,ML-CL with <25% coarse-grained particles	50	200	400	1,000
Fine-grained soils (LL < 50) with no plasticity. CL,ML,ML-CL with >25% coarse-grained particles	100	400	1,000	2,000
Coarse-grained soils with fines. GM,GC,SM,SC contain >12% fines	100	400	1,000	2,000
Coarse-grained soils with little or no fines. GW,SW,GP,SP contain <12% fines (or any borderline soil beginning with one of these symbols GM-GC,GC-SC)	200	1,000	2,000	3,000
Crushed rock	1,000		3,000	

[a] See Table 5.4 for Unified Soil Classification Group Symbols.

TABLE 5.4 Unified Soil Classification Group Symbols[a]

GW	Well-graded gravels, gravel–sand mixtures, little or no fines
GP	Poorly graded gravels, gravel–sand mixtures, little or no fines
GM	Silty gravels, poorly graded gravel–sand-silt mixtures
GO	Clayey gravels, poorly graded gravel–sand–clay mixtures
SW	Well-graded sands, gravelly sands, little or no fines
SP	Poorly graded sands, gravelly sands, little or no fines
SM	Silty sands, poorly graded sand–silt mixtures
SC	Clayey sands, poorly graded sand–clay mixtures
ML	Inorganic silts and very fine sand, silty or clayey fine sands
CL	Inorganic clays of low to medium plasticity
MH	Inorganic silts, micaeous or diatomaceous fine sandy or silty soils, elastic silts
CH	Inorganic clays of high plasticity, fat clays
OL	Organic silts and organic silt clays of low plasticity
OH	Organic clays of medium to high plasticity
Pt	Peat and other highly organic soils

[a]ASTM-D-2487-83 Classification of Soils for Engineering Purposes.

of 4 inches of sand or tamped earth beneath the pipe to act as a cushion and to protect the pipe from damage.

Before installing any section of pipe, the pipe should be carefully inspected for cuts, scratches, gouges, nicks, buckling, or any other imperfection. Any section of pipe or fitting that has an imperfection should not be used.

Thermal expansion and contraction must be taken into account, particularly in hot weather. On a hot summer day, the sun's rays could cause the wall of the pipe to reach a temperature of 150°F (66°C). If this pipe were to be installed, contraction would occur overnight, which would cause weakening or failure of

TABLE 5.5 Values of Deflection Lag Factor D

Installation condition	D
Burial depth <5 ft with moderate to high degree of compaction (85% or greater Proctor, ASTM D-698 or 50% or greater relative density ASTM 2049)	2.0
Burial depth <5 ft with dumped or slight degrees of compaction (Proctor, ASTM D-698, less than 85% relative density; ASTM D-2049, less than 40%)	1.5
Burial depth >5 ft with moderate to high degree of compaction	1.5
Burial depth > 5 ft with dumped or slight degree of compaction	1.25

TABLE 5.6 Bedding Factor K_x

Type of installation	K_x
Shaped bottom with tamped backfill material placed at the sides of the pipe, 95% Proctor density or greater	0.083
Compacted coarse-grained bedding and backfill material placed at the side of the pipe, 70–100% relative density	0.083
Shaped bottom, moderately compacted backfill material placed at the sides of the pipe, 85–95% Proctor density	0.103
Coarse-grained bedding, lightly compacted backfill material placed at the sides of the pipe, 40–70% relative density	0.103
Flat bottom, loose material placed at the sides of the pipe (not recommended); <35% Proctor density, <40% relative density	0.110

TABLE 5.7 Minimum Trench Widths

Pipe size (in.)	Trench width (in.)
≤3	8
4 and 6	12
8	16

the joints. To avoid this, the pipe is "snaked" along in the trench. The loops formed will compensate for the contraction of the pipe, thus protecting the joints. For example, 100 feet of PVC pipe will contract or expand $\frac{3}{4}$ inch for every 20°F temperature change. Should the pipe wall reach 150°F (66°C) during the day and cool off to 70°F (21°C) during the night, the pipe would contract 3 inches. $\Delta T = 80°F$ (44°C). Snaking of the pipe will allow this to happen without imposing any stress on the piping. Table 5.8 gives the required offset and loop length when snaking the pipe for each 10°F (5°C) temperature change.

TABLE 5.8 Offset and Loop Length Requirements

	Maximum temperature variation (°F/°C)									
Loop length	10	20	30	40	50	60	70	80	90	100
(ft)	5.6	11.2	16.8	22.4	28	39.6	39.2	44.8	50.4	56
20	3	4	5	5	6	6	7	7	8	8
50	7	9	11	13	14	16	17	18	19	20
100	13	18	22	26	29	32	35	37	40	42

It is preferable to backfill the trench with the installed pipe during the early morning hours before the pipe has had an opportunity to expand as a result of the sun. The pipe should be continuously and uniformly supported over its entire length on firm stable material. Do not use blocks to support the pipe or to change grade. Backfill materials should be free of rocks with a particle size of $\frac{1}{2}$ inch or less. After 6–8 inches of cover over the pipe has been placed, it should be compacted by hand or with a mechanical tamper. The remainder of the backfill can be placed in uniform layers until the trench is completely filled, leaving no voids under or around, rocks, or clumps of earth. Heavy tampers should be used only on the final backfill.

Additional details for installation of underground piping can be found in (1) ASTM D2774, *Underground Installation of Thermoplastic Pressure Piping* and (2) ASTM D2321, *Underground Installation of Flexible Thermoplastic Sewer Pipe.*

E. Supporting Thermoplastic Pipe

The general principles of pipe support used for metallic pipe can be applied to plastic pipe with a few special considerations. Spacings of supports for thermoplastic pipe is closer than for metallic pipe and varies for each specific system. The support spacing for each plastic marterial is given under the chapter for that specific piping system. It is critical that the support spacings given be followed to provide a piping system that will give long uninterrupted service. Care should be taken not to install plastic pipes near or on steam lines or other hot surfaces.

Supports and hangers can be by means of clamps, saddle, angle, spring, or other standard types. The most commonly used are clevis hangers. It is advisable to select those with broad, smooth bearing surfaces, which will minimize the danger of stress concentrations. Relatively narrow or sharp support surfaces can damage the pipe or impose stresses on the pipe. If desired, protective sleeves may be installed between the hanger and pipe (see Figure 5.4). When the recommended support spacing is relatively close (2–3 feet or less), it will be more economical to provide a continuous support. For this purpose channel iron or inverted angle iron can be used (see Figure 5.5). This procedure should also be followed when the line is operating at elevated temperatures.

The support hangers should not rigidly grip the pipe but should permit axial movement to allow for thermal expansion. All valves should be individually supported and braced against operating torques. All fittings except couplings should also be individually supported. See Figure 5.6 for a typical valve support. Rigid clamping or anchor points should be located at valves and fittings close to a pipeline change in direction (see Figures 5.7, 5.8 and 5.9.) Figure 5.10 shows a means of supporting a horizontal line from the bottom.

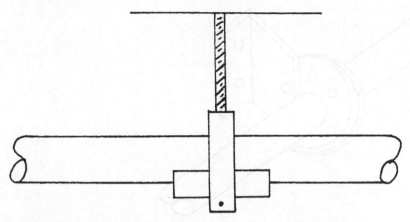

FIGURE 5.4 Typical clevis hander support with protective sleeve.

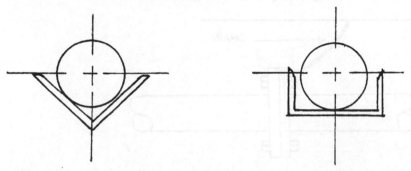

FIGURE 5.5 Continuous support of pipe using angle or channel iron.

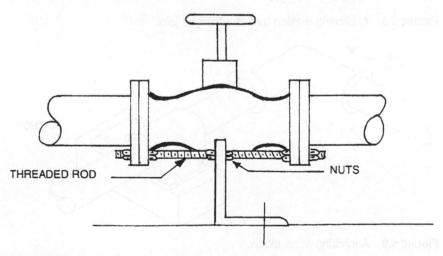

THREADED ROD ——————————— NUTS

FIGURE 5.6 Typical valve support.

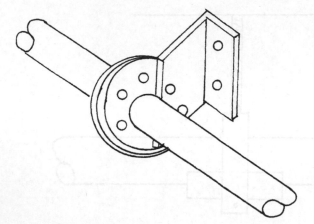

FIGURE 5.7 Anchoring making use of a flanged joint.

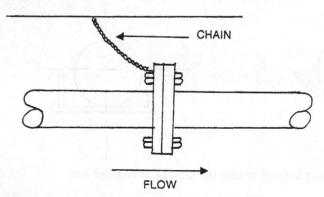

FIGURE 5.8 Anchoring making use of a flanged joint.

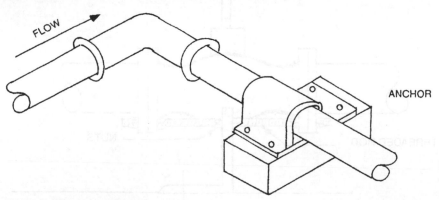

FIGURE 5.9 Anchoring at an elbow.

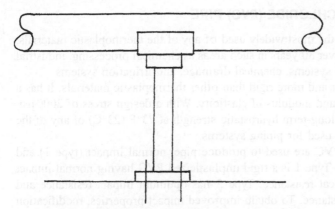

FIGURE 5.10 Typical support of a pipe from the bottom.

Vertical runs should be supported by means of a saddle at the bottom and the use of long U-bolts as guides on the vertical. These guides should not grip the pipe but be loose enough to permit axial movement but not traverse movement.

It is important that vibration from pumps or other equipment not be transmitted to the pipe. These vibrations can be isolated by the use of flexible couplings installed at the junction of the pipe and pump or equipment.

F. Fire Hazards

Potential fire hazards should also be considered when selecting the specific plastic material to be used. Although the fluid being conveyed may not be combustible, there is always the danger of a fire from other sources. In the physical and mechanical properties table located under the heading of each specific thermoplastic material are two entries relating to this topic. The first is the limiting oxygen index percentage. This is a measure of the minimum oxygen level required to support combustion of the thermoplast. The second is the flame spread classification. These ratings are based on common tests as outlined by the Underwriters Laboratories and are defined as follows:

Flame Spread Rating	Classification
0–25	Noncombustible
25–50	Fire retardant
50–75	Slow burning
75–200	Combustible
over 200	Highly combustible

5.2 POLYVINYL CHLORIDE (PVC) PIPE

Polyvinyl chloride is the most widely used of any of the thermoplastic materials. It has been used for over 30 years in such areas as chemical processing, industrial plating, water supply systems, chemical drainage, and irrigation systems.

PVC is stronger and more rigid than other thermoplastic materials. It has a high tensile strength and modulus of elasticity. With a design stress of 2000 psi, PVC has the highest long-term hydrostatic strength at 73°F (23°C) of any of the major thermoplastics used for piping systems.

Two types of PVC are used to produce pipe, normal impact (type 1) and high impact (type 2). Type 1 is a rigid unplasticized PVC having normal impact with optimum chemical resistance. Type 2 has optimum impact resistance and normal chemical resistance. To obtain improved impact properties, modification of the basic polymer is required, thus compromising chemical resistance. This is characteristic of many thermoplastics. The high-impact pipe is designed to take greater shock loads than normal-impact pipe. Type 1 pipe has a maximum allowable temperature rating of 150°F (66°C). Most PVC pipe is of the high-impact variety.

A. Pipe Data

PVC piping is available in $\frac{1}{4}$ inch to 16 inches nominal diameter in schedule 40 and schedule 80 wall thicknesses. Dimensions are shown in Table 5.1. the maximum allowable working pressures are given in Table 5.9 for schedule 40 and schedule 80. Note from the table that the operating pressure decreases as the nominal diameter increases. To overcome this, PVC piping is also available in Standard Dimension Ratings (SDR). For any one SDR specification, the allowable operating pressure is constant at a given temperature, regardless of pipe diameter. The dimensional ratio is derived by dividing the outside diameter by the minimum wall thickness of the pipe.

There are six SDR standards for PVC pipe:

SDR 13.5
SDR 18
SDR 21
SDR 25
SDR 26
SDR 32.5

Dimensions of these SDR-rated pipe systems are given in Tables 5.10A and 5.10B. The range of pipe sizes for each SDR system varies, as seen in the tables. This is because of the purpose for which each rating was designated. Table 5.11 supplies the maximum allowable operating pressure for each rating.

TABLE 5.9 Allowable Working Pressure (psi) of Schedule 40 and Schedule 80 PVC Pipe at 73.4°F/23°C

Nominal pipe size (in.)	Schedule 40, solvent-welded		Schedule 80			
			Solvent-welded		Threaded	
	Type 1	Type 2	Type 1	Type 2	Type 1	Type 2
$\frac{1}{4}$	780	390	1130	565	575	280
$\frac{3}{8}$	620	310	920	460	460	230
$\frac{1}{2}$	600	300	850	425	420	210
$\frac{3}{4}$	480	240	690	345	340	170
1	450	225	630	315	320	160
$1\frac{1}{4}$	370	185	520	260	260	130
$1\frac{1}{2}$	330	165	470	235	240	120
2	280	140	400	200	235	120
$2\frac{1}{2}$	300	150	420	210	200	100
3	260	130	370	185	190	95
$3\frac{1}{2}$	240	120	345	170	210	105
4	220	110	320	160	185	90
5	195	100	290	145	175	85
6	180	90	280	140	NR[a]	NR[a]
8	160	80	250	120	NR[a]	NR[a]
10	140	70	230	115	NR[a]	
12	130	65	230	115	NR[a]	
14	130	65				
16	130	65				

[a] NR = Not recommended.

A complete range of pipe sizes is not available in all SDR ratings. SDR 13.5, which has the highest operating pressure, is only available in sizes of $\frac{1}{2}$ inch through 4 inches. SDR 14 and SDR 21 both have the same pressure rating. However, wall thicknesses vary for the same size pipe. SDR 14 is available in sizes of 4–12 inches and has a heavier wall. This system is recommended for underground work. SDR 21 is available in sizes of $\frac{3}{4}$ inch through 12 inches and is usually recommended for above-ground installation. For the same reasons, SDR 18 and SDR 25 pipe are recommended for burial, and the latter is available from 4 through 24 inches. Most of the larger diameter buried pipe is used to handle potable water and meets either AWWA C900 or AWWA C905 specifications.

TABLE 5.10A Dimensions of PVC Pipe Made to SDR Standards

Nominal pipe size (in.)	Outside diameter (in.)	SDR 13.5		SDR 21		SDR 26		SDR 32.5	
		Inside diameter (in.)	Wall thickness (in.)	Inside diameter (in.)	Wall thickness (in.)	Inside diameter (in.)	Wall thickness (in.)	Inside diameter (in.)	Wall thickness (in.)
1/2	0.840	0.716	0.062						
3/4	1.050	0.894	0.078	0.930	0.060				
1	1.315	1.121	0.097	1.189	0.063				
1 1/4	1.660	1.414	0.123	1.502	0.079	1.532	0.064		
1 1/2	1.900	1.618	0.141	1.720	0.090	1.754	0.073	1.78	0.060
2	2.375	2.023	0.176	2.149	0.113	2.193	0.091	2.229	0.073
2 1/2	2.875	2.449	0.213	2.601	0.137	2.655	0.110	2.699	0.088
3	3.500	2.982	0.259	3.166	0.167	3.230	0.135	3.284	0.108
4	4.500	3.834	0.333	4.072	0.214	4.154	0.173	4.224	0.138
6	6.625			5.993	0.316	6.115	0.255	6.217	0.204
8	8.625			7.803	0.411	7.961	0.332	8.095	0.265
10	10.750			9.728	0.511	9.924	0.413	10.088	0.331
12	12.750			11.538	0.606	11.770	0.490	11.966	0.392

TABLE 5.10B   Dimensions of PVC Pipe Made to SDR Standards

Nominal pipe size (in.)	Outside diameter (in.)	SDR 14		SDR 18		SDR 25	
		Inside diameter (in.)	Wall thickness (in.)	Inside diameter (in.)	Wall thickness (in.)	Inside diameter (in.)	Wall thickness (in.)
4	4.80	4.07	0.343	4.23	0.267	4.39	0.192
6	6.90	5.86	0.493	6.09	0.383	6.30	0.276
8	9.05	7.68	0.646	7.98	0.503	8.28	0.362
10	11.10	9.42	0.793	9.79	0.617	10.16	0.444
12	13.20	11.20	0.943	11.65	0.733	12.08	0.528
14	15.3					14.08	0.612
16	17.4					16.01	0.696
18	19.5					17.94	0.780
20	21.6					19.87	0.864
24	25.8					23.74	1.032

A transparent schedule 40 PVC pipe system is also available in sizes of $\frac{1}{4}$ inch through 6 inches. It is joined by solvent cementing. Maximum operating pressures are the same as for regular PVC pipe. This transparent material is manufactured by Thermoplastic Processes Inc. and is sold under the tradename of Excelon R-4000.

Flanged PVC piping systems have a somewhat different pressure rating than other schedule 40 and schedule 80 pipes. The rating is constant through 8-inch diameters. However, the pressure rating decreases with increasing temperatures (see Table 5.12).

As with all thermoplatic piping systems, the maximum allowable pressure rating decreases with increasing temperature. Table 5.13 supplies the temperature

TABLE 5.11   Maximum Allowable Pressure of SDR-Rated Pipe at 73°F/23°C

SDR Rating	Maximum pressure (psi)
13.5	315
14	200
18	150
21	200
25	165
26	160
32.5	125

TABLE 5.12 Maximum Allowable Operating Pressure of Flanged PVC Systems Based on Operating Temperature[a]

Operating temperature (°F/°C)	Maximum allowable operating pressure (psi)
100/38	150
110/43	135
120/49	110
130/54	75
140/60	50

[a] Threaded flanges $> 2\frac{1}{2}$ inches must be backwelded in order to operate at ratings shown.

correction factors to be used with PVC pipe in schedule 40, schedule 80, and the SDR ratings. They are applied in the following manner:

Assume that a 2-inch schedule 80 solvent-welded, type 1 line is to operate at 110°F (43°C); at 73.4°F (23°C) the maximum allowable operating pressure from Table 5.9 is 400 psi. The temperature correction factor from Table 5.13 is

TABLE 5.13 Temperature Correction Factors for PVC Pipe Referred to 73°F/23°C

Operating temperature (°F/°C)	Factor	
	Type 1	Type 2
50/10	1.20	1.20
60/16	1.10	1.10
70/21	1.05	1.05
73.4/23	1.00	1.00
80/27	0.95	0.95
90/32	0.90	0.90
100/38	0.85	0.85
110/43	0.80	0.70
115/46	0.75	0.55
120/49	0.70	0.45
125/52	0.70	0.35
130/54	0.65	0.25
140/60	0.60	NR[a]
150/66	0.55	NR[a]

[a] NR = Not recommended.

TABLE 5.14 Collapse Ratings of Type 1 PVC Pipe at
73.4°F/23°C

Pipe size (in.)	Schedule 40	Schedule 80
$\frac{1}{2}$	450	575
$\frac{3}{4}$	285	499
1	245	469
$1\frac{1}{4}$	160	340
$1\frac{1}{2}$	120	270
2	75	190
$2\frac{1}{2}$	100	220
3	70	155
4	45	115
6	25	80
8	16	50
10	12	43
12	9	39

0.80. Therefore the maximum allowable operating pressure at 110°F (43°C) is
400(0.80) = 320 psi.

On occasion PVC pipe is used under such conditions that it is operating
under a negative internal pressure, such as in pump suction lines and vacuum
lines. For this type of service, solvent-cemented joints should be used. Under
continuous pumping, PVC lines can be evacuated to 5 μm, but when the system
is shut off, the pressure will rise to and stabilize at approximately 10,000 μm or
10 mmHg at 73°F (23°C). For this type of service, the collapse pressure rating of
the pipe must be taken into account (see Table 5.14). These values determine the
allowable pressure differential between internal and external pressure. If a
temperature other than 73.4°F (23°C) is to be used, the correction factors in
Table 5.13 must be applied.

B. Joining

The preferred method of joining PVC pipe is by means of solvent cementing.
Other methods that can be used are

Threaded joints on schedule 80 pipe
Flanged joints

TABLE 5.15 Temperature Correction Factors for Underground Fiberglass-Armored PVC Pipe

Operating temperature (°F/°C)	Pressure class (psi)	
	250	350
73/23	1.0	1.0
80/27	0.92	0.93
90/32	0.85	0.85
100/38	0.80	0.80
110/43	0.70	0.70
120/49	0.60	0.60
130/54	0.55	0.55
140/60	0.50	0.50

Bell and ring gasket joints for underground piping
Grooved-end mechanical joints
Compression insert joints

The physical properties and allowable operating pressures of PVC pipe at the temperatures shown must also be taken into account. PVC has found wide application as process piping, laboratory drainage piping, potable water supply piping, and vent line piping.

5.3 FIBERGLASS-ARMORED PVC PIPE

The dual construction of a fiberglass outer covering and a PVC inner liner has definite advantages, particularly for buried piping systems. PVC has relatively low physical properties as compared to a fiberglass pipe. Fiberglass-armored PVC pipe is essentially a fiberglass pipe with a PVC liner. In mechanical design, the PVC liner is ignored for structural purposes. All mechanical strength comes from the fiberglass pipe. The PVC liner provides the corrosion resistance, which is superior to that of the fiberglass. Other liners can also be used.

A. Pipe Data

This piping is produced for underground as well as above-ground applications, and the former is the predominant usage for water distribution systems, fire lines, forced sewer lines, and industrial pipe lines.

Underground piping is available in diameters of 4–12 inches in two standard pressure ratings of 250 and 350 psi. These pressure ratings are sustained working pressures including surge pressure at 73°F (23°C). As operating

TABLE 5.16 Maximum Operating Pressure and Temperature of Above-Ground Fiberglass-Armored PVC Pipe

Nominal pipe size (in.)	Operating temperature (°F/°C)	Operating pressure (psi)
2	200/93	150
$2\frac{1}{2}$	200/93	150
3	200/93	150
4	200/93	100
6	200/93	100

temperatures increase, the allowable operating pressure decreases. Temperature correction factors can be found in Table 5.15. To use these factors, multiply the SDR rating at 70°F (21°C) by the appropriate factor for the actual operating temperature. For example, if the operating temperature is 100°F (38°C), the correction factor is 0.80. Assuming that pressure class 250 pipe is to be used, then the allowable operating pressure at 100°F (38°C) will be 250 (0.80) = 200 psi.

Above-ground piping is manufactured in sizes of 2–6 inches. Maximum operating pressures and temperatures are given in Table 5.16.

B. Joining

Underground piping is manufactured with an integral bell and spigot end. An elastomeric ring is "locked" in the ring groove at the time of manufacture. The purpose of the ring is twofold: It aids in the centering of the pipe, and it is used to make a tight seal. Unfortunately, fittings are not available, but the pipe can be used directly with cast iron dimensional fittings, both slip joint and mechanical joint. When using mechanical joint fittings, caution should be excercised when tightening the bolt. Overtorquing can damage the pipe. A maximum torque of 50 ft-lb should be used. Installation is in accordance with the details of buried fiberglass pipe.

Above-ground piping is joined in a two-step procedure. The inner PVC liner is solvent welded, after which the entire joint is wrapped with fiberglass tape saturated with the proper epoxy adhesive, which is supplied by the pipe manufacturer.

5.4 CHLORINATED POLYVINYL CHLORIDE (CPVC) PIPE

PVC is chlorinated to increase the chlorine content to approximately 67% from 56.8% to produce CPVC. This additional chlorine increases the heat deflection temperature from 160°F (71°C) to 212°F (100°C). This also permits a higher allowable operating temperature.

CPVC has been used for over 25 years for handling corrosive materials in the chemical process industry. It is also widely used as condensate return lines to convey hot water, particularly in areas where external corrosion is present.

A. Pipe Data

CPVC pipe is available in diameters of $\frac{1}{4}$ inch through 12 inches in both schedule 40 and schedule 80. Dimensions are shown in Table 5.1. The maximum allowable operating pressures at 73°F (23°C) are given in Tables 5.17 and 5.18. As with all thermoplastic piping systems, the allowable operating pressure is decreased as the temperature increases. Table 5.19 provides the factors by which the operating pressures must be reduced at elevated temperatures. To use the temperature correction factors in Table 5.19, multiply the allowable working pressure at

TABLE 5.17 Maximum Allowable Operating Pressure of CPVC Pipe at 73°F/23°C

Nominal pipe size (in.)	Schedule 40	Schedule 80	
	Solvent-welded	Solvent-welded	Threaded
$\frac{1}{4}$	640	900	450
$\frac{3}{8}$	620	800	365
$\frac{1}{2}$	600	850	340
$\frac{3}{4}$	480	690	275
1	450	630	250
$1\frac{1}{4}$	370	520	205
$1\frac{1}{2}$	330	470	185
2	280	400	160
$2\frac{1}{2}$	300	420	170
3	260	370	150
4	220	320	130
5	180	290	115
6	180	280	115
8	160	250	NR[a]
10	140	230	NR[a]
12	130	230	NR[a]

[a] NR = Not recommended.

TABLE 5.18 Maximum Allowable Operating Pressure of Flanged CPVC Systems Based on Operating Temperature[a]

Operating temperature (°F/°C)	Maximum allowable operating pressure (psi)
100/38	150
110/43	145
120/49	135
130/54	125
140/60	110
150/66	100
160/71	90
170/77	80
180/82	70
190/88	60
200/93	50
210/99	40

[a] Threaded flange sizes $2\frac{1}{2}$ through 6 inches must be backwelded in order to operate at the ratings shown above.

TABLE 5.19 Temperature Correction Factors for CPVC Pipe

Operating temperature		
°F	°C	Factor
70	21	1.00
80	27	1.00
90	32	0.91
100	38	0.82
120	49	0.65
140	60	0.50
160	71	0.40
180	82	0.25
200	93	0.20

Source: Courtesy of B. F. Goodrich Specialty Polymers Chemical Division.

TABLE 5.20 Collapse Pressure
Rating of Schedule 80 CPVC Pipe at
73.4°F/23°C

Nominal pipe size (in.)	Pressure rating (psi)
$\frac{1}{2}$	575
$\frac{3}{4}$	499
1	469
$1\frac{1}{4}$	340
$1\frac{1}{2}$	270
2	190
$2\frac{1}{2}$	220
3	155
4	115
6	80
8	50

70°F (21°C) by the factor for the actual operating temperature. For example, if the operating temperature were 140°F (60°C) for a 2-inch schedule 80 solvent-welded line, the maximum allowable operating pressure at 140°F (60°C) would be (400)(0.5) = 200 psi.

Table 5.20 provides the collapse pressure rating of schedule 80 pipe. This is useful when the piping will be operating under negative pressure such as in a pump suction line. The pressure shown is the differential between the external and internal pressures.

B. Joining

The preferred method of joining CPVC pipe is by means of solvent cementing. However, schedule 80 pipe may be threaded, and flanges are also available. Threaded joints should not be used at temperatures above 150°F (66°C). Flanges should be limited to making connections to pumps, equipment, or accessories installed in the line or where frequent disassembly of the line may be necessary.

5.5 POLYPROPYLENE (PP) PIPE

Polypropylene is one of the most common and versatile thermoplastic used. Polypropylene is closely related to polyethylene. They are both members of a group known as "polyolefins," composed only of carbon and hydrogen. Other common thermoplasts, such as PVC and PTFE, also contain chlorine and fluorine, respectively.

If unmodified, polypropylene is the lightest of the common thermoplasts, having a specific gravity of approximately 0.91. In addition to the light weight, its major advantages are high heat resistance, stiffness, and chemical resistance. In some cases PP is actually a combination of PE and PP formed during a second stage of the polymerization. The form is designated copolymer PP and produces a product that is less brittle than the homopolymer. The copolymer is able to withstand impact forces down to $-20°F$ ($-29°C$), while the homopolymer is extremely brittle below $40°F$ ($4°C$).

The homopolymers are generally long-chain, high molecular weight molecules with a minimum of random molecular orientation, thus optimizing their chemical, thermal, and physical properties. Homopolymers are preferred over copolymer materials for demanding chemical, thermal, and physical conditions. Increased impact resistance is characteristic of most copolymers. However, tensile strength and stiffness are substantially reduced, increasing the potential for distortion and cold flow, particularly at elevated temperatures.

Piping systems produced from the homopolymer have an upper temperature limit of $180°F$ ($82°C$), while those produced from the copolymer have an upper temperature limit of $200°F$ ($93°C$). The minimum operating temperatures are also affected, with the homopolymer limited to $50°F$ ($10°C$) and the copolymer to $15°F$ ($4°C$).

A. Pipe Data

Polypropylene pipe is produced from both the copolymer and the homopolymer in schedule 40, schedule 80, and pressure-rated systems. The maximum allowable operating pressures of schedule 40 and schedule 80 fusion-welded homopolymer PP pipe at $74°F$ ($23°C$) are shown in Table 5.21, while the maximum allowable operating pressure for threaded schedule 80 homopolymer pipe at $70°F$ ($21°C$) is given in Table 5.22. These pressure ratings must be reduced if the operating temperature exceeds $70°F$ ($21°C$). Table 5.23 provides the temperature correction factors to be used at elevated temperatures. Threaded PP pipe is not to be used at elevated temperatures. The temperature correction factor for homopolymer PP pipe at $120°F$ ($49°C$) is 0.75. To determine the allowable operating pressure of 2-inch schedule 40 pipe at $120°F$ ($49°C$), which has an allowable operating pressure of 140 psi at $70°F$ ($21°C$), multiply 140 by 0.75 to given an allowable operating pressure of 105 psi at $120°F$ ($49°C$).

Table 5.24 provides the maximum allowable operating pressure of flanged homopolymer PP systems based on operating temperature.

TABLE 5.21 Maximum Allowable Operating Pressure of Homopolymer PP Fusion-Welded Pipe at 73.4°F/23°C

Nominal pipe Size (in.)	Maximum allowable working pressure, psi	
	Schedule 40	Schedule 80
$\frac{1}{2}$	290	410
$\frac{3}{4}$	240	330
1	225	300
$1\frac{1}{4}$	185	250
$1\frac{1}{2}$	165	230
2	140	200
$2\frac{1}{2}$	130	185
3	130	190
4	115	165
6	90	140
8	80	130
10	75	115
12	60	110

TABLE 5.22 Maximum Operating Pressure for Threaded Schedule 80 Homopolymer PP Pipe at 73.4°F/23°C

Nominal pipe size (in.)	Maximum operating pressure (psi)
$\frac{1}{2}$	20
$\frac{3}{4}$	20
1	20
$1\frac{1}{4}$	20
$1\frac{1}{2}$	20
2	20
3	20
4	20

TABLE 5.23 Temperature Correction Factors for Schedule 40 and Schedule 80 Fusion-Welded Homopolymer PP Pipe Referred to 73.4°C/23°C

Operating temperature (°F/°C)	Pipe size	
	$\frac{1}{2}$–8 in.	10 and 12 in.
40/4	1.220	1.230
50/10	1.150	1.150
60/16	1.080	1.090
70/23	1.000	1.000
80/27	0.925	0.945
90/32	0.875	0.870
100/38	0.850	0.800
110/43	0.800	0.727
120/49	0.750	0.654
130/54	0.700	0.579
140/60	0.650	0.505
150/66	0.575	0.430
160/71	0.500	0.356

TABLE 5.24 Maximum Allowable Operating Pressure of Flanged Homopolymer PP Systems Based on Operating Temperature

Operating temperature (°F/°C)	Maximum allowable operating pressure (psi)
100/38	150
110/93	140
120/49	130
130/54	118
140/60	105
150/66	93
160/71	80
170/77	70
180/82	50

TABLE 5.25 Collapse Pressure Rating of Schedule 80 Homopolymer PP Pipe at 73.4°F/23°C

Nominal pipe size (in.)	Pressure rating (psi)
$\frac{1}{2}$	230
$\frac{3}{4}$	200
1	188
$1\frac{1}{4}$	136
$1\frac{1}{2}$	108
2	76
3	62
4	46
6	32

When a piping system is used under negative pressure, such as a pump suction line, care should be taken not to exceed the collapse pressure rating of the pipe. These values are given in Table 5.25 for schedule 80 homopolymer PP pipe.

Homopolymer polypropylene pipe is also available in standard pressure ratings of 45, 90, and 150 psi. Pipe sizes available at each of these ratings are as follows:

Rating	Size (in.)
45 psi	4–32
90 psi	2–32
150 psi	$\frac{1}{2}$–20

Copolymer PP pipe is available in SDRs of 32, with an operating pressure of 45 psi at 73.4°F (23°C) and SDR 11, which is rated at 150 psi at 73.4°F (23°C). Temperature correction factors for the SDR-rated pipe may be found in Table 5.26. These factors are used in the same way as those for the schedule 40 and schedule 80 pipe.

B. Joining

Polypropylene pipe can be joined by thermal fusion, threading, and flanging. The preferred method is thermal fusion. Threading is restricted to schedule 80 pipe. Flanged connections are used primarily for connecting to pumps, instruments, valves, or other flanged accessories.

TABLE 5.26 Temperature Correction Factors for SDR-Rated Copolymer PP Pipe Referred to 73.4°F/23°C

Temperature (°F/°C)	Correction factor
73.4/23	1.00
100/38	0.64
140/60	0.40
180/82	0.28
200/93	0.10

5.6 POLYETHYLENE (PE) PIPE

Polyethylene is a thermoplastic material which varies from type to type depending upon the molecular structure, its crystallinity, molecular weight, and molecular weight distribution. These variations are made possible through the changes in the polymerization conditions used in the manufacturing process. PE is formed by polymerizing ethylene gas obtained from petroleum hydrocarbons.

The terms low, high, and medium density refer to the ASTM designations based on unmodified polyethylene. Low-density PE has a specific gravity of 0.91–0.925; medium-density PE has a specific gravity of 0.926–0.940; high-density PE has a specific gravity of 0.941–0.959. The densities, being related to the molecular structure, are indicators of the properties of the final product.

Ultra high molecular weight (UHMW) polyethylene has an extremely long molecular chain. According to ASTM specifications, UHMW PE must have a molecular weight of at least 3.1 million. The molecular weight of pipe-grade PE is in the range of 500,000.

A. Pipe Data

Polyethylene pipe is available in several categories. An industrial grade such as Plexico 3408 EHMW pipe is stabilized with both UV stabilizers and a minimum of 2% carbon black to protect against degradation due to ultraviolet light. This stabilized system may be used above ground without fear of loss of pressure rating or other important properties due to ultraviolet light degradation. Pipe is available in sizes of $\frac{1}{2}$ inch through 36 inches in standard pressure ratings from 255 psi to 50 psi at 73°F (23°C) (see Table 5.27). In pressure operations, it has a maximum allowable temperature rating of 140°F (60°C). For nonpressure applications, it may be used to 180°F (82°C). As with other thermoplastic piping systems, as the temperature increases above 73°F (23°C), the pressure rating must

TABLE 5.27 Pressure Rating of Plexico 3408 EHMW PE Pipe at 73.4°F/23°C

SDR	Pressure rating (psi)
7.3	255
9.0	200
11.0	160
13.5	130
15.5	110
17	100
21	80
26	65
32.5	50

Source: Courtesy of Plexico Inc., subsidiary of Chevron Chemical Company, Franklin Park, IL.

be reduced. Table 5.28 provides the temperature correction factors to be used at elevated temperatures. These factors can be used in the following manner. Assume that SDR 9 pipe rated for 200 psi at 73°F (23°C) is to be used at a temperature of 120°F (49°C). From Table 5.28 we find the correction factor of 0.63. Therefore the maximum allowable operating pressure at 120°F/49°C is (200) (0.63) = 126 psi.

A high-temperature pipe, Plexico HTPE, is also available. Its pressure/temperature ratings are given in Table 5.29.

TABLE 5.28 Temperature Correction Factors for 73.4°F/23°C Rating of Plexico 3408 EHMW Pipe

Operating temperature (°F/°C)	Correction factors
40/4	1.2
60/16	1.80
73/23	1.00
100/38	0.78
120/49	0.63
140/60	0.50

Source: Courtesy of Plexico Inc., subsidiary of Chevron Chemical Company, Franklin Park, IL.

TABLE 5.29 Pressure Rating of Plexico High-Temperature PE Pipe[a]

SDR	Rating (psi) at °F/°C			
	120 (49)	140 (60)	160 (71)	176 (80)
7.3	175	160	125	100
9.0	140	125	100	80
11.0	110	100	80	65
13.5	90	80	65	50
15.5	75	70	55	45
17.0	70	60	50	40
21.0	55	50	40	30
26.0	45	40	30	25
32.5	35	30	25	20

[a]Pressure ratings based on water service for an estimated 10-year service life for temperatures above 140°F (60°C).
Source: Courtesy of Plexico Inc., subsidiary of Chevron Chemical Company, Franklin Park, IL.

Pipe for burial underground is furnished color coded by means of longitudinal stripes in blue, red, or yellow. These are designated by Plexico as Bluestripe for potable water, Redstripe for underground fire main systems, and Yellowstripe for gas distribution systems. This pipe is manufactured from the same 3408 EHMW resin. The colored stripes form an integral part of the pipe and have the same properties as the pipe itself. The Redstripe piping system has Factory Mutual approval for use as underground fire mains. Pipe is designed to meet class 150 (150 psi at 73°F [23°C]) and class 200 (200 psi at 73°F [23°C]) with a temperature rating of from −50°F (−46°C) to 140°F (60°C). Pressure–temperature ratings are give in Table 5.30.

TABLE 5.30 Pressure–Temperature Ratings of Redstripe EHMW PE Pipe

Class	Service Pressure Ratings (psi)			
	73°F/23°C	100°F/38°C	120°F/49°C	140°F/60°C
150	150	125	100	80
200	200	170	135	110

Source: Courtesy of Plexico Inc., subsidiary of Chevron Chemical Company, Franklin Park, IL.

Ryerson Plastics also produces a line of polyethylene in two grades under the trade name Driscopipe. Grade 8600 is recommended for use wherever environmental stress cracking or abrasion resistance are major concerns. It has superior toughness and fatigue strength and high impact strength. It is also protected from ultraviolet degradation and is available in the following ratings:

SDR	Allowable pressure (psi)	Size range available (in.)
32.5	65	5–42
25.3	65	10–48
15.5	110	2–24
11.0	160	3/4–27
9.33	190	3/4–6
8.3	220	8
7.3	254	14, 18

For less demanding applications, grade 1000 is recommended and is available in the following ratings:

SDR	Allowable pressure (psi)	Size range available (in.)
32.5	51	3–54
26.0	64	3–54
21.0	80	3–36
19.0	89	3–36
17.0	100	2–36
15.5	110	2–32
13.5	128	2–30
11.0	160	1–24
9.0	200	2–18
7.0	267	2–14

B. Joining

Polyethylene pipe is joined by thermal fusion techniques. Specific instructions should be requested from the manufacturer.

5.7 ETHYLENE CHLOROTRIFLUOROETHYLENE (ECTFE) PIPE

Ethylene chlorotrifluoroethylene is a 1 : 1 alternating copolymer of ethylene and chlorotrifluoroethylene. This chemical structure gives the polymer a unique combination of properties. It possesses excellent chemical resistance, a broad-use temperature range from cryogenic to 340°F (171°C) with continuous service to 300°F (149°C), and has excellent abrasion resistance.

ECTFE is a tough material with excellent impact strength over its entire operating temperature range. Outstanding in this respect are properties related to impact at low temperature. In addition to these excellent impact properties, ECTFE also possesses good tensile, flexural, and wear-related properties. It is also one of the most radiation-resistant polymers. Other important properties include a low coefficient of friction and the ability to be pigmented. Table 2.19 gives the physical and mechanical properties. ECTFE is sold by Ausimont under the trade name Halar.

A. Pipe Data

ECTFE piping is available in sizes of 1 inch through 3 inches in an SDR pressure-rated system of 160 psi at 68°F(20°C). Table 5.31 show the correction factors to be used to determine the maximum allowable operating pressure at elevated temperatures.

TABLE 5.31 Temperature Correction Factors for ECTFE Pipe

Temperature		Correction factor
°F	°C	
68	20	1.0
83	30	0.90
104	40	0.82
121	50	0.73
140	60	0.65
158	70	0.54
176	80	0.39
194	90	0.27
212	100	0.20
256	125	0.10
292	150	Drainage
340	170	pressure only

From Table 5.31, the temperature correction factor for 140°F (60°C) is found to be 0.65. Applying this factor to the allowale 150 psi at 73°F (23°C) gives 150 (0.65) = 97.5 psi at 140°F (60°C). Note that above 256°F (125°C) the piping should be used only for gravity flow without any positive pressure.

B. Joining

All pipe and fittings are joined by the butt fusion method.

5.8 VINYLIDENE FLUORIDE (PVDF) PIPE

PVDF is a crystalline, high molecular weight polymer of vinylidene fluoride containing 50% fluorine. It has a very linear chemical structure and is similar to PTFE, with the exception that it is not fully fluorinated. It has high tensile strength, a high heat deflection temperature, and is resistant to permeation of gases. Approval has been granted by the Food and Drug Administration for its repeated use on contact with food in food-handling and processing equipment. Much of the strength and chemical resistance of PVDF is retained within a temperature range of −40 to 320°F (−40 to 160°C).

PVDF is manufactured under the trade name Kynar by Elf Atochem, Solef by Solvay, Hylar by Ausimont USA, and Super Pro 230 & ISO by Asahi/America.

A. Pipe Data

PVDF pipe is available in schedule 40, schedule 80, and two pressure-rated systems: 150 and 230 psi at 73.4°F (23°C). These piping systems can be operated continuously at 280°F (138°C). Table 5.32 supplies the maximum allowable operating pressures for schedule 40 and schedule 80 pipe. This pipe is available in sizes of $\frac{1}{2}$ inch through 6 inches. Table 5.33 provides the temperature correction factors when the pipe is used at elevated temperatures. These factors are to be applied to the maximum allowable operating pressure at 70°F (21°C). The correction factor is used in the following manner. A 2-inch schedule 80 solvent-welded pipe at 70°F (21°C) has an allowable operating pressure of 275 psi. the temperature correction factor from Table 5.33 at 140°F (60°C) is 0.670. Therefore, the allowable operating pressure of 2-inch solvent-welded pipe at 140°F (60°C) is (275) (0.670) = 184 psi. Table 5.34 supplies the maximum allowable operating pressure of flanged PVDF systems based on operating temperatures.

If pipe is to be used in negative pressure applications, care should be exercised that the collapse pressure rating not be exceeded. Table 5.35 gives the collapse pressure rating for schedule 80 pipe.

TABLE 5.32 Maximum Allowable Operating Pressure of Schedule 40 and Schedule 80 PVDF at 70°F/21°C

Nominal pipe size (in.)	Maximum Working Pressure (psi)		
	Schedule 40, fusion-welded	Schedule 80, threaded	Schedule 80, fusion-welded
$\frac{1}{2}$	360	290	580
$\frac{3}{4}$	300	235	470
1	280	215	430
$1\frac{1}{4}$	—	215	320
$1\frac{1}{2}$	220	160	320
2	180	135	275
3	NR[a]	NR[a]	250
4	NR[a]	NR[a]	220
6	NR[a]	NR[a]	190

[a] NR = Not recommended.

TABLE 5.33 Temperature Correction Factors for Schedule 40 and Schedule 80 PVDF Pipe

Operating temperature (°F/°C)	Factor	Operating temperature (°F/°C)	Factor
30/−1	1.210	160/70	0.585
40/4	1.180	170/77	0.543
50/10	1.110	180/82	0.500
60/16	1.060	190/88	0.458
70/21	1.000	200/93	0.423
80/27	0.950	210/99	0.388
90/32	0.900	220/104	0.352
100/38	0.850	230/110	0.317
110/43	0.804	240/116	0.282
120/49	0.760	250/121	0.246
130/54	0.712	260/127	0.212
140/60	0.670	270/132	0.183
150/66	0.627	280/138	0.155

TABLE 5.34 Maximum Allowable Operating Pressure of Flanged PVDF Systems Based on Operating Temperature

Operating temperature (°F/°C)	Maximum allowable operating pressure (psi)
100/38	150
110/43	150
120/49	150
130/54	150
140/60	150
150/66	140
160/71	133
170/77	125
180/82	115
190/88	106
200/93	97
210/99	90
240/116	60
280/138	25

The 230-psi pressure-rated pipe system manufactured by Asahi/America is available in sizes of $\frac{3}{8}$ inch through 4 inches as standard with sizes through 12 inches available on special order. The 150-psi rated system is available in sizes $2\frac{1}{2}$ through 12 inches. These systems are marketed under the trade names of Super Pro 230 and Super Pro 150. Table 5.36 provides the pressure rating correction factors when the piping is to be operated at elevated temperatures.

TABLE 5.35 Collapse Pressure Rating of Schedule 80 PVDF Pipe at 73°F/23°C

Nominal pipe size (in.)	Collapse pressure rating (psi)
$\frac{1}{2}$	391
$\frac{3}{4}$	339
1	319
$1\frac{1}{2}$	183
2	129

TABLE 5.36 Pressure Rating Correction Factors for Elevated Operating Temperatures for Super Pro[a] 230 and Super Pro[a] 150 Pipe

Temperature		Correction factor
°F	°C	
70	21	1.00
80	27	0.95
90	32	0.87
100	38	0.80
120	49	0.68
140	60	0.58
160	71	0.49
180	82	0.42
200	93	0.36
240	115	0.25
260	138	0.18

[a] Super Pro 230 and Super Pro 150 are trademarks of Asahi/America.

B. Joining

PVDF pipe can be joined by threading, fusion welding, and flanging. Of these methods, fusion welding is the preferred technique. Threading can be done only using schedule 80 pipe. It is recommended that flanging be limited to connections to equipment, pumps, instruments, etc.

5.9 POLYVINYLIDENE CHLORIDE (SARAN) PIPE

Polyvinylidene chloride pipe is manufactured by Dow Chemical and sold under the trade name Saran. Because of the poor physical properties of Saran, applications are somewhat limited. The pipe system has a relatively low allowable operating pressure, which decreases rapidly as operating temperatures increase above ambient. Consequently, Saran-lined steel pipe is used more commonly than solid Saran pipe.

Saran pipe meets FDA regulations for use in food processing and for potable water, as well as regulations prescribed by the meat inspection division of

TABLE 5.37 Dimensions of Schedule 80 Saran Pipe

Nominal pipe size (in.)	Outside diameter (in.)	Inside diameter (in.)	Wall thickness (in.)
$\frac{1}{2}$	0.840	0.546	0.147
$\frac{3}{4}$	1.050	0.742	0.154
1	1.315	0.957	0.179
$1\frac{1}{4}$	1.660	1.278	0.191
$1\frac{1}{2}$	1.900	1.500	0.200
2	2.375	1.939	0.218
$2\frac{1}{2}$	2.875	2.277	0.299
3	3.500	2.842	0.329
4	4.500	3.749	0.376
6	6.625	5.875	0.375

the U.S. Department of Agriculture for transporting fluids used in meat production.

A. Pipe Data

Saran pipe is available only in schedule 80 diameters of $\frac{1}{2}$ inch through 6 inches. The dimensions of Saran pipe differ from those of other thermoplastic piping

TABLE 5.38 Maximum Operating Pressures of Saran Pipe at Various Temperatures

Nominal pipe size (in.)	Maximum operating pressure (psi) at °F (°C):						
	50(10)	68(20)	77(25)	86(30)	104(40)	140(60)	176(80)
$\frac{1}{2}$	270	235	220	200	170	120	100
$\frac{3}{4}$	230	200	190	180	150	110	80
1	210	180	170	155	145	100	70
$1\frac{1}{4}$	180	155	140	135	110	80	60
$1\frac{1}{2}$	160	140	130	120	100	70	50
2	140	120	110	105	90	65	45
$2\frac{1}{2}$	125	110	105	100	80	60	40
3	115	100	90	85	70	50	35
4	100	85	80	70	60	40	30
6	80	70	65	60	50	30	20

systems (see Table 5.37). The maximum allowable operating pressures at various operating temperatures are given in Table 5.38.

B. Joining

Saran pipe is joined only by threading. No other method has proven successful.

6

Thermoset Piping Systems

6.1 INTRODUCTION

A. Design Information

Thermoset piping materials are the result of a careful selection of base resin backbone reactive end group, catalyst, accelerator, fillers, and other additives. By varying these ingredients, specific properties can be imparted to the finished product. For thermoset piping systems, the four primary resin groups used are vinyl esters, unsaturated polyesters, epoxies, and furans.

Vinyl esters have a wide range of corrosion resistance, particularly to strong corrosive acids, bases, and salt solutions, up to temperatures of 200°F (93°C). These are produced from the reaction of epoxy resins with ethylene-unsaturated carboxylic acids. The epoxy resin backbone imparts toughness and superior tensile elongation properties. The vinyl ester generally used for chemical process piping systems is derived from a diglycidyl ether of bisphenol-A epoxy resin with the addition of methacrylate end groups to provide maximum corrosion resistance.

Unsaturated polyesters are alkyd thermosetting resins with vinyl unsaturation on the polyester backbone. Because they are simple, versatile, and economical, they are the most widely used resin family. They can be compounded to resist most chemicals at temperatures of 75°F (24°C) and many chemicals to 160°F (71°C) or higher.

As with the vinyl esters, the unsaturated polyesters can be formulated to yield specific properties. A typical unsaturated polyester resin includes a base polyester resin, reactive diluents, catalyst, accelerator, fillers and inhibitors. Fillers are selected to improve flame retardancy, shrinkage control, and impact resistance. Catalyst systems are formulated to extend resin shelf life and control reaction and cure times.

Epoxies are thermosetting matrix resins that cure to cross-linked, insoluble, infusible matrix resins with or without the addition of heat. These resins are specified when continuous operation at elevated temperatures [up to 437°F (225°C)] is anticipated. They can be specially formulated through the selection of a wide variety of base resins, curing agents, catalysts, and additives to meet specific applications.

The furan polymer is a derivative of furfuryl alcohol and furfural. Although these resins cost approximately 30% more than other thermosetting resins, in many cases they are the most economical choice when

1. Solvents are present in a combination of acids and bases.
2. They are an intrinsic choice to high nickel alloys.
3. Process changes may occur that would result in exposure of solvents in oxidizing atmospheres.

The greatest single advantage of the furan resins is their extremely good resistance to solvents in combination with acids and bases.

B. Reinforcing Materials

Because of its low cost, high tensile and impact strength, light weight, and good corrosion resistance, fiberglass is the most widely used reinforcement. The predominant fiberglass used is E-glass (aluminum borosilicate), while S-glass (magnesium aluminoborosilicate) is selected for higher tensile strength, modulus, and temperature requirements. For extremely corrosive appications, C or ECR glass is used.

Although resistant to most chemicals, fiberglass can be attacked by alkalies and a few acids. To overcome this, synthetic veils, most commonly thermplastic or polyester, are used. For increased resistance to chemical attack, a veil of C glass may be specified.

When chemical resistance to hydrofluoric acid is required, graphite (carbon) fibers are supplied. These fibers also possess high tensile strengths and moduli, low density, and excellent fatigue and creep resistance and can be engineered to yield an almost zero coefficient of thermal expansion. These fibers will also impart some degree of electrical conductivity to the pipe.

Other fibers are finding increasing applications. Among these are aramid, polyester, and oriented polyethylene. Although these fibers have excellent corrosion resistance and high strength-to-weight ratios, they have some drawbacks. Polyethylene has reduced fiber-resin adhesion properties, and aramid has much lower compressive properties than tensile properties.

Extremely effective is a synthetic nexus surfacing veil. This material has been used in particularly severe applications.

C. Working Pressure

Thermoset piping systems are not manufactured to a specific standard. Therefore there is no generalization as to allowable operating pressures. Some manufacturers supply their piping systems to a true inside diameter, for example, a 1-inch pipe would have a 1-inch inside diameter. Other manufacturers furnish pipe with outside diameters equal to IPS diameters and varying inside diameters depending upon pressure ratings. Therefore, each manufacturer's system must be treated independently as to operating pressures, operating temperatures, and dimensions.

D. Burial of Thermoset Pipe

This section deals with the design and installation of fiberglass-reinforced polyester, epoxy ester, and vinyl ester piping. Formulas are provided for the calculation of pipe stiffness and allowable deflection. For operating pressures below 100 psi, normally the vertical pressure on the pipe from ground cover and vehicle line loading will dictate the required wall thickness. At operating pressures above 100 psi, the internal pressure usually dictates the required wall thickness. In all cases the wall thickness should be checked for both conditions.

Soil conditions should be checked prior to design since these conditions affect the design calculations. Once these are determined, the design calculations can proceed.

The minimum pipe stiffness can be calculated using the following equation:

$$PS = \frac{E_F I}{0.149 \, r_M{}^3}$$

where

PS = pipe stiffness (psi)
E_F = hoop flexural modulus of elasticity (psi)
I = moment of inertia of structural wall (in.^{4}/in. = t^{-3}/12, where t = wall thickness)
r_M = mean radius of wall = D_M/2 where D_M = mean diameter of wall (in.).

Table 6.1 shows the minimum stiffness requirements by pipe diameter. If the PS calculated is less than that shown in Table 6.1, a thicker pipe wall must be used. The wall thickness must also be checked to be sure that it is adequate for the internal pressure.

The allowable deflection is calculated by

$$\Delta Y = \frac{[DW_C + W_L]K_X r^3}{E_F I + 0.061\ K_a E' r^3} + a$$

where

ΔY = vertical pipe deflection (in.) (ΔY should not exceed a value of 0.05 D_M; $Y = D_M/2$.)

D = deflection lag factor (see Table 5.1)

W_C = vertical soil load

r_M = mean radius wall (in.)

K_X = bedding factor (see Table 5.6)

$E_F I$ = stiffness factor of structural wall (in.^{2}lb/in.)

K_a = deflection coefficient (see Table 6.2)

E' = soil modulus (see Table 5.3)

a = deflection coefficient (see Table 6.2)

and

$$W_C = \frac{Y_S H D_o}{144}$$

where

Y_S = specific weight of soil (lb/ft^3) (If a specific weight of soil is unknown, use a soil density of 120 lb/ft^3.)

H = burial depth to top of pipe (ft)

D_o = outside diameter of pipe

W_E = live load on pipe (lb in.)

and

$$W_L = \frac{C_L P(1 + I_F)}{12}$$

TABLE 6.1 Minimum Pipe Stiffness Requirements

Nominal diameter (in.)	Minimum pipe stiffness at 5% deflection (psi)
1–8	35
10	20
12–144	10

TABLE 6.2 Deflection Coefficients

Installed condition	K_a	Δa
For all installation conditions with buried depths of 16 ft or less	0.75	0
For burial depths greater than 16 ft and installation conditions as follows:		
Dumped or slight degree of compaction (Proctor less than 85% or 1.00 relative density less than 40%)	1.00	$0.02\ D_M$
Moderate degree of compaction (Proctor of 85–95% or relative density of 40–70%)	1.00	$0.01\ D_M$
High degree of compaction (Proctor greater than 95% or relative 1.00 density greater than 70%)	1.00	$0.005\ D_M$

where

$\quad C_L$ = live load coefficient (see Table 6.3A and 6.3B)
$\quad P$ = wheel load (lb)
$\quad L_F$ = impact factor = $0.766 - 0.133H$ $(0 \leq I_F \leq 0.50)$

The trench should be excavated to a depth of approximately 12 inches below the pipe grade, which will permit a 6-inch foundation and a 6-inch layer of pipe bedding material. The bottom of the trench should be as uniform and

TABLE 6.3A Live Load Coefficient C_L—Single-Wheel Load

Pipe diameter (in.)	C_L				
	4[a]	6	8	12	16
8	0.020	0.010	0.006	0.003	0.001
10	0.025	0.012	0.007	0.002	0.002
12	0.029	0.014	0.008	0.004	0.002
14	0.034	0.016	0.009	0.004	0.002
16	0.038	0.018	0.010	0.005	0.003
18	0.042	0.020	0.012	0.005	0.003
20	0.046	0.922	0.013	0.006	0.003
24	0.055	0.026	0.015	0.007	0.004
30	0.066	0.032	0.019	0.007	0.005
36	0.076	0.038	0.022	0.010	0.006
42	0.085	0.044	0.026	0.012	0.007
48	0.094	0.049	0.029	0.014	0.008
54	0.101	0.053	0.032	0.015	0.009
60	0.104	0.055	0.033	0.016	0.009

[a] Height of cover over pipe in feet.

TABLE 6.3B Live Load Coefficient C_L for Two Passing Trucks

Pipe diameter (in.)	C_L				
	4^a	6	8	12	16
8	0.0294	0.0169	0.0112	0.0062	0.0039
10	0.0367	0.0210	0.0139	0.0077	0.0049
12	0.0443	0.0253	0.0167	0.0092	0.0059
14	0.0517	0.0295	0.0195	0.0108	0.0069
16	0.0589	0.0336	0.0223	0.0123	0.0078
18	0.0664	0.0379	0.0251	0.0139	0.0088
20	0.0740	0.0422	0.0280	0.0155	0.0098
24	0.0886	0.0506	0.0335	0.0185	0.0118
30	0.1108	0.0632	0.0419	0.0232	0.0147
36	0.1329	0.0759	0.0503	0.0278	0.0177
42	0.1551	0.0886	0.0586	0.0325	0.0207
48	0.1773	0.1012	0.0670	0.0371	0.0236
54	0.1994	0.1139	0.0754	0.0417	0.0266
60	0.2216	0.1265	0.0838	0.0464	0.0295

^{a} Height of cover over pipe in feet.

continuous as possible. High spots in the trench will cause uneven bearing on the pipe, which will cause stress on the pipe during backfill and unnecessary wear at these points. Sharp bends and changes in elevation of the line should be avoided.

All backfill material should be free of stones above $\frac{3}{4}$ inch in size, vegetation, and hard clods of earth. An ideal type of backfill is pea gravel or crushed stone corresponding to ASTM C33 graduation 67 (grain size $\frac{3}{4}-\frac{3}{16}$ inch).

Pea gravel has some advantages: When poured directly into the trench it will compact to 90% or more of its maximum density, eliminating the need for equipment and labor for compaction. The pea gravel will not retain water, eliminating the problem of clumping. Above the 70% height of the pipe, good native soil, free of clods, stones over $\frac{3}{4}$ inch, organic matter, and other foreign material can be used. This layer should be placed 6–16 inches above the pipe using any standard compaction method other than hydraulic compaction. Above this level, the trench can be backfilled without compaction as long as there are no voids present. The backfilling should be done as soon as possible after testing of the pipe to eliminate the possibility of damage to the pipe, floating of the pipe due to flooding, and shifting of the line due to cave-ins.

If the pipe is to be laid under a road crossing, it is good practice to lay the pipe in a conduit, taking care to see that the pipe is properly bedded at the entrance and exit of the conduit. If not properly bedded, excessive wear and/or stress can be imposed on the pipe.

E. Joining of Thermoset Pipe

As with thermoplastic piping systems, there are several methods available for joining thermoset piping systems. If properly made, the joint will be as strong as the pipe itself, but when improperly made, it becomes the weak link in the system. The methods of joining are

1. butt and strap joint
2. socket-type adhesive joint
3. flanged joints
4. bell and spigot joint
5. threaded joint

1. Butt and Strap Joint

This joint is made by butting together two sections of pipe and/or fittings and overwrapping the joint with successive layers of resin impregnated mat or mat and roving (see Fig. 6.1). It is the standard method used with polyester and vinyl ester piping systems and is so stated in the commercial standards, which stipulate that all pipe 20 inches in diameter and larger shall be overlaid both inside and outside. Pipe having diameters of less than 20 inches shall be overlaid only on the outside.

The butt joint provides a mechanical or adhesive bond and not a chemical bond. Therefore it is important that the surface be free from any contamination. In addition, the width of the strapping materials is also important since they must be long enough to withstand the shear stress of the pipe. As the strapping material shrinks around the pipe during the cure, it forms a very tight joint.

The width of the strapping material will vary as successive layers are applied. It is important that each layer be long enough to surround the pipe completely and provide approximately a 2-inch overlap. Table 6.4 indicates the material requirements for a butt and strap joint on 50 psi rated pipe, while Table 6.5 provides the same information for 100 psi rated pipe and Table 6.6 for 150 psi rated pipe.

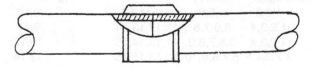

FIGURE 6.1 Butt and strap joint.

TABLE 6.4 Strapping Material Required for Butt and Strap Joint of 50 psi Rated Pipe

Nominal pipe size (in.)	Mat width (in.)			Roving width (in.), layer sequence 4
	3[a]	4[a]	6[a]	
2	1,2	3	4	4
3	1,2	3	4	4
4	1,2	3	4	4
6	1,2	3	4	4
8	1,2	3,5,6	4	4
10	1,2	3,5,6	4	4
12	1,2	3,5,6	4	4
14		1,2,3	4,6,7	5
16		1,2,3	4,6,7	5
18		1,2,3	5,7,8	4,6
20		1,2,3	5,7,8	4,6
24		1,2,3	4,6,8,9	5,7

[a] Layer sequence.

TABLE 6.5 Strapping Material Required for Butt and Strap Joint of 100 psi Rated Pipe

Nominal pipe size (in.)	Mat width (in.)					Roving width (in.)		
	3[a]	4[a]	6[a]	8[a]	12[a]	4[a]	6[a]	8[a]
2	1,2	3				4		
3	1,2	3				4		
4	1,2	3,5,6				4		
6	1,2	3,5,6				4		
8		1,2,3	4,6,7			5		
10		1,2,3	5,7,8			4,6		
12		1,2,3	4,6,8			5,7		
14		1,2,3	4,5,6	8,10,11			7,9	
16			1,2,3	4,5,6	7,9,11,12			8,10
18			1,2,3,4	5,6,7,8	10,12,13,14			9,11
20			1,2,3,4	5,6,7,8,9	11,13,14,15			10,12
24			1,2,3,4,5	6,7,8,9,10	11,13,15,16			12,14

[a] Layer sequence.

TABLE 6.6 Strapping Material Required for Butt and Strap Joint of 150 psi Rated Pipe

Nominal pipe size (in.)	Mat width (in.)					Roving width (in.)		
	3[a]	4[a]	6[a]	8[a]	12[a]	4[a]	6[a]	8[a]
2	1,2	3				4		
3	1,2	3,5,6				4		
4	1,2	3,5,6				4		
6		1,2,3	5,7,8			4,6		
8		1,2,3	4,6,8,9			5,7		
10		1,2,3	4,5,6	8,10,11			7,9	
12			1,2,3,4	5,6,8	10,12,13,14			9,11
14			1,2,3,4,5	6,7,8,9,10	11,13,15,16			12,14

[a] Layer sequence.

Detailed procedures for making a butt and strap joint can be obtained from any reliable vendor of the pipe. Listed below are the general steps that must be followed:

1. Clean all pipe surfaces in the areas of the joint.
2. Roughen the surface of the pipe or fitting with a file or sander.
3. Coat all raw edges of the pipe with resin to prevent penetration by the fluid to be handled.
4. Cut all ends of pipe straight
5. Align the two edges of the pipe and check to see that they are square.

Cure time for each joint is approximately 45 minutes.

2. Socket-Type Adhesive Joint

These joints are used with polyester and vinyl ester piping systems, as well as reinforced epoxy systems. The joint is made by utilizing a coupling into which the fitting end or pipe end is placed and cemented. Figure 6.2 illlustrates a typical joint.

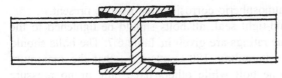

FIGURE 6.2 Socket-type adhesive joint.

The adhesive for these joints is furnished in either liquid or paste form in two parts—the cement and the catalyst—which must be mixed together. Since the adhesive has a pot life of only 15–30 minutes, the pipe and fitting surfaces should be cleaned and ready to be joined before the adhesive is mixed. The general steps to follow are

1. Prepare the pipe and fittings by cleaning and sanding.
2. Mix the cement and catalyst.
3. Apply the prepared adhesive to the pipe and fitting to be joined.
4. Assemble the fitting to the pipe, and rotate 180° to distribute the cement and to eliminate air pockets.
5. Wipe a fillet of cement around the fitting and remove any excess.
6. To speed up the hardening of an epoxy cemented joint, apply heat but do not exceed 180°F (82°C). Heat must be applied slowly. Hold the joint at this temperature for 10 minutes or until the adhesive fillet is no longer tackly. Then slowly increase the temperature to 200–230°F (93–110°C) for 10 minutes. If bubbling occurs at the edges of the joint, remove the heat for a few minutes, then reapply.
7. Let the joint cool to 100°F (38°C) before handling.

The application of heat in no way affects the strength or corrosion resistance—it only shortens the curing time. It is not necessary to apply heat to effect a cure. A hot air gun can be used to apply the heat.

3. Flanged Joints

Thermoset pipe systems can be joined by means of flanged joints. However, a complete system is not normally flanged. Flanges are used when connecting to equipment, e.g., pumps, tanks, etc., or accessory items such as valves or instruments. They are also used when sections of the piping system may have to be removed.

Flanges are generally provided press molded or hand-laid-up stub ends. The press-molded flanges can be furnished with a short length of pipe as a stub end.

Full-face gaskets having a Shore durometer of 50 to 70 should be used. Care must be taken in the selection of the gasket material to ensure its compatibility with the material being handled. When connecting flanged joints, a washer should be placed under both the nut and bolt head. Nuts and bolts must also be selected to resist any atmospheric corrosion that may be present.

In order to effect a liquid-tight seal, all bolts should be tightened to the proper torque. Maximum torque ratings are given in Table 6.7. The bolts should be gradually tightened in an alternating 180° pattern so that the maximum pressure is not achieved on one bolt while others have little or no pressure applied.

TABLE 6.7 Maximum Bolt Torque for Flanged Joints on Reinforced Thermoset Pipe at Different Pressure Ratings

Nominal pipe Size (in.)	Torque (ft-lb)					
	25 psi	50 psi	75 psi	100 psi	125 psi	150 psi
2	25	25	25	25	25	25
3	25	25	25	25	25	25
4	25	25	25	25	25	25
6	25	25	25	25	25	25
8	25	25	30	40	35	40
10	25	25	30	40	50	60
12	25	25	35	45	50	70
14	25	30	40	60	60	80
16	25	30	50	70	75	100
18	30	35	50	80	80	
20	30	35	60	90	100	
24	35	40	70			

4. Bell and Spigot O-Ring Joints

The bell and spigot joint is widely used for buried lines and long straight runs. Sealing of the joint is effected by means of an elastomeric O-ring. Care must be taken in the selection of the O-ring to ensure that it is compatible with the fluid being handled. (Appendix 1 provides the fluid compatibility of various types of elastomeric materials in contact with a variety of corrodents.) Figure 6.3 illustrates a typical joint.

The advantages of this type of joint lie in its ease and speed of assembly regardless of weather conditions, which results in the lowest installation cost for long straight runs. By virtue of the nature of the joint, several degrees of

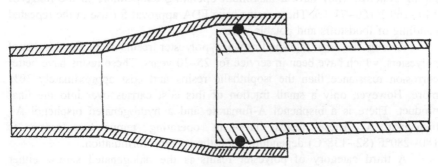

FIGURE 6.3 Bell and spigot O-ring joint.

misalignment can be tolerated. The joint also will act as an expansion joint to a limited degree. If installed above ground, provision must be made to restrain thrust.

F. Fire Hazards

Potential fire hazards should also be considered when selecting the specific thermoset to be used. Although the fluid being conveyed may not be combustible, there is always the potential danger of an external fire.

In the Physical and Mechanical Properties Table found under the heading of each specific resin system are two entries relating to this topic. The first is the limiting oxygen index percentage. This is a measure of the minimum oxygen level required to support combustion of the thermoset resin. The second is the flame spread classification. These ratings are based on common tests as outlined by the Underwriters Laboratories and are defined as follows:

Flame Spread Rating	Classification
0–25	Noncombustible
25–50	Fire retardant
50–200	Combustible
over 200	Highly combustible

6.2 REINFORCED POLYESTER PIPING SYSTEMS

Polyester does not describe a specific resin, but rather a family of resins. There are three basic types of polyester resins which are used for corrosion-resistant piping. With each of these types there are subtypes.

The isophthalic polyesters are the least expensive of the resins used for piping systems. They have a maximum operating temperature in the range of 140–160°F (60–71°C). These resins have FDA approval for use in the repeated handling of foodstuffs and potable water.

Among the first high-performance polyester resins were the bisphenol polyesters, which have been in service for 25–30 years. These resins have better corrosion resistance than the isophthalic resins and cost approximately 30% more. However, only a small fraction of this cost carries over into the final product. There is a bisphenol A-fumarate and a hydrogenated bisphenol A–bisphenol A. These resins have an operating temperature range of 180–280°F (82–138°C) depending upon the specific formulation.

A third category of polyester resins is the halogenated series—either chlorinated or brominated. These resins can be used at temperatures up to

250°F (121°C). Their cost is approximately the same as the bisphenol polyesters. Chlorinated polyesters have a very high heat distortion point, and laminates show a very high retention of physical strength at elevated temperatures. These resins are inherently fire retardant. If antimony trioxide is added to these resins, they will burn with extreme difficulty. The other polyesters will support combustion, but burn slowly. Since specific physical properties are determined by the type of laminate construction, resin selection, selection of reinforcing material, and ratio of resin to reinforcing material, values can only be averages and ranges, since these factors will vary between manufacturers.

A. Pipe Data

Reinforced polyester pipe systems are available in diameters of 1 inch to 24 inches, with larger diameters being available on special order. Standard lengths are 10 and 20 feet. Unlike other piping systems, polyester pipe systems have not been standardized dimensionally. Some manufacturers designate their pipe with a nominal pipe size. For example, a nominal 6-inch-diameter pipe would have an inside diameter of 6.129–6.295 inches depending upon the pressure rating. Other manufacturers would supply a pipe with a true 6-inch inside diameter. Usually, pipe manufactured by the filament-wound process will have a true inside

TABLE 6.8 Reinforced Polyester Pipe Wall Thicknesses at Different Pressure Ratings

Pipe size (in.)	Minimum pipe wall thickness (in.)					
	25 psi	50 psi	75 psi	100 psi	125 psi	150 psi
2	0.1875	0.1875	0.1875	0.1875	0.1875	0.1875
3	0.1875	0.1875	0.1875	0.1875	0.25	0.25
4	0.1875	0.1875	0.1875	0.25	0.25	0.25
6	0.1875	0.1875	0.25	0.25	0.3125	0.375
8	0.1875	0.25	0.25	0.3125	0.375	0.4375
10	0.1875	0.25	0.3125	0.375	0.4375	0.50
12	0.1875	0.25	0.375	0.4375	0.50	0.625
14	0.25	0.3125	0.375	0.50	0.625	0.75
16	0.25	0.3125	0.4375	0.5625	0.6875	
18	0.25	0.375	0.50	0.625	0.75	
20	0.25	0.375	0.50	0.6875		
24	0.25	0.4375	0.625	0.8125		
30	0.3125	0.50	0.75			
36	0.375	0.625				
42	0.375	0.75				

TABLE 6.9 Collapsing Pressure of Reinforced Polyester Pipe Based on 20-Foot Lengths at Different Wall Thicknesses (in.)[a]

Pipe size (in.)	Collapsing pressure (psig)									
	0.1875	0.25	0.3125	0.375	0.4375	0.50	0.5625	0.625	0.6875	0.75
6	38.5	100[b]	207	378	570	800	1040	1410	1780	2200
8	17.3	44.7	94	173	264	378	515	680	865	1040
10	9.23	23.7	50	93	143	206	284	378	485	610
12	5.36	14.2	30	56	86	125	173	230	298	370
14	3.42	9	19.3	36	56	81	113	151	196	249
16	2.3	6.1	13	21.6	38	56	78	104	136	173
18	1.62	4.3	9.3	17.4	27	39.7	56	75	98	125
20	1.2	3.2	6.88	13	20.7	29.7	41.5	56	73	99
24	0.89	2.1	4.1	7.65	12	17.6	24.6	33.4	43.6	56
30	0.63	1.5	2.97	5.2	7.6	10.4	13.8	18	23	29.7
36	0.48	1.15	2.28	3.99	5.78	8	10.7	13.8	17.4	21.6
42	0.39	0.91	1.8	3.17	4.6	6.4	8.5	11	13.9	17.4
48	0.31	0.75	1.48	2.59	3.75	5.2	7	9.18	11.4	14.3
54	0.26	0.63	1.24	2.18	3.16	4.4	5.9	7.7	9.7	12
60	0.22	0.54	1.06	1.85	2.07	3.8	5.06	6.6	8.3	10.3

[a] Any reading above 14.7 will withstand full vacuum. If stiffeners are added every 10 feet, the above figures will increase.
[b] Wall thicknesses above the horizontal lines are capable of withstanding full vacuum with a 5:1 safety factor, minimum.

diameter, while pipe manufactured by the centrifugally cast process will have a constant outside diameter and a varying inside diameter.

Table 6.8 shows the minimum wall thicknesses for reinforced polyester pipe at various operating pressures, and Table 6.9 shows the external collapsing pressure of reinforced polyester pipe. This latter set of data is of particular importance when the piping is to be used under negative pressure, such as in vacuum systems or as pump sections.

Piping is usually furnished as a standard with pressure ratings of 50, 100, and 150 psig. Other pressure ratings are easily obtained because of the method of manufacture. Table 6.10 shows the minimum flange thicknesses required for reinforced polyester pipe at various operating pressures.

The life of the pipe in service is affected by more than the type of resin and type of reinforcement. Equally important is the method of manufacture employed, the resin cure, and the reliability and experience of the manufacturer. If the piping is purchased from a reliable and experienced polyester pipe manufacturer, these conditions should be met. In order to guarantee good performance, it must be

TABLE 6.10 Minimum Flange Thicknesses Required for Reinforced Polyester
Pipe at Various Operating Pressures

Pipe size (in.)	Minimum flange thickness (in.)					
	25 psi	50 psi	75 psi	100 psi	125 psi	150 psi
2	0.5	0.5	0.5	0.5625	0.625	0.6875
3	0.5	0.5	0.625	0.6875	0.75	0.8125
4	0.5	0.5625	0.6875	1.1875	0.875	1.3125
6	0.5	0.625	0.75	0.875	1.0	1.0625
8	0.5625	0.75	0.875	1.0	1.125	1.25
10	0.6875	0.875	1.0625	1.1875	1.3125	1.4375
12	0.75	1.0	1.25	1.4375	1.625	1.75
14	0.8125	1.0625	1.3125	1.5	1.75	1.875
16	0.875	1.1875	1.4375	1.625	1.875	
18	0.9375	1.25	1.5	1.75	2.0	
20	1.0	1.3125	1.625	1.875		
24	1.125	1.5	1.875			
30	1.375	1.875				
36	1.75					
42	2.0					

determined that the material being used and the method of manufacture are
satisfactory for the specific application.

B. Joining

Reinforced polyester pipe can be joined by any of the methods given on pages
443–448. Of the methods given, the butt and strap technique is most widely used.
Pipe less than 20 inches in diameter is overlaid on the outside only, while pipe
exceeding 20 inches in diameter should be overlaid both inside and outside.

6.3 REINFORCED VINYL ESTER PIPING SYSTEMS

The vinyl ester resins became commercially available in the mid-1960s. There are
several basic advantages to the vinyl esters:

1. The basic structure of the vinyl ester molecule is such that it is more
   resistant to some types of chemical attack, such as hydrolysis and
   oxidation or halogenation.
2. The vinyl esters have better impact resistance and greater tolerance to
   temperature and pressure fluctuations and mechanical shock than the
   polyesters.

3. Laminate tests have shown that the vinyl ester resins have strengths somewhat higher than the polyesters but not equal to the heat-cured epoxies.
4. Because of their molecular structure, they cure rapidly and give high early strength and superior creep resistance.
5. They provide excellent fiber wet-out and good adhesion to the glass fiber, in many cases similar to the amine-cured epoxies but less than the heat-cured epoxies.

At elevated temperatures the flexural modulus must be checked, since at 225°F (107°C) the vinyl esters have lost up to half of their flexural strength. This is an important consideration in negative pressure applications.

In general, this family of resins possesses good physical and mechanical properties.

A. Pipe Data

Vinyl ester piping systems vary in allowable operating pressures and maximum allowable operating temperatures from manufacturer to manufacturer. This is due to the particular resin used and the method of manufacture. Because of this, it is necessary to check with the supplier whose pipe is to be used as to the allowable design conditions.

The Fibercast Company produces pipe in sizes of $1\frac{1}{2}$ through 14 inches with a maximum allowable operating temperature of 175°F (80°C). Table 6.11 gives the properties of Cl-2030 pipe, including the recommended operating external pressure at 75°F (24°C). As the operating temperature increases, the

TABLE 6.11 Recommended Operating Data of Fibercast Cl-2030 Pipe

Nominal pipe size (in.)	Internal pressure at 175°F/80°C (psi)	External pressure at 75°F/24°C (psi)	Axial load at 75°F/24°C (psi)
$1\frac{1}{2}$	875	432	3,000
2	800	511	3,700
3	525	302	6,500
4	500	206	10,200
6	500	141	19,400
8	375	80	25,500
10	300	26	32,000
12	250	16	38,100
14	175	12	41,900

Source: Courtesy of Fibercast Company.

TABLE 6.12 Operating External Pressure for Cl-2030 Pipe at Various Temperatures

Nominal pipe size (in.)	Maximum allowable external pressure (psi) at various temperatures (°F/°C)			
	75/24	150/66	175/80	200/93
1½	432	384	327	261
2	511	455	387	309
3	302	269	229	183
4	206	183	156	125
6	141	125	107	85
8	80	71	61	48
10	26	23	20	16
12	16	14	12	9
14	12	11	9	7

Source: Courtesy of Fibercast Company.

allowable external pressure decreases (see Table 6.12). Standard fittings to be used with the Cl-2030 pipe have a somewhat lower pressure rating than the pipe (see Table 6.13).

Smith Fiberglass Products produces a line of vinyl ester piping in sizes 2 through 8 inches rated for 150 psi at 225°F (107°C). There are numerous other manufacturers, but these two examples have been chosen to illustrate the differences between manufacturers.

TABLE 6.13 Pressure Ratings (psi) of Fibercast Vinyl Ester Fittings at 175°F/80°C Used Weldfast 200 Adhesive

Nominal pipe size (in.)	Elbows, tees, reducers, couplings, flanges, socket and threaded nipples		Laterals, crosses, saddles, and grooved nipples
	Socket fittings	Flanged fittings	
1½	300	300	
2	275	200	125
3	200	150	125
4	150	150	100
6	150	150	100
8	150	150	100
10	150	150	75
12	150	150	75
14	125	150	

Source: Courtesy of Fibercast Company.

B. Joining

Reinforced vinyl ester piping can be joined by butt and strap joints, socket-type adhesive joints, flanged joints, or bell and spigot joints (see pages 443–448). The specific joining method to be used will depend upon the manufacturer. It is important that the manufacturer's recommendations be strictly adhered to.

6.4 REINFORCED EPOXY PIPING SYSTEMS

The family of epoxy resins dominated the pipe market for many years until the introduction of the vinyl esters. They are still a major factor. In small-diameter high-pressure pipe (through 12 inches), the epoxy resins predominate, although vinyl esters are available. The epoxy resins can provide outstanding service in the chemical processing industry. Virtually all pipe is produced by the heat-cured process, which uses aromatic amines and acid anhydrides, in order to obtain the best physical properties.

There are many epoxy formulations. A formulation can be tailor-made to meet a specific corrosive and/or mechanical requirement. The heat-cured epoxies have higher strengths than the vinyl esters and provide superior fiber wet-out and adhesion to the glass fiber than the vinyl esters.

A. Pipe Data

Reinforced epoxy pipe systems are not standardized as to dimensions and/or pressure ratings. They vary depending upon the resin used, the reinforcing material, and the manufacturing techniques employed. The fibercast Company produces an RB-2530 pipe system in sizes 1 inch through 14 inches, which is nominally rated up to 450 psi in the smaller sizes up to 250°F (121°C). Table 6.14 lists the recommended operating data for the pipe, and Table 6.15 provides the maximum allowable pressure ratings for epoxy fittings. Since these ratings are lower than the ratings for straight pipe, the fitting ratings will control the overall rating of the system.

Table 6.16 supplies the maximum allowable external pressure for Fibercast RB-2530 epoxy pipe systems at different temperatures.

Reinforced Plastic Systems, Inc. manufactures two epoxy piping systems designated Techstrand 1000 and Techstrand 2000. The Techstrand 1000 system is available in sizes 2 through 16 inches and is rated for 150 psi at 210°F (99°C). Table 6.17 lists the pressure ratings of the Techstrand 1000 piping system.

Techstrand 2000 epoxy piping system is available in sizes 2 through 16 inches and is rated for 150 psi at 225°F (107°C). Table 6.18 lists the pressure ratings of the Techstrand 1000 piping system.

These three piping systems are typical of those available. Many other manufacturers also produce epoxy piping systems.

TABLE 6.14 Recommended Operating Data for Fibercast RB-2530 Epoxy Pipe

Nominal pipe size (in.)	Internal pressure at 270°F/132°C (psi)	Internal pressure at 250°F/121°C (psi)	Internal pressure at 75°F/24°C (psi)	Axial load at 75°F/24°C (lb)
1	200	300	1,469	1,680
$1\frac{1}{2}$	850	1,275	1,454	3,700
2	670	1,000	837	4,700
3	430	650	678	7,100
4	370	500	274	9,200
6	370	500	160	17,700
8	250	375	90	23,200
10	200	300	34	29,100
12	165	250	20	34,600
14	150	225	15	38,000

Source: Courtesy of Fibercast Company.

TABLE 6.15 Pressure Rating of Fibercast Epoxy Fittings Up to 225°F/107°C Using Weldfast 440 Adhesive

Nominal pipe size (in.)	Maximum allowable operating pressure (psi)			
	Elbows, tees, reducers, couplings flanges, sockets, and threaded nipples			Laterals, crosses, saddles and grooved nipples
	Socket flanges	Flanged fitting	Flanges	
1	300	300	300	
$1\frac{1}{2}$	450	150	450	
2	450	150	450	125
3	300	150	300	125
4	225	150	225	100
6	225	150	225	100
8	225	150	225	100
10	225	150	225	75
12	225	150	225	75
14	125	150	150	

Source: Courtesy of Fibercast Company.

TABLE 6.16 Allowable Operating External Pressures for Fibercast RB-2530 Epoxy Pipe at Various Temperatures

Nominal pipe size (in.)	Allowable pressue (psi) at °F/°C						
	75/24	150/66	175/80	200/93	225/107	250/121	270/132
1	1,469	1,439	1,378	1,343	1,160	940	812
$1\frac{1}{2}$	1,454	1,425	1,364	1,329	1,149	931	804
2	837	820	785	765	661	536	463
3	678	654	636	619	535	434	375
4	274	269	257	251	217	176	152
6	160	157	150	146	126	102	89
8	90	88	84	82	71	58	50
10	34	33	32	31	27	22	19
12	20	20	19	19	16	13	11
14	15	15	14	14	12	10	9

Source: Courtesy of Fibercast Company.

B. Joining

Reinforced epoxy piping systems can be joined by butt and strap joints, socket-type adhesive joints, and flanged joints as described on pages 443–448. The specific system to be used will be based on the manufacturer's design.

TABLE 6.17 Pressure Rating of Techstrand 1000 Epoxy Piping System

Nominal pipe size (in.)	Short-term burst pressure[a] 210°F/99°C (psi)	External pressure at 75°F/24°C (psi)
2	2,200	100
3	1,800	70
4	1,600	40
6	1,500	30
8	1,300	17
10	1,300	17
12	1,300	17
14	1,300	17
16	1,300	17

[a] Failure by weeping.

TABLE 6.18 Pressure Rating of Techstrand 2000 Epoxy Piping System

Nominal pipe size (in.)	Short-term burst pressure[a] 225°F/107°C (psi)	External pressure at 75°F/24°C (psi)
2	3,000	260
3	2,000	120
4	1,600	75
6	1,400	31
8	1,300	18
10	1,300	18
12	1,300	18
14	1,300	18
16	1,300	18

[a] Failure by weeping.

6.5 REINFORCED PHENOL-FORMALDEHYDE PIPING SYSTEMS

Reinforced phenol-formaldehyde pipe is manufactured from a proprietary resin produced by the Haveg Division of Ametek, Inc. and sold under the trade name of Haveg 41NA. It is reinforced with silicate fibers. When necessary to provide corrosion resistance against hydrofluoric acid or certain fluoride salts, graphite is substituted for the silicate fillers. This is one of the oldest synthetic piping materials available, having been in existence for more than 50 years. In general, it does not have the impact resistance of the polyesters or epoxies.

A. Pipe Data

Pipe is available in sizes $\frac{1}{2}$ inch through 12 inches in lengths of 4 feet in the $\frac{1}{2}$-inch and $\frac{3}{4}$-inch sizes and in lengths of 10 feet in all other sizes. The pressure rating varies with diameter and operating temperature. Refer to Table 6.19 for operating pressures of each size pipe at different operating temperatures.

Insulation is seldom required on Haveg 41NA because of its low thermal conductivity. The pipe may be steam traced providing the steam temperature does not exceed 300°F (149°C). All sizes of pipe are suitable for use under full vacuum.

B. Joining

Haveg 41NA pipe may be joined by means of either flanged joints or cemented joints. The cemented joint eliminates metal flanges, bolts, and gaskets, as well as

TABLE 6.19 Operating Pressure Versus Temperature for Haveg 41NA Pipe Systems

Nominal pipe size (in.)	Maximum operating pressure (psi) at °F/°C					
	50/10	100/38	150/66	200/93	250/121	300/149
$\frac{1}{2}$	100	100	100	100	100	100
$\frac{3}{4}$	100	100	100	100	100	100
1	100	100	100	100	100	100
$1\frac{1}{2}$	100	100	100	100	100	100
2	100	100	100	100	100	100
3	95	90	85	82	80	75
4	67	65	61	59	55	52
6	41	40	39	37	35	30
8	35	34	32	31	30	29
10	29	28	27	26	25	24
12	24	23	22	20	19	18

Source: Courtesy of Ametek Inc.

the maintenance they entail. The use of flanges is best restricted to make connections to equipment, pumps, expansion joints, or valves that are constructed of a material other than Haveg 41NA. Bell and spigot or threaded joints are available by special order. Flanged connections are made by using cast iron split flanges set in tapered grooves which are machined near the pipe ends.

A cemented joint consists of a machined sleeve that has been split longitudinally and cemented onto the butted ends of two sanded pipes. It is important that the manufacturer's directions for the mixing of the cement, preparation of the pipe ends, and installation of the sleeve be followed exactly if a leak-free joint is to be obtained.

6.6 REINFORCED FURAN RESIN PIPING SYSTEMS

Furan resins are produced from furfuryl alcohol and furfural. These resins are more expensive than other thermoset resins but are the most economical choice when the presence of solvents exists in a combination with acids and bases or when process changes may occur that result in exposure to solvents in oxidizing atmospheres.

Furan laminates have the ability to retain their physical properties at elevated temperatures. Pipe produced from these resins can be used at a temperature of 300°F/149°C.

Reinforced furfuryl alcohol-formaldehyde pipe is produced from a proprietary resin with silicate fillers manufactured by the Haveg Division of Ametek, Inc. and sold under the trade name of Haveg 61NA. It is considered a furan resin. Graphite filler is substituted when the pipe is to be used to convey hydrofluoric acid or certain fluoride salts. The material is tough, durable, light weight, and has been available for over 50 years.

A. Pipe Data

Haveg 61NA is available in sizes $\frac{1}{2}$ inch through 12 inches. Standard lengths of the $\frac{1}{2}$-inch and $\frac{3}{4}$-inch sizes are 4 feet. All other diameters are available in 10-foot lengths. The pressure rating varies with temperature and pipe diameter. Table 6.20 provides the maximum allowable operating pressure at different temperatures for each diameter of pipe. All sizes are suitable for full vacuum at 300°F (149°C).

Because of the low thermal conductivity of the pipe, insulation is seldom required. The pipe may be steam traced providing the steam temperature does not exceed 300°F (149°C).

B. Joining

Haveg 61NA pipe may be connected by either flanged or cemented joints. The cemented joint eliminates metal flanges, bolts, and gaskets, as well as the maintenance they require. It is a good policy to limit the use of flanges to making connections to pumps, equipment, expansion joints, valves, or other

TABLE 6.20 Operating Pressure versus Temperature for Haveg 61NA Pipe Systems

Nominal pipe size (in.)	Maximum operating pressure (psi) at °F/°C					
	50/10	100/38	150/66	200/93	250/121	300/149
$\frac{1}{2}$	100	100	100	100	100	100
$\frac{3}{4}$	100	100	100	100	100	100
1	100	100	100	100	100	100
$1\frac{1}{2}$	100	100	100	100	100	100
2	100	100	100	100	100	100
3	95	90	85	82	80	75
4	67	65	61	59	55	52
6	41	40	39	37	35	30
8	35	34	32	31	30	29
10	29	28	27	26	25	24
12	24	23	22	20	19	18

Source: Courtesy of Ametek Inc.

accessory equipment to be installed in the pipe line. Flanges can also be used to connect Haveg 61NA to dissimilar materials. Bell and spigot of threaded connections are available by special order.

Flanged connections are made by using cast iron split flanges set in tapered grooves, which are machined near the end of the pipe.

Cemented joints are made by using a machined sleeve, which has been split longitudinally and cemented on to the butted, sanded ends of two pipe lengths. It is important that the manufacturer's directions for the mixing of the cement, preparation of the pipe ends, and installation of the sleeve be followed exactly if a leak-free joint is to be obtained.

6.7 REINFORCED PHENOLIC PIPING SYSTEM

Reinforced phenolic pipe consists of a proprietary phenolic resin produced by Haveg Division of Ametek, Inc. and silica filaments and fillers. It is sold under the trade name Haveg SP.

A. Pipe Data

Haveg SP pipe and fittings are made to nominal IPS schedule 40 OD and are available in sizes of 1 inch through 8 inches. Piping in diameters of 1, $1\frac{1}{2}$, 2, 3, and 4 inches are designed for pressure applications and are furnished with threaded ends in nominal 10-foot lengths. The 6- and 8-inch-diameter pipe is designed for low-pressure drainage operations. This pipe is supplied with plain nonthreaded ends. The maximum operating temperature for all sizes is 300°F (149°C). Refer to Table 6.21 for maximum allowable operating pressures at 70°F (21°C) and 300°F (149°C).

TABLE 6.21   Maximum Allowable Working Pressure of Haveg SP Pipe

Nominal pipe size (in.)	Maximum allowable working pressure (psi)	
	70°F/21°C	300°F/149°C
1	150	100
$1\frac{1}{2}$	150	100
2	150	100
3	100	75
4	100	75
6	25	15
8	15	10

Source: Courtesy of Haveg Div. of Ametek Inc.

B. Joining

The pressure-rated piping system, sizes 1 inch through 4 inches, is assembled using threaded joints and cement. The purpose of the threads is to align and immobilize the joint until the cement hardens.

Low-pressure drain piping, sizes 6 and 8 inches, are assembled by flanging and/or cementing. Since it is somewhat difficult to field thread 6- and 8-inch pipe, a spigot-and-socket cemented joint is used instead of a threaded joint. If a flange connection is desired, it can be made with a slip-on cemented construction rather than a threaded cemented construction.

7

Miscellaneous Applications

7.1 LINED PIPING SYSTEMS

Nonmetallic materials such as the thermoplasts, thermosets, and elastomers possess a decided advantage over metallics in resisting corrosion. However, the thermoplasts do not have the physical strength or ruggedness of metals. In addition, special pipe supports are often required and safety considerations may limit allowable pressures and temperatures and also system locations. The majority of the thermoplasts are affected by ultraviolet light and oxygen, requiring UV stabilizers and antioxidants to be added to the formulation.

The thermosets increase the temperature range and pressures but still require special supporting techniques, protection from external loading, and allowances for expansion and construction. As operating temperatures increase, the allowable operating pressure decreases.

Most of the elastomeric materials do not have sufficient mechanical strength to permit the manufacture of entire piping systems but they can be utilized as liners, as can thermoplastic and thermoset resins.

In general, lining thicknesses range from 0.10 to 0.20 inch in the most widely used pipe sizes. The outer shell is usually schedule 40 or schedule 80 carbon steel pipe.

Materials used for linings are as follows:

Thermoplasts		Elastomers
PVC	PTFE	Natural rubber
PE	FEP	Neoprene
PP	PFA	Butyl rubber
ETFE	Saran	Chlorobutyl rubber
PVDF		Nitrile (Buna N) rubber
ECTFE		EPDM rubber
		Hypalon

As with any piping system, the system needs to be engineered properly in order to meet specified process conditions. Proper design of the system requires more than sizing the pipe to meet the flow rate and allowable pressure drop. Other factors must be taken into account including:

Permeation
Absorption
Environmental stress cracking
Fluid velocity
Bolt torquing
Piping support.

A discussion of these and other factors can be found in: Schweitzer PA. Corrosion Resistant Piping Systems. New York: Marcel Dekker, 1994.

7.2   REINFORCED THERMOSET PLASTIC TANKS (RTP)

RTP tanks are available in a great many different shapes and sizes from 1 gallon bottles to 250,000-gallon field-erected tanks. One important use has been in the use of RTP tanks for underground storage of gasoline.

Many manufacturers have a line of standard sizes and shapes. If one of these meets your requirements, this is the most economical approach. However, there are companies who will custom-design tanks.

Although polyester resins are the most widely used, tankage is available with other resins. Selection of the proper resin is important if corrosion resistance is to be achieved.

The design principles for RTP tanks are the same for metallic tanks. Wall thicknesses must be calculated based on internal and external pressures, operating temperatures, and service conditions. A detailed discussion of the design techniques for RTP tankage can be found in Mallinson, J.H. Corrosion-Resistant Plastic Composites in Chemical Plant Design. New York: Marcel Dekker, 1988.

7.3 DUCT SYSTEMS

Many chemical processes involve handling of contaminated exhaust air or gases, sometimes in large volumes, from process vats, tanks, or equipment. Individual systems may range from small units handling approximately 1000 cfm to extremely large systems handling approximately 1,000,000 cfm or more. Between these two extremes are many installations in the 10,000 to 35,000-cfm range. These are ideally suited to be handled in RTP duct systems including ductwork, fans, discharge stacks and dampers. There are excellent fire retardant resins available, provided that the chemicals in the proposed system are compatible.

Fiberglass reinforced plastic systems of this type have definite advantages over metal systems which may be subject to corrosion, and over elastomeric lined systems which are being replaced by FRP systems. As a result of air pollution control regulations, the need for ductwork, scrubbers, and accessory equipment to handle the pollutants has increased. The application of FRP units is the most economical approach for both interior and exterior applications.

Details of ductwork design can be found in Mallinson, JH. Corrosion-Resistant Plastic Composites in Chemical Plant Design. New York: Marcel Dekker, 1988.

7.4 STRUCTURAL APPLICATIONS

The use of fiber-reinforced plastics in structural applications has greatly increased, not only in the chemical industry but in industry in general. Applications include use of thermoplastic and thermoset resins. Products produced using thermoset resins include

Car bodies	Bus bar insulators
Airplane sections and interiors	Bus duct insulators
Truck hoods	ladder cages
Fenders	Ladder rails and rungs
Roof paneling	Motor wedges
Golf clubs	Piping supports
Tennis racquets	Platforms
Fishing poles	Railings
Pole vaulting poles	Scrubber packing supports
Boats	Separators and dividers
Antennas	Cable trays
Automobile springs	Cell partitions
Base and floor plates	Columns
Beams	Concrete forms
Benches	Cooling towers
Braces	Door and window frames

Bridge supports	Ducts
Electrical enclosures	Electrostatic precipitators
Hay poles	Hockey sticks
Hoods	Instrument trays
Streetlight poles	Supports
Tank covers	Tool handles
Trench covers	Troughs
Wind charger blades	Windmill blades

In producing these products, a variety of reinforcement materials are used, including glass fiber (E glass), aramid fiber, and carbon fibers.

The selection of the resin system is critical when the finished product is to be used in a corrosive environment. Quite often, general purpose and isophthalic resins are used. These are suitable for mildly corrosive services. If a more severe atmosphere is present, then a vinyl ester should be considered.

In addition to reinforcements, fillers and additives are also used in the formulation to impart certain specific properties. Some of the more common materials used are

Aluminum silicate: to improve chemical resistance, opacity, insulation properties, surface finish.

Calcium carbonate: to reduce costs, improve whiteness and finish.

Alumina trihydrate: to improve fire retardancy (do not use in severe chemical exposure because of heavy loading).

Antimony trioxide: to improve fire retardancy; use at a 3–5% loading level with chlorinated or brominated polyesters or vinyl ester.

Sand or grit: to impart nonskid properties to walking surfaces, such as grating, platforms, and stair treads.

Alumina: to improve insulation properties, provide better abrasion resistance.

Carbon or graphite: in certain applications, the normally non-conductive grating or flooring must be made conductive by adding carbon or graphite to the resin; carbon veil may also be used.

Appendix 1

Comparative Corrosion Resistance of
Selected Elastomers

Chemical	Butyl °F	Butyl °C	Hypalon °F	Hypalon °C	EPDM °F	EPDM °C	EPT °F	EPT °C	Viton A °F	Viton A °C	Kalrez °F	Kalrez °C	Natural rubber °F	Natural rubber °C	Neoprene °F	Neoprene °C	Buna-N °F	Buna-N °C
Acetaldehyde	80	27	60	16	200	93	210	99	x	x	x	x	x	x	200	93	x	x
Acetamide	150	66	x	x	200	93	200	93	210	99	x	x	x	x	200	93	180	82
Acetic acid 10%	110	43	200	93	140	60	x	x	190	88	200	93	150	66	160	71	200	93
Acetic acid 50%	110	43	200	93	140	60	x	x	180	82	200	93	x	x	160	71	200	93
Acetic acid 80%	90	32	200	93	140	60	x	x	180	82	90	32	x	x	160	71	210	99
Acetic acid, glacial	150	66	x	x	140	60	x	x	x	x	80	27	x	x	x	x	100	38
Acetic anhydride	160	71	220	93	x	x	x	x	x	x	210	99	x	x	90	32	200	93
Acetone			x	x	300	148	x	x	x	x	210	99	x	x	x	x	x	x
Acetyle chloride			x	x	x	x	x	x	x	x	210	99	x	x	x	x	x	x
Acrylic acid									x	x			x	x	x	x	x	x
Acrylonitrile	x	x	140	60	140	60	100	38	x	x	110	43	90	32	160	71	x	x
Adipic acid	x	x	140	60	200	93	140	60	180	82	210	99	80	27	160	71	180	82
Allyl alcohol	190	88	200	93	300	148	80	27	190	88			80	27	120	49	180	82
Allyl chloride	x	x			x	x	x	x	100	38			x	x	x	x	x	x
Alum	190	88	200	93	200	93	140	60	190	88	210	99	150	66	200	93	200	93
Aluminum acetate			x	x	200	93	180	82	180	82	210	99	x	x	x	x	200	93
Aluminum chloride, aqueous	150	66	250	121	210	99	180	82	190	88	210	99	140	60	200	93	200	93
Aluminum chloride, dry											190	88						
Aluminum fluoride	180	82	200	93	210	99	180	82	180	82	210	99	150	66	200	93	190	88
Aluminum hydroxide	100	38	250	121	210	99	140	60	190	88	210	99	150	66	180	82	180	82
Aluminum nitrate	190	88	250	121	210	99	180	82	190	88	210	99	150	66	200	93	200	93
Aluminum oxychloride									x	x								
Aluminum sulfate	190	88	200	93	210	99	210	99	190	88	210	99	160	71	200	93	210	99
Ammonia gas			140	60	140	60	140	60	x	x	210	99	x	x	140	60	190	88

Each cell below is given as °F/°C (x = not resistant; blank = no data). The nine numeric column-pairs read left-to-right as printed.

Chemical	1	2	3	4	5	6	7	8	9
Ammonium bifluoride	190/x	60/60	300/300	140/60	140/60	60	x	x	180/82
Ammonium carbonate	140/88	140/93	300/148	180/82	190/88	210/99	150/66	200/93	200/93
Ammonium chloride 10%	200/88	200/93	210/148	180/82	190/88	210/99	150/66	200/93	200/93
Ammonium chloride 50%	200/88	200/93	210/99	180/82	190/88	210/99	150/66	190/88	200/93
Ammonium chloride, sat.	200/88	200/93	300/148	180/82	190/88	210/99	150/66	200/93	200/93
Ammonium fluoride 10%	200/66	200/93	210/99	210/99	140/60	210/99	160/71	100/38	200/93
Ammonium fluoride 25%	200/66	250/121	300/148	140/60	140/60	210/99	80/27	200/93	120/49
Ammonium hydroxide 25%	250/88	250/121	100/38	140/60	190/88	210/99	x	200/93	200/93
Ammonium hydroxide, sat.	250/88	200/93	100/38	140/60	190/88	210/99	90/32	210/99	200/93
Ammonium nitrate	200/82	80/27	250/121	180/82	x	300/148	170/77	200/93	180/82
Ammonium persulfate	80/88	140/60	300/148	210/99	140/60	210/99	150/66	200/93	200/93
Ammonium phosphate	140/82	200/93	300/148	180/82	180/82	210/99	150/66	200/93	200/93
Ammonium sulfate 10–40%	200/88	200/93	300/148	180/82	180/82	210/99	150/66	200/93	200/93
Ammonium sulfide	200	60/16	300/148	210/99	x	210/99		160/71	180/82
Ammonium sulfite		200/93	210/99			210/99			160/71
Amyl acetate	60/x	x	210/99	x	x	210/99	x	x	x
Amyl alcohol	200/82	200/93	x	180/82	200/93	210/99	150/66	200/93	180/82
Amyl chloride	x	140/60	140/60	x	190/88	210/99	x	x	x
Aniline	140/66	140/60	300/x	x	230/110	250/121	x	x	x
Antimony trichloride	140/66	200/93	x	x	190/88	210/99	x	140/60	x
Aqua regia 3:1		250/121	300/148	x	190/88	210/99	x	x	180/82
Barium carbonate	200/88	250/121	250/121	180/82	250/121	210/99	180/82	160/71	200/93
Barium chloride	250/88	250/121	250/121	180/82	190/88	210/99	150/66	200/93	200/93
Barium hydroxide	250/88	200/93	300/148	180/82	190/88	210/99	150/66	200/93	200/93
Barium sulfate	200	200/93	140/60	180/82	190/88	210/99	180/82	160/71	180/82
Barium sulfide	200/88	x	150/66	140/60	190/88	210/99	150/66	200/93	200/93
Benzaldehyde	150/32	x	x	x	x	210/99	x	x	x
Benzene	x	x	x	x	190/88	210/99	x	x	x

(Continued)

Chemical	Butyl °F	Butyl °C	Hypalon °F	Hypalon °C	EPDM °F	EPDM °C	EPT °F	EPT °C	Viton A °F	Viton A °C	Kalrez °F	Kalrez °C	Natural rubber °F	Natural rubber °C	Neoprene °F	Neoprene °C	Buna-N °F	Buna-N °C
Benzenesulfonic acid 10%	90	32	x	x	x	x	x	x	170	77	210	99	x	x	100	38	x	x
Benzoic acid	150	66	200	93	x	x	140	60	190	88	310	154	150	66	200	93	x	x
Benzyl alcohol	190	88	140	60	x	x	x	x	350	177	210	99	x	x	x	x	140	60
Benzyl chloride	x	x	x	x	x	x	x	x	110	43	210	99	x	x	x	x	x	x
Borax	190	88	200	93	300	148	210	99	190	88	210	99	150	66	200	93	180	82
Boric acid	190	88	290	143	190	88	140	60	190	88	210	99	150	66	200	93	180	82
Bromine gas, dry			60	16	x	x	x	x			210	99			x	x	x	x
Bromine gas, moist			60	16	x	x									x	x	x	x
Bromine, liquid			60	16	x	x			350	177	140	60			x	x	x	x
Butadiene			x	x	x	x	x	x	190	88	210	99			140	60	200	93
Butyl acetate	x	x	60	16	140	60	x	x	x	x	210	99	x	x	60	16	x	x
Butyl alcohol	140	60	250	121	200	93	180	82	250	121	240	116	150	66	200	93	x	x
r-Butylamine							x	x	x	x	210	99					80	27
Butyric acid	x	x	x	x	140	60	x	x	120	49	x	x	x	x	x	x	x	x
Calcium bisulfide	120	49	250	121					190	88	210	99	120	49	180	82	180	82
Calcium bisulfite	150	66	90	32	x	x	x	x	190	88	210	99	180	82	200	93	200	93
Calcium carbonate	190	88	90	32	210	99	140	60	190	88	200	93	150	66	60	16	180	82
Calcium chlorate	190	88	200	93	140	60	140	60	190	88	210	99	150	66	200	93	200	93
Calcium chloride	190	88	200	93	210	99	180	82	190	88	210	99	200	93	200	93	180	82
Calcium hydroxide 10%	190	88	200	93	210	99	180	82	190	88	210	99	200	93	220	104	180	82
Calcium hydroxide, sat.	190	88	250	121	220	104	180	82	190	88	210	99	200	93	220	104	180	82
Calcium hypochlorite	190	88	250	121	210	99	180	82	190	88	210	99	200	93	220	104	180	82
Calcium nitrate	190	88	100	38	300	148	180	82	190	88	210	99	150	66	200	93	80	27
Calcium oxide	190	88	200	93	210	99			190	88	210	99			160	71	180	82

	1		2		3		4		5		6		7		8		9	
	°C	°F	°C	°F	°C	°F	°C	°F	°C	°F	°C	°F	°C	°F	°C	°F	°C	°F
Calcium sulfate	82	180	71	160	82	180	99	210	93	200	82	180	148	300	121	250	38	100
Caprylic acid	x	x	x	x	x	x	99	210	88	190	82	180	x	x	x	x	x	x
Carbon bisulfide	93	200	93	200	66	150	99	210	x	x	82	180	121	250	93	250	88	190
Carbon dioxide, dry	93	200	93	200	66	150	99	210	x	x	82	180	121	250	93	250	88	190
Carbon dioxide, wet	93	200	x	x	66	150	99	210	x	x	82	180	121	250	110	230	88	190
Carbon disulfide	x	x	x	x	x	x	99	210	88	190	82	180	x	x	x	x	x	x
Carbon monoxide	82	180	93	200	x	x	99	210	88	190	x	x	121	250	93	250	32	90
Carbon tetrachloride	x	x	x	x	x	x	99	210	88	190	82	180	x	x	x	x	x	x
Carbonic acid	x	x	x	x	x	x	99	210	177	350	x	x	148	300	x	x	66	150
Cellosolve	x	x	x	x	x	x	99	210	x	x	x	x	71	160	x	x	66	150
Chloroacetic acid, 50% water	x	x	x	x	x	x	99	210	x	x	x	x	x	x	x	x	71	160
Chloroacetic acid	x	x	x	x	x	x	99	210	x	x	x	x	x	x	32	90	x	x
Chlorine gas, dry	x	x	x	x	x	x			88	190	x	x	x	x	66	160	x	x
Chlorine gas, wet	x	x	x	x	x	x	99	210	88	190	x	x	x	x	71		x	x
Chlorine, liquid	x	x	x	x	x	x	99	210	88	190	x	x	x	x	x	x	x	x
Chlorobenzene	x	x	x	x	x	x	99	210	88	190	x	x	x	x	x	x	x	x
Chloroform	x	x	x	x	x	x	99	210	88	190	x	x	x	x	x	x	x	x
Chlorosulfonic acid	x	x	x	x	x	x	99	210	x	x	x	x	x	x	x	x	x	x
Chromic acid 10%	88	190	60	140	x	x	99	210	177	350	x	x	99	210	66	150	38	100
Chromic acid 50%	88	190	38	100	x	x	99	210	177	350	x	x	99	210	71	160	x	x
Chromyl chloride							99	210	x	x	x	x	38	100	x	x	99	210
Citric acid 15%	82	180	93	200	43	110	99	210	88	190	82	180	99	210	66	150	88	190
Citric acid, concd	82	180	93	200	66	160	99	210	88	190	82	180	99	210	71	160	88	190
Copper acetate	82	180	71	160			99	210	x	x	38	100	99	210	x	x	88	190
Copper carbonate	x	x	x	x	x	x	99	210	99	210	99	210	99	210	x	x	88	190
Copper chloride	93	200	93	200	66	150	99	210	88	190	82	180	99	210	93	200	88	190
Copper cyanide	82	180	71	160	71	160	99	210	88	190	99	210	99	210	121	210	88	190
Copper sulfate	93	200	93	200	66	150	99	210	88	190	82	180	99	210	121	210	88	190

(Continued)

Chemical	Butyl °F	Butyl °C	Hypalon °F	Hypalon °C	EPDM °F	EPDM °C	EPT °F	EPT °C	Viton A °F	Viton A °C	Kalrez °F	Kalrez °C	Natural rubber °F	Natural rubber °C	Neoprene °F	Neoprene °C	Buna-N °F	Buna-N °C
Cresol	x	x	x	x	x	x	100	38	x	x	210	99	x	x	x	x	x	x
Cupric chloride 5%			200	93	210	99	210	99	180	82					210	99	210	99
Cupric chloride 50%			200	93	210	99	210	99	180	82					160	71	180	82
Cyclohexane	x	x	x	x	x	x	x	x	190	88	210	99	x	x	x	x	180	82
Cyclohexanol			x	x	x	x	x	x	190	88	210	99			x	x	x	x
Dibutyl phthalate							x	x	80	27	210	99	x	x	x	x	x	x
Dichloroacetic acid											210	99			x	x	x	x
Dichloroethane	x	x	x	x	x	x	x	x	190	88	210	99	x	x	x	x	x	x
Ethylene glycol	190	88	200	93	200	93	180	82	350	177	210	99	150	66	160	71	200	93
Ferric chloride	190	88	250	121	220	104	180	82	190	88	210	99	150	66	160	71	200	93
Ferric chloride 50% in water	160	71	250	121	210	99	180	82	180	82	210	99	150	66	160	71	180	82
Ferric nitrate 10–50%	190	88	250	121	210	99	180	82	190	88	210	99	150	66	200	93	200	93
Ferrous chloride	190	88	250	121	200	93	180	82	180	82	210	99	150	66	90	32	200	93
Ferrous nitrate	190	88			210	99	180	82	210	99			150	66	200	93	200	93
Fluorine gas, dry	x	x	140	60	60	16	100	38	x	x	x	x	x	x	x	x	x	x
Fluorine gas, moist					60	16					x	x	x	x	x	x	x	x
Hydrobromic acid, dil	150	66	90	32	90	32	140	60	190	88	210	99	100	38	x	x	x	x
Hydrobromic acid 20%	160	71	100	38	140	60	140	60	190	88	210	99	110	43	x	x	x	x
Hydrobromic acid 50%	110	43	100	38	140	60	140	60	190	88	210	99	150	66	x	x	x	x
Hydrochloric acid 20%	x	x	160	71	100	38	x	x	350	177	210	99	150	66	90	32	130	54
Hydrochloric acid 38%	x	x	140	60	90	32	x	x	350	177	210	99	160	71	90	32	x	x
Hydrocyanic acid 10%	140	60	90	32	200	93	x	x	190	88	210	99	90	32	x	x	200	93
Hydrofluoric acid 30%	350	177	90	32	60	16	140	60	210	99	210	99	100	38	200	93	x	x
Hydrofluoric acid 70%	150	66	90	32	x	x	x	x	350	177	210	99	x	x	200	93	x	x

This table lists the maximum resistance temperature (°F / °C) of nine elastomers to each reagent. An "x" indicates the elastomer is not resistant.

Reagent	1 °F	1 °C	2 °F	2 °C	3 °F	3 °C	4 °F	4 °C	5 °F	5 °C	6 °F	6 °C	7 °F	7 °C	8 °F	8 °C	9 °F	9 °C
Hydrofluoric acid 100%	x	x	x	x	x	x	x	x	x	x	60	16	x	x	x	x	x	x
Hypochlorous acid	x	x	x	x	x	x	x	x	x	x	190	88	x	x	x	x	x	x
Iodine solution 10%	90	x	250	x	x	x	x	x	x	x	190	88	66	x	x	x	80	27
Ketones, general	x	x	x	x	x	x	x	x	x	x	x	x	x	x	x	x	x	x
Lactic acid 25%	120	49	140	60	140	60	210	99	210	99	190	88	80	x	140	60	80	x
Lactic acid, concd	120	49	80	27	60	27	210	99	210	99	150	66	x	x	90	32	90	x
Magnesium chloride	200	93	250	121	250	121	180	82	180	82	180	82	80	210	99	180	82	—
Malic acid	x	x	121	82	x	x	80	26	80	26	160	82	80	150	180	180	180	82
Manganese chloride	90	32	180	x	x	210	210	99	210	99	180	82	80	27	200	93	100	38
Methyl chloride	x	x	x	x	x	x	x	x	x	x	190	88	x	x	x	x	x	x
Methyl ethyl ketone	x	x	80	80	27	x	x	x	x	x	190	88	x	x	x	x	x	x
Methyl isobutyl ketone	x	x	60	60	16	x	x	x	x	x	160	88	x	x	x	x	x	x
Muriatic acid	140	60	x	x	x	x	x	x	x	x	350	177	x	x	x	x	x	x
Nitric acid 5%	100	71	60	38	16	16	x	x	x	x	190	88	x	x	x	x	x	x
Nitric acid 20%	100	71	60	38	16	16	x	x	x	x	190	88	x	x	x	x	x	x
Nitric acid 70%	x	32	x	x	x	x	x	x	x	x	190	88	x	x	x	x	x	x
Nitric acid, anhydrous	x	x	x	x	x	x	x	x	x	x	190	88	x	x	x	x	x	x
Nitrous acid, concd	120	49	x	x	x	x	x	x	x	x	100	38	99	x	210	88	x	x
Oleum	x	x	x	x	x	x	x	x	x	x	190	88	x	x	x	x	x	x
Perchloric acid 10%	150	66	100	38	140	60	190	88	190	88	190	88	99	150	210	210	x	x
Perchloric acid 70%	90	32	140	60	140	60	88	210	88	60	190	88	66	150	210	140	x	x
Phenol	150	66	x	x	80	x	80	27	210	x	210	99	x	110	210	80	x	x
Phosphoric acid 50–80%	150	66	200	93	140	60	180	82	180	82	210	99	43	x	210	180	200	82
Picric acid	80	80	27	300	148	140	60	82	60	190	88	x	160	200	200	93	130	54
Potassium bromide 30%	250	121	250	121	210	99	180	82	180	82	190	88	71	160	160	200	180	82
Salicylic acid	80	27	x	x	x	x	x	x	x	x	190	88	x	x	x	x	x	x
Silver bromide 10%	x	x	x	x	x	x	x	x	x	x	210	x	x	x	x	x	x	x
Sodium carbonate	180	82	250	121	300	148	190	88	190	88	210	99	82	180	200	200	93	93

(Continued)

Chemical	Butyl °F	°C	Hypalon °F	°C	EPDM °F	°C	EPT °F	°C	Viton A °F	°C	Kalrez °F	°C	Natural rubber °F	°C	Neoprene °F	°C	Buna-N °F	°C
Sodium chloride	180	82	240	116	140	60	180	82	190	88	210	99	130	54	200	93	180	82
Sodium hydroxide 10%	180	82	250	121	210	99	210	99	x	x	210	99	150	66	200	93	160	71
Sodium hydroxide 50%	190	88	250	121	180	82	200	93	x	x	210	99	150	66	200	93	150	66
Sodium hydroxide, concd	180	82	250	121	180	82	80	27	x	x	210	99	150	66	200	93	150	66
Sodium hypochlorite 20%	130	54	250	121	300	148	x	x	190	88			90	32	x	x	x	x
Sodium hypochlorite	90	32			300	148	x	x	190	88			90	32	x	x	x	x
Sodium sulfide to 50%	150	66	250	121	300	148	210	99	190	88	210	99	150	66	200	93	180	82
Stannic chloride	150	66	90	32	300	148	210	99	180	82	210	99	150	66	210	99	180	82
Stannous chloride	150	66	200	93	280	138	210	99	190	88	210	99	150	66	160	71	180	82
Sulfuric acid 10%	150	66	250	121	150	66	210	99	350	177	240	116	150	66	200	93	150	66
Sulfuric acid 50%	150	66	250	121	150	66	210	99	350	177	210	99	100	38	200	93	200	93
Sulfuric acid 70%	100	38	160	71	140	60	210	99	350	177	150	66	x	x	200	93	x	x
Sulfuric acid 90%	x	x	x	x	x	x	80	27	350	177	150	66	x	x	x	x	x	x
Sulfuric acid 98%	x	x	110	43	x	x	x	x	350	177			x	x	x	x	x	x
Sulfuric acid 100%	x	x	x	x	x	x	x	x	190	88			x	x	x	x	x	x
Sulfuric acid, fuming			x	x	x	x	x	x							x	x	x	x
Sulfurous acid	150	66	160	71	x	x	180	82	190	88	210	99	x	x	x	x	x	x
Thionyl chloride	x	x							x	x	210	99	x	x	x	x	x	x
Toluene	x	x	x	x	80	27	x	x	190	88	80	27	x	x	x	x	150	66
Trichloroacetic acid	x	x	x	x	x	x	x	x	190	88	210	99	x	x	x	x	x	x
White liquor					300	148	180	82	190	88	210	99	x	x	140	60	140	60
Zinc chloride	190	88	250	121	300	148	180	82	210	99	210	99	150	66	160	71	190	88

[a] The chemicals listed are in the pure state or in a saturated solution unless otherwise indicated. Compatibility is shown to the maximum allowable temperature for which data are available. Incompatibility is shown by an x. A blank space indicates that data are unavailable.

Source: Schweitzer PA. Corrosion Resistance Tables, 4th ed., Vols. 1–3. New York: Marcel Dekker, 1995.

Appendix 2

Chemical Synonyms

Chemical	Synonym
Acetic acid, crude	Pyroligneous acid
Acetic acid amide	Acetamide
Acetic ether	Ethyl acetate
Acetol	Diacetone alcohol
Acetylbenzene	Acetophenone
Acetylene tetrachloride	Tetrachloroethane
Almond oil	Benzaldehyde
Aluminum hydrate	Aluminum hydroxide
Aluminum potassium chrome	Chrome alum
Alum potash	Aluminum potassium sulfate
Aminobenzene	Aniline
Ammonium fluoride, acid	Ammonium bifluoride
Baking soda	Sodium carbonate
Benzene carbonal	Benzaldehyde
Benzenecarboxylic acid	Benzoic acid
Benzol	Benzene
Boracic acid	Boric acid
Bromomethane	Methyl bromide
Butanoic acid	Butyric acid
Butanol-1	Butyl alcohol
Butanone	Methyl ethyl ketone
Butter of antimony	Antimony trichloride

(*Continued*)

Chemical	Synonym
Butyl phthalate	Dibutyl phthalate
Calcium sulfide	Lime sulfur
Carbamide	Urea
Carbolic acid	Phenol
Carbonyl chloride	Phosgene
Caustic potash	Potassium hydroxide
Caustic soda	Sodium hydroxide
Chlorobenzene	Monochlorobenzene
1-Chlorobutane	Butyl chloride
Chloroethane	Ethyl chloride
Chloroethanoic acid	Chloroacetic acid
Chloromethane	Methyl chloride
Chloropentane	Amyl chloride
3-Chloropropene-1	Allyl chloride
Chlorotoluene	Benzyl chloride
Chromium trioxide	Chromic acid
Cupric acetate	Copper acetate
Cupric carbonate	Copper carbonate
Cupric fluoride	Copper fluoride
Cupric nitrate	Copper nitrate
Cupric sulfate	Copper sulfate
Cuprous chloride	Copper chloride
Diacetone	Diacetone alcohol
Dibromomethane	Ethylene bromide
Dibutyl ether	Butyl ether
Dichloroethane	Ethylene dichloride
Dichloroethane	Dichloroethylene
Dichloromethane	Methylene chloride
Diethyl	Butane
Diethylene dioxide	Dioxane
Diethylenimide oxide	Morpholine
Dihydroxyethane	Ethylene glycol
Dimethylbenzene	Xylene
Dimethyl polysiloxane	Silicone oil
Dipropyl	Hexane
Dipropyl ether	Isopropyl ether
Dowtherm	Diphenyl
Epsom salt	Magnesium sulfate
Ethanal	Acetaldehyde
Ethanamide	Acetamide
Ethanoic acid	Acetic acid
Ethanoic anhydride	Acetic anhydride
Ethanol	Ethyl alcohol

Chemical	Synonym
Ethanonitrile	Acetonitrile
Ethanoyl chloride	Acetyl chloride
Ethanoxy ethanol	Cellosolve
Ethylene chloride	Ethylene dichloride
Ethyl ether	Diethyl ether
Formalin	Formaldehyde
Furfuraldehyde	Furfural
Furfurol	Furfural
Glucose	Dextrose
Glycerol	Glycerine
Glycol	Ethylene glycol
Glycol ether	Diethylene glycol
Glycol methyl ether	Methyl cellosolve
Hexamethylene	Cyclohexane
Hexanedioic acid	Adipic acid
Hexose	Dextrose
Hexyl alcohol	Hexanol
Hydroxybenzoic acid	Salicylic acid
Hypo photographic solution	Sodium bisulfate
Lime	Calcium oxide
Marsh gas	Methane
Methanal	Formaldehyde
Methanoic acid	Formic acid
Methanol	Methyl alcohol
Methylbenzene	Toluene
Methyl chloroform	Trichloroethane
Methyl cyanide	Acetonitrile
Methyl phenol	Cresol
Methyl phenol ketone	Acetophenone
Methyl phthalate	Dimethyl phthalate
Methyl propane-2	Butyl alcohol, tertiary
Muriatic acid	Hydrochloric acid (crude)
Nitrogen trioxide	Nitrous acid
Oil of mirbane	Nitrobenzene
Oil of wintergreen	Methyl salicylate
Oxalic nitrile	Cyanogen
Phenylamine	Aniline
Phenyl bromide	Bromobenzene
Phenyl carbinol	Benzyl alcohol
Phenol chloride	Chlorobenzene
Phenyl ethane	Ethylbenzene
Pimelic ketone	Cyclohexanone

(Continued)

Chemical	Synonym
Propanoic acid	Propionic acid
Propanol	Propyl alcohol
Propanone	Acetone
Propenyl alcohol	Allyl alcohol
Propenoic acid	Acrylic acid
Propyl acetate	Isopropyl acetate
Prussic acid	Hydrocyanic acid
Pyrogallol	Progallic acid
Red oil	Oleic acid
Sal ammoniac	Ammonium chloride
Sodium borate, tetra	Borax
Sodium phosphate, dibasic	Disodium phosphate
Starch gum	Dextrin
Sugar of lead	Lead acetate
Sulfuric chlorohydrin	Chlorosulfonic acid
Tannin	Tannic acid
Tetrachloroethylene	Perchloroethylene
Tetrachloromethane	Carbon tetrachloride
Trichloromethane	Chloroform
Trihydroxybenzene	Pyrogallic acid
Trihydroxybenzoic acid	Gallic acid
Trinitrophenol	Picric acid
Vinyl cyanide	Acrylonitrile
Water glass	Sodium silicate

Index

479